Health Information Tec

M000247782

Series Editor
Tim Benson, R-Outcomes Ltd, Newbury, UK

Health information technology is one of the fastest growing industry sectors. The purpose of this book series is to provide monographs covering the rationale, content and use of these and other standards to help bridge the gap between the need for and availability of qualified and knowledgeable staff. This series will be focused on health informatics technology standards and the technology driving change in health IT. It will appeal to the traditional informatics market, but also cross over into more technical disciplines, but without leaving the remit that this is to expand knowledge in healthcare IT. It will comprise a set of single-author, practically focused, academically driven concise reference monographs on the leading standards and their application. Each volume will focus on one or more specific standards and explain how to use each one individually or in combination. This provides a tight focus for each book. The aim is to offer a set of "must have" references on the widely used standards, and in particular those mandated by the ONC.

More information about this series at http://www.springer.com/series/10471

Tim Benson • Grahame Grieve

Principles of Health Interoperability

SNOMED CT, HL7 and FHIR

Third Edition

 Springer

Tim Benson
R-Outcomes Ltd
Newbury, UK

Grahame Grieve
Health Intersections Pty Ltd
Melbourne, Australia

ISSN 2199-2517   ISSN 2199-2525   (electronic)
Health Information Technology Standards
ISBN 978-3-319-30368-0   ISBN 978-3-319-30370-3   (eBook)
DOI 10.1007/978-3-319-30370-3

Library of Congress Control Number: 2016938106

This Springer imprint is published by Springer Nature
The registered company is Springer International Publishing AG Switzerland

Tim Benson dedicates this book to his sons Laurence, Oliver, Alex and Jamie.

Grahame Grieve dedicates this book to his family, who lit the fire in the first place.

Foreword to the Third Edition

Recent US Government reports have included statements such as:

> *The apparent inability of the private sector to achieve interoperable systems suggests the need for national leadership to support their creation.*

> *Information blocking occurs when persons or entities knowingly and unreasonably interfere with the exchange or use of electronic health information.*

> *Health lacks a common language to share data.*

Each of these points oversimplifies the real issues facing healthcare information exchange. A combination of technology, policy and alignment of incentives has worked in every industry to enable data liquidity. If stakeholders understand all the issues, the same thing will happen in healthcare.

Unfortunately, domain expertise in interoperability is rare. The standards are esoteric and detailed. Politics and emotion can cloud the objective evaluation of standards that are suitable for purpose, well documented and mature enough for adoption.

Principles of Health Interoperability: SNOMED CT, HL7 and FHIR (3rd edition) by Tim Benson and Grahame Grieve provides an accessible, well-organized primer that is objective and clear. It clarifies that interoperability is not just as simple as pushing HL7 transactions from point to point.

When I was 2 years old in 1964, my mother gave me ampicillin and I developed two red dots on my stomach. She declared me allergic to penicillin. For 50 years my medical record has said "penicillin allergy" and not:

Substance: Pencillins and Cephalosporins
Reaction: Urticaria
Observer: Mother
Level of Certainty: Very Uncertain
Date of observation: January 1, 1964

If we are to share data among stakeholders, we need easy to implement technologies that provide a structure for the information (such as the five components of an allergy above), appropriate vocabularies (how do we describe the nature of the reaction in a uniform fashion) and a secure means of transmitting that information over the wire. If I was diagnosed with a live threatening strep infection, for which Penicillin is the most effective drug, would a clinician make a different decision on treatment knowing that my allergy is uncertain and minor? Certainly.

Principles of Health Interoperability is a must read for policymakers, technology leaders and industry implementers. The book distills thousands of pages of standards into the essential information you need to know. The addition of the *Fast Healthcare Interoperability Resources* (FHIR) makes the 3rd edition even better than the 2nd edition. FHIR will enable an ecosystem of apps, which layer on top of existing EHRs, reduce the cost of interfacing and accelerate innovation.

If you are looking for the definitive resources on the latest techniques to implement content, transport and vocabulary interoperability, look no further than this book. It will be a centerpiece of my own bookshelf.

Beth Israel Deaconess Health System   John D. Halamka
Boston, MA, USA
Harvard Medical School
Boston, MA, USA

Foreword to the First Edition

Health data standards are a necessary component of interoperability in healthcare. Aggregation of health-related data mandates the use of standards, and aggregation is necessary to support safe and quality care. The American Recovery and Reinvestment Act (ARRA) includes $19 billion dollars in direct funding and an additional $18.5 billion in returned savings tagged to the use of health information technology (HIT). The resulting expanding use of HIT has engaged a growing number of stakeholders, many of whom now realize the value of standards.

All aspects of creating and "meaningful use" of electronic health records (EHRs) require standards. With the increasing demand for individuals knowledgeable in what standards are available and how and when to use those standards, this book is most welcome. The author, Tim Benson, has been engaged in the creation of standards since the beginning. His experiences span organizations – including HL7, CEN and ISO and terminologies such as SNOMED and LOINC. He has engaged the global community and understands similarities as well as differences among the global community. He has a top reputation as a teacher and writer within the international community. I know no other individual more qualified to write this book than Tim Benson.

In *Principles of Health Interoperability HL7 and SNOMED*, Tim focuses on major contributors to the set of required standards. In the first section, he lays out a framework for why interoperability is important and what is needed to accomplish that interoperability. Health Level Seven (HL7) is pre-eminent among the several contributing Standards Developing Organizations (SDOs) in the global community. HL7 standards are widely used and cover the full spectrum of applications. Its membership is international (currently including over 35 countries) and includes the major HIT vendors and representatives of the full set of stakeholders. The International Healthcare Technology Standards Developing Organization (IHTSDO) is rapidly promoting SNOMED CT as the preferred terminology in healthcare. While focusing on HL7 and SNOMED CT, Tim has included much useful information on other standards and other organizations.

Readers will find this book easy to read, even if it is their first exposure to standards. In this rapidly changing field, this book is a must for anyone who is involved or has interest in the use of health information technology – and who isn't.

Duke Centre for Health Informatics   W. Ed Hammond
Duke Translational Medicine Institute,
Duke University, Durham, NC, USA
Biomedical Informatics Core
Duke Translational Medicine Institute
Durham, NC, USA
Community and Family Medicine
Duke University
Durham, NC, USA
Founding Member of HL7 (1987),
Chair HL7 (1991, 1996–97, 2008–09)
Durham, NC, USA

Foreword to the Second Edition

The success of this book validates the above remarks. Interoperability and the focus of the broad community on this topic and the implementation of systems and standards that support interoperability have grown at an exponential rate. As the implementation of Health Information Interchange systems grows, more and more people join the workforce to support this growth. They need to be taught and learn about standards supporting interoperability. A number of colleagues and I use this book as a text. The students love it – it is clear and easy to read and understand. Technology and the ensuing standards to support standards change rapidly. In this second addition, Tim has astutely addressed this challenge. In some sections, he expanded the material; in others, he reorganized the material; and, most importantly, he added new sections to increase the comprehension and coverage of the topic. The second edition is even better than the first.

Duke Centre for Health Informatics   W. Ed Hammond
Duke Translational Medicine Institute,
Duke University, Durham, NC, USA
Biomedical Informatics Core
Duke Translational Medicine Institute
Durham, NC, USA
Community and Family Medicine
Duke University
Durham, NC, USA
Founding Member of HL7 (1987),
Chair HL7 (1991, 1996–97, 2008–09)
Durham, NC, USA

Preface

Interoperability is one of the hottest topics in healthcare, yet one of the least well understood. Successful interoperability offers great opportunities to improve quality and outcomes while reducing waste and costs. The task of interoperability is to deliver the right information at the right time to the right place. Everybody (patient, clinician, manager and payer) stands to benefit from more soundly based decisions, safer care and less waste, errors, delays and duplication.

Interoperability needs appropriate standards to link computer systems, and to share information in a way that meets security and privacy needs. SNOMED CT and HL7 (including FHIR) provide key standards that underpin efforts to improve healthcare interoperability. HL7 provides the structure, rather like English grammar, while SNOMED CT provides the words that computers understand.

This book gives a broad introduction to healthcare interoperability in general, and the main standards, setting out the core principles in a clear readable way for analysts, students and clinicians.

The third edition of this book is fully revised, reorganized and extended. There are five new chapters on FHIR (Fast Healthcare Interoperability Resources), written by Grahame Grieve, the father of FHIR. This is the first comprehensive introduction to FHIR in any book.

FHIR APIs are likely to have a massive disruptive impact on healthcare interoperability, being an order of magnitude less expensive to implement than previous standards. FHIR will also support an explosion of patient-centric apps that can interoperate with legacy systems.

To accommodate these changes, we have changed the order of the chapters, so that clinical terminology and SNOMED CT come before HL7 interchange formats, v2, v3, CDA and FHIR. The introductory chapters have also been revised and updated.

The book is organized in four parts. The first part covers the principles of health-care interoperability, why it matters, why it is hard and why modeling is an important part of the solution. The second part covers clinical terminology and SNOMED CT. The third part covers the longer established HL7 standards, v2, v3, CDA and IHE XDS. The final part covers FHIR.

Newbury, UK   Tim Benson
Melbourne, Australia   Grahame Grieve
January 2016

Principles of Health Interoperability: SNOMED CT, HL7 and FHIR (3rd Edition)

Healthcare interoperability delivers information when and where it is needed. Everybody stands to gain from safer more soundly based decisions and less duplication, delays, waste and errors. This book provides an introduction to healthcare interoperability and the main standards used.

The third edition includes a new part on FHIR (Fast Healthcare Interoperability Resources), the most important new health interoperability standard for a generation. FHIR combines the best features of HL7's v2, v3 and CDA, while leveraging the latest web standards and a tight focus on implementation. FHIR can be implemented at a fraction of the price of existing alternatives and is well suited for mobile phone apps, cloud communications and EHRs.

The book is organized into four parts. The first part covers the principles of health interoperability, why it matters, why it is hard and why models are an important part of the solution. The second part covers clinical terminology and SNOMED CT. The third part covers the main HL7 standards: v2, v3, CDA and IHE XDS. The new fourth part covers FHIR and has been contributed by Grahame Grieve, the original FHIR chief.

Newbury, UK   Tim Benson
Melbourne, Australia   Grahame Grieve

Acknowledgements

Tim Benson: Many people have contributed to my understanding of this healthcare interoperability. In particular, I want to thank Ed Hammond, David Markwell, Roddy Neame, Abdul-Malik Shakir, Alan Rector, Bob Dolin, Charlie McCay, Charlie Mead, Clem McDonald, David Ingram, Ed Cheetham, Ed Conley, Georges de Moor, Jack Harrington, James Read, Kent Spackman, Larry Weed, Leo Fogarty, Mark Schafarman, Mike Henderson, Réné Spronk, Sigurd From, Tom Marley and Woody Beeler. Finally, I wish to thank my family for their forbearance and all the great people who have created SNOMED, HL7 and FHIR.

Grahame Grieve: FHIR is a community, a collective accomplishment, and many people have contributed, too many to list. But a few deserve mention: Ewout Kramer, Lloyd McKenzie, Josh Mandel, James Agnew, Brian Postlethwaite and David Hay for contributing the most to the community and the specification. More personally, Kevin Moynihan, David Rowlands, Thomas Beale, Kim Clohessy, Chuck Jaffe, Gunther Schadow, Charlie McCay, Andy Bond and Woody Beeler have contributed enormously to my understanding of healthcare, integration and the business environment in which it thrives. Also thanks to Mel Grieve for editing the FHIR part, and to my family for sharing their holidays with this book.

Contents

About the Authors

Tim Benson graduated from the University of Nottingham as a mechanical engineer. He was introduced to healthcare computing at the Charing Cross Hospital, London, where he evaluated the socio-economic benefits of medical computing systems. He founded one of the first GP computer suppliers (Abies Informatics Ltd). There, with James Read and David Markwell, he helped develop the Read Codes, which became the national standard for NHS primary care and one of the two sources of SNOMED CT. Tim led the first European project team on open standards for health interoperability, which led to CEN/TC251 and collaboration with HL7, where he was a co-chair of the Education Committee for several years. He has also developed a family of short generic patient-reported outcome measures (http://www.r-outcomes.com).

Grahame Grieve graduated from the University of Auckland as a biochemist and worked as a clinical diagnostic scientist at St Vincent's Hospital, Melbourne, before spending 4 years performing medical research in diabetes, lipid metabolism and oxidation. He then switched focus and joined Kestral Computing P/L, a Laboratory and Imaging Information Systems vendor, where he ended up as Chief Technology Officer, before leaving to establish his own consulting business (http://www.healthintersections.com.au). A growing involvement in integration, and interoperability, lead him to the HL7 community where he has led committees and edited standards for HL7 v2, v3 and CDA. The outcome of this was the recognition that something new was needed, and this led to the creation of the FHIR specification, which now consumes his life.

Part I
Principles of Health Interoperability

Chapter 1
The Health Information Revolution

Abstract This chapter sets out some of the core problems and opportunities facing the digital healthcare sector. Healthcare is all about communication. Large investments in digital health have failed to live up to expectations, partly due to poor interoperability. Patient centered care requires a new approach, organized primarily for patient benefit, not just for provider organizations. What matters most is the point of care, which is inevitably complex. Many lessons can be learnt from past experience, successes and failures.

Keywords EHR • Communication • Information • Patient-centered care • Outcomes • Key performance indicators • Quality • Waste • Clinical decisions • Clinical specialty • El Camino hospital • POMR • GP computing • Prescription form • NHS National Programme for IT • Summary care record • Detailed care record • Infoway standards collaborative • MedCom • Meaningful use

Healthcare is Communication

Modern healthcare depends on teamwork and communication. Interoperability is needed to provide information when and where required, facilitate quicker and more soundly based decision making, reduce waste by cutting out repeated work and improve safety with fewer errors.

Convergence of digital health interoperability, wireless sensors, imaging technology and genomics will transform the way that healthcare is practiced, its efficiency and effectiveness. Patients using their own mobile devices are leading this revolution. Patients won't wait, even if it takes years for physicians to adopt new medical advances [1].

Most healthcare processes involve communication within the system. Billions of documents are generated mostly using pen and paper. Healthcare remains the largest remaining market for pens, paper and fax-machines. The long-promised digital health revolution has been slow to arrive and is still characterized by "hope, hype and harm" [2]. Large initiatives such as the $30Bn *Meaningful Use* scheme have failed to improve efficiency as much as was hoped, in large part due to failure to address interoperability at the clinical level.

© Springer-Verlag London 2016
T. Benson, G. Grieve, *Principles of Health Interoperability*,
Health Information Technology Standards, DOI 10.1007/978-3-319-30370-3_1

Paper-based patient records are widely recognised as unfit for purpose. What Bleich complained of more than 20 years ago is still common:

> *The medical record is an abomination ... it is a disgrace to the profession that created it. More often than not the chart is thick, tattered, disorganized and illegible; progress notes, consultants notes, radiology reports and nurses notes are all co-mingled in accession sequence. The charts confuse rather than enlighten; they provide a forbidding challenge to anyone who tries to understand what is happening to the patient* (Bleich 1993) [3].

Paper records can only be used by one person at a time, and are often not where they are needed. Once to hand, it is hard to find what you want in a disorganized, illegible, inconsistent, incomplete, badly sorted collection. The user has to work hard just to glean any useful information. An enormous amount of staff time is spent locating, transporting and reviewing these paper repositories.

On the other hand, it is easy to overlook just how flexible and durable paper-based patient records are in spite of these deficiencies. EHRs need to become just as flexible, reliable and easy to use.

Traditionally, healthcare information systems have been organised hierarchically on the basis of the flow of money and authority, flowing from payer to provider organizations and down to departments, clinicians and finally patients. This model is way out of alignment with the natural flow of information needed to care for individual patients, which is more like a social network, with each patient at the center of his or her own net.

All people want the same things from health and social care. They want to feel better physically and mentally, to do more and be independent. They want this now and in the future, with a long healthy life followed by a quick peaceful death, not a slow demise. Every patient also wants excellent care and service, to be treated kindly, to be listened to and have issues fully explained, be seen promptly and for systems to perform reliably and safely.

More confident and engaged patients tend to report better outcomes and experience and have lower costs. These patients are typically more empowered, knowledgeable, confident to manage their own health, able to get help when they need it and participate in shared decision-making,

Given that patients are the sole reason for healthcare activity, health services increasingly need to focus on the outcomes that matter to patients. Great organizations have always used a small number of key performance indicators (KPIs).

> *What matters is ... settling upon a consistent and intelligent method of assessing your outcome results and then tracking your trajectory with rigor (Collins 2006)* [4].

Efforts to improve quality often lead to lower costs, while efforts to cut costs invariably lead to lower quality. The primary focus needs to be on quality improvement not cost cutting.

In a person-centered model, care is based on continuous clinical relationships, customized to individual patient needs, with the patient ultimately in control. Knowledge is shared, information flows freely and decisions are based on evidence. Transparency and collaboration are virtues, patient needs are anticipated and effort is devoted to eliminating waste, which is any activity that costs money but delivers no benefit to patients.

However, today's healthcare information systems were designed mainly to support the traditional medical model, based around discrete conditions, visits and episodes. Each clinician decides independently on what investigations and treatment to order based on their training and experience. This has led to large variation in treatment, much of which is unwarranted (Wennberg 2010) [5]. The patient record is often just a log of what happened, incentivized to maximize fee income and kept secret from the patient. The system defends professional demarcation and reacts to patient needs only as and when they arise.

However, when so much of what we do is performed over the Internet, there are no technical barriers to sharing information and providing joined-up patient-centered care, yet good interoperability remains a rarity.

Back in 2001 the Institute of Medicine in *Crossing the Quality Chasm* set out rules for a person-centered healthcare system [6] (see also Table 1.1).

1. Care should be based on continuous healing relationships, not based on payment for discrete episodes. This means continuous access, taking full advantage of modern information technology, 24-h a day, 7 days a week and 365 days a year.
2. Customization based on patients' needs and values. Variation should be based on patients' informed needs and wishes, not professional autonomy.
3. The patient should be in control over decisions, access and information sharing – "no decision about me without me".
4. Knowledge and information should be shared with patients as a right, without restriction, delay or the need for anyone else's permission.
5. The best care results come from the conscientious explicit and judicious use of current best evidence and knowledge of patient values by well-trained experienced clinicians.
6. Safety should be a system property, not be regarded as an individual responsibility. Systems should prevent error when possible, detect any errors that occur and mitigate the harm done if an error does reach the patient.

Table 1.1 Contrast between traditional and patient-centered healthcare models (Based on Institute of Medicine 2001)

Aspect	Traditional	Patient-centered
Focus of care	Discrete visits/episodes	On-going care relationship
Variation mainly due to	Professional autonomy	Patient needs and values
Control	Professionals in control	Patient in control
Decisions based on	Training and experience	Evidence
Safety	Individual responsibility	A system property
Openness	Secrecy	Transparency subject to patient privacy
Reactivity	React to patient needs	Anticipate patient needs
Economic focus	Cut costs	Eliminate waste
Collaboration	Demarcation	Cooperation
Information technology	Silos	Interoperability

7. Use knowledge of individual patients, local conditions, and the natural history of illness to predict and anticipate needs not simply react to events.
8. Economies by reducing all types of waste, not by cutting costs. Improving quality can save money, but cutting costs reduces quality. Examples of waste in the US health system include [7]:

  - Service overuse ($210 billion)
  - Inefficiency ($130 billion)
  - Excess administrative costs ($190 billion)
  - Prices that are too high ($105 billion)
  - Missed prevention opportunities ($55 billion)
  - Fraud ($75 billion)

9. Teamwork, cooperation, collaboration, communication and coordination are more important than professional prerogatives and roles.

Information Handling

Information handling has evolved over several thousand years through the four stages originally set out by Marshall McLuhan (1962) [8].

In the first stage, information and knowledge was held only in the human brain and transferred from one person to another by speech. Oral tribe culture provides an example. Access depends on the person with the knowledge being present and this is lost forever when they die. Much of medicine still relies on this model of communication and the clinician's memory.

The second stage began with the invention of handwriting. Hand-written records are formatted at the time of writing, cannot be replicated without transcription and may be hard to read. However, modern healthcare, involving teams of doctors and nurses, each doing a specialised task, would be impossible without written records. Hospital medical records are still largely hand-written.

Gutenberg

The third stage was triggered by the invention of printing by Johannes Gutenberg around 1455, which provided the means to replicate and broadcast information widely. This led to the Renaissance, the Age of Enlightenment, the Industrial Revolution and the Information Society. The impact of top-down broadcasting and dissemination of knowledge on medical education has been massive, but there has been little impact on how people perform routine consultations or maintain records.

The fourth and last stage, the electronic age, has its origins in the electronic computers and information science developed during the Second World War and has gathered pace exponentially ever since. The digital revolution has led to explosive development of the Internet, the Web, mobile phones and social networking, following Moore's and Metcalfe's laws.

Moore's Law is the prediction made in 1965 that the power of computer devices would continue to double every 2 years; this has held good for 50 years and shows few signs of stopping yet. Two to the power 25 is over 33 million. Metcalfe's Law says that as networks grow, the value to each user increases linearly but the total value of the network increases exponentially.

Topol has drawn close parallels between the transformative effects of Gutenberg's press and those of the smartphone. These include explosion of knowledge, spurring innovation, promoting individualization, promoting revolution and wars, fostering social networks, reducing interpersonal interaction, spreading ideas and creativity, promoting do-it-yourself, flattening the Earth, reducing costs, archiving and reducing boredom. He suggests that just as Gutenberg democratized reading, smartphones will democratize medicine by giving individuals unfettered direct access to all of their health data and information [9].

We are moving towards new relationships between patients and citizens, their clinicians and smartphone apps and supporting algorithms, sharing the same health and care information. This co-production triangle can reduce data-action latency, the delay between information being available and its being acted upon [10].

Use of Information

Healthcare is the quintessential information-based industry, yet has singularly failed to harness these forces. The electronic health record (EHR) lies at the heart of digital health. The wide range of uses, clinical and non-clinical, are shown in Fig. 1.1.

Clinical care is task-oriented. At any moment a clinician is performing one of a number of well-defined tasks, but every clinical microsystem is different. Clinical care is made up of thousands of discrete tasks, each with its own information and communication needs and requiring systems, terms and classifications tailored to the needs of the task.

These tasks are ultimately determined by the complexity and variety of the natural history of disease processes and their corresponding diagnostic, treatment and administrative procedures. Automating these tasks, which include everything needed to support clinical decision-making, to order tests and treatment, to correspond with all those involved in the care of individuals (patients, hospital specialists, GPs, community and social care services), is the core task of digital health.

Managers cannot and do not need to understand every detail of clinical care; their focus is to provide a safe, efficient and courteous service, smooth administration of each patient's visit, and to ensure that everything is done in order to get paid.

Their focus is on service management. These tasks are far more homogeneous than clinical uses, focused on meeting the contractual obligations imposed by regulators and payers. However, such regulations and contracts change frequently and are ultimately determined politically.

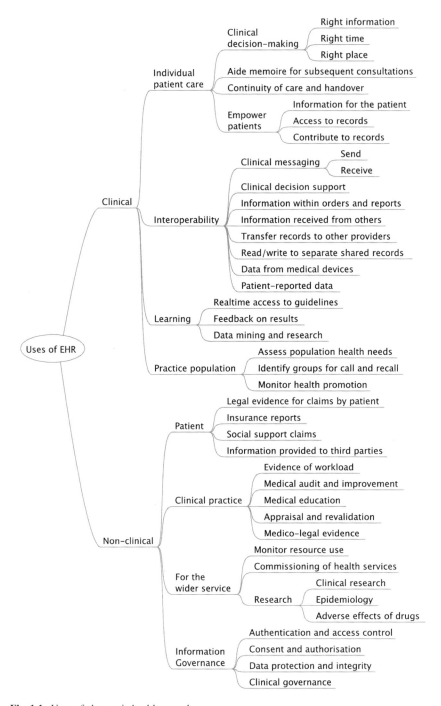

Fig. 1.1 Uses of electronic health records

One of the problems is that health professionals can be overwhelmed by information and by demands for information. Herbert Simon noted that:

> *Information consumes the attention of its recipients; a wealth of information creates a poverty of attention and a need to allocate that attention amongst the overabundance of information sources that might consume it* (Simon 1971) [11].

In medicine, as in art, the value of information is often related to its rarity. The key to medical decision-making is Bayes law, which is based on how much a new piece of information changes the prior probabilities.

A combination of EHR features, such as auto-population, templates and cut-and-paste, which were conceived to save data entry effort and maximize income, often generate voluminous notes where it is hard to find what you are seeking.

Clinical Decisions

Differences in treatment and investigation patterns of individual doctors lead directly to different costs and outcomes. Doctors spend the money. It is always important to do tests efficiently, but if a test or procedure is inappropriate, it is waste irrespective of how efficiently it is done. Don Berwick has written:

> *The ultimate measure by which to judge the quality of a medical effort is whether it helps patients (and their families) as they see it. Anything done in healthcare that does not help a patient or family is, by definition, waste whether or not the professions and their associations traditionally hallow it* (Berwick 1997) [12].

What principally determines cost is doing the right things. Only a small proportion of cost variance is down to service efficiency – doing things right.

Electronic patient records are key to improved clinical decision-making. Computer-based records are legible and, in theory, information can be displayed in the best way for the task at hand. Several people can work on the same record at the same time in different places, saving the delays and effort required to locate, retrieve and transport paper. Prompts can improve quality and safety, prevent key data being omitted, and save time by not needing to record the same data time and again.

Healthcare communication and information flow patterns involve large numbers of people over a wide geographical area and diverse subject matter. For example, each primary care doctor refers patients to many specialists and each specialist receives referrals from many referrers. Each doctor communicates with a multitude of specialised investigation and treatment services, community care agencies, administrative and funding bodies. These highly complex many-to-many communication patterns are found throughout the health and social care services (Fig. 1.2).

The half-life of information (how long a piece of information has much value) differs enormously between contexts, such as outpatient clinics, wards, intensive care units and operating theatres. There is little benefit in showing information well after its half-life is over, even if it has to be preserved for medical legal purposes.

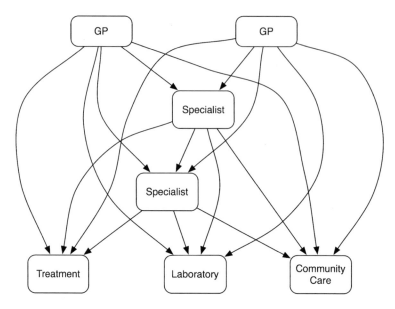

Fig. 1.2 Complex information flow patterns

Every specialty has its own needs. Doctors are organised into 60 or more specialties and there are similar numbers for nursing, treatment and investigation professions. Each specialty has its own governance, education and quality assurance
requirements, speaks its own dialect and has its own ways of working. Specialists
understand their own specialty very well, but not other specialties.

This helps explain why successful electronic patient record systems are often
limited to a single specialty, such as general practice, maternity care or renal medicine, where the needs are relatively homogeneous and well understood.

Clinicians in hospitals are very mobile; they are found on any ward where they
have patients, in clinics, at any one of several hospitals, on domiciliary visits in the
community, in laboratories or in their own office. Mobility just adds to the problems
of computerization.

The concept of the one-size-fits-all patient record has seldom been successful
except where enormous efforts and adequate resources have been devoted to tailor
the system to individual specialty needs and where management has been able to
mandate its use.

Lessons from History

El Camino Hospital

These issues have been well known for at least 40 years. The first hospital to implement a comprehensive EHR was the El Camino Hospital, Mountain View, California,
which went live in 1971. This project was subjected to a detailed 6-year evaluation,

which compared its costs and other outcomes with control hospitals. Such detailed long-term evaluations are surprisingly rare and when done are not always published, but almost all other studies have shown similar findings.

The following quotes come from an account of the experience by Melville Hodge, who led the project for the supplier (Hodge 1990) [13]. The project at El Camino was initially met with:

Massive resistance from important segments of the medical staff, spreading quickly to ... national newspaper headlines. This resistance, initially justified in part by early system shortcomings, seemed intractable.

Healthcare is a complex socio-technical system, involving the interaction of both people and technology. You cannot design organizational and technical systems independently of each other, nor expect to re-engineer healthcare systems successfully without a thorough understanding of both the human and technology requirements to make all the parts work smoothly together (Coiera 2004) [14]. Hodge warned:

Never forget that introduction of [EHR] into a hospital impacts a human organization to an unparalleled degree. If the need to manage the change process is ignored, resistance and even rebellion may be reasonably predicted.

The initial resistance was overcome by learning these lessons and:

By effective leadership of the more visionary El Camino physicians.

The outcome was that 10 years later, in 1981, the hospital chief executive could claim a triumph:

The hospital inpatient cost per case is 40% less than the county average for 13 similar community hospitals.

To summarize:

Success has repeatedly been demonstrated to be the consequence of each doctor, one at a time, coming to see how his performance is enhanced by investing his always-scarce time in learning how to use the system efficiently. Similarly, hospital managers must participate in and buy into a carefully designed benefits realization program before they can be reasonably expected to act.

These problems and risks, and the knowledge of how to mitigate them, were first understood almost 40 years ago, yet the same things still happen (Wachter 2015) [2].

Success in GP Surgeries

In the UK all GPs (family physicians) use EHRs in their consulting room and almost all work paper-free – they rely entirely on electronic records while consulting. All primary care prescriptions are printed by computer or sent electronically [15]. This all happened by the mid 1990s, more than 20 years ago.

Leadership and incentives played a big part in why GPs use computers (Benson 2002) [16]. Over a 30-year period, the leaders of the GP profession worked hand-in-hand with the government to encourage and remove barriers to computerizing practices.

The story of the NHS computer-printed prescription form provides a good example of how governments can remove barriers to computerization. Computer-assisted repeat prescribing saves writing out prescriptions by hand and improves legibility and safety. The computer-printed FP10 (comp) form is twice the width of a standard prescription, with a blank area on the right hand side. The original reason for the blank space was that narrow tractor-feed printers were not available when the form was developed in the 1970s. The blank right hand side was used to provide each patient with a record of his or her medication; this was so useful that no one then considered doing away with it.

In spite of reservations that the wider form would be more expensive and computers would make it easier to prescribe more, hence increase the NHS drugs bill, the Department of Health approved the national use of the form in 1981. This single regulatory change was critical in stimulating the spread of GP computing. In other countries computer printed prescriptions remained illegal for decades longer, slowing uptake there.

Failure in Hospitals

The story in hospitals is very different. Attempts to replicate the success of GP computing in hospitals have failed repeatedly. There are several reasons.

You cannot shoehorn a system that works well in one specialty into another, yet the information systems used by different specialties need to work together, which requires interoperability.

GPs work as individuals mainly from a single consulting room, but hospital clinicians work as teams and are very mobile; their work is individually specialised, there are many specialties and each works in a different way. No one understands everything that goes on in a hospital.

Hospital clinicians need excellent communication within their work-group (the clinical micro-system) between doctors, nurses and other professions. An Australian study of hospital doctors found that they spent about 33 % of their time in communicating with other professionals, compared with 15 % of their time in direct care, including communication with the patient and their family. 70 % of the tasks performed by junior hospital doctors were with another member of staff, usually another doctor. Interns spent twice as much time on documenting (22 %) as on direct care (11 %) [17].

Hospital doctors have been offered few incentives or career encouragement to become involved, leading to alienation. Hospital computing has usually been treated as an administration overhead, reporting to the finance director, who is usually concerned with maximizing revenue and cutting costs.

NHS National Programme of IT

The NHS National Programme for IT (NPfIT) was described as the biggest computer programme in the world (Brennan 2005) [18] and turned into one of the biggest failures. It set out to provide detailed electronic health records for everyone in England, but this central objective was abandoned. How did this come about?

Conceived in 2002 during the period between 9/11 and the invasion of Iraq, things went badly wrong from the start. The central recommendation of the report, which led to the creation of the project, was for:

> A doubling of spending on ICT to fund ambitious targets of the kind set out in the NHS Information Strategy. To avoid duplication of effort and resources and to ensure that the benefits of ICT integration across health and social services are achieved, the Review recommends that stringent standards are set from the center to ensure that systems across the UK are fully compatible with each other [19].

More detail was provided in a strategy document 3 months later, which stated:

> The core of our strategy is to take greater control over the specification, procurement, resource management, performance management and delivery of the information and IT agenda. We will improve the leadership and direction given to IT, and combine it with national and local implementation that are based on ruthless standardization (DH 2002) [20].

Note two important differences between these quotations. The vision of integration across health and social services and cross-UK compatibility was dropped. Then the recommendation to set stringent standards was changed to one of ruthless standardization (an odd term as standards are usually based on consensus). The revised focus was to provide a centrally procured set of one-size-fits-all systems, and to rip-and-replace every system in the country. However, many local managers simply refused to replace working systems with those that were procured, which many did not consider to be fit for purpose.

The Strategy had ten key elements, the final one being to:

> Create national standards for data quality and data interchange between systems at local, regional and national levels (paragraph 2.3.2) [20].

From the outset, the challenges of developing and deploying the necessary standards were greatly underestimated. The strategy document published in June 2002 strongly and wrongly implied that the relevant standards were already available.

> Work is already underway on a strategy for electronic Clinical Communications and a report is due at the end of March 2002 (sic) (paragraph 4.2.2) [20].

The first phase of the project, between April 2002 and March 2003, was to be used to:

> Define the data and data interchange standards we will require in the future (paragraph 1.2.3) [20].

Responsibility for standards development was spread across four separate organizations for strategic direction, defining standards, ratifying standards and certification testing. No one had overall control of the whole picture. These national functions were eventually brought together April 2005 under NHS Connecting for Health. By then the key decisions on scope, technology and budgets had all been set in stone.

A central team developed specifications for national services using HL7 v3, but the specification and deployment of local services was left to local providers who adopted different releases of HL7 v2.

Two key parts of the program were the summary care record (SCR) and the detailed care records (DCR). The SCR is a nationally stored summary of patient's medical records in England, for use out of hours and emergency care. It contains details of medication, allergies and adverse drug reactions.

The evaluation of the SCR identified damaging conflicts between separate but interacting socio-technical networks [21]:

• The design network – policy makers, advisers, software developers and those involved in the technical infrastructure
• The implementation network – involved in implementation
• The governance network, responsible for privacy and security
• The front-line user network – users
• The evaluation network – evaluators.

Early use of the SCR was lower than expected, although this has now been turned into a success after several more years of effort. DCRs were even less successful.

Canada

In Canada, the Health Infoway project established a centrally funded Infoway Standards Collaborative, to:

> Support and sustain health information standards and foster collaboration to accelerate the implementation of pan-Canadian standards-based solutions [22].

The scope of the Infoway Standards Collaborative covers the interoperability standards that are required to meet the needs of the program, including their establishment, promotion, support and maintenance, and liaison with international standards development organizations.

The process used engages all stakeholders, stimulates market demand for these standards and seeks to reduce the risks and barriers to adoption. An open governance structure and long-term funding support it.

Denmark

Denmark has been uniquely successful in linking primary care doctors with laboratories, hospitals and pharmacies. In 1994 the Danish Government established MedCom as a national public project collaborating with public authorities, healthcare organizations and private firms. A small group of experts developed a set of standards for referrals, discharge letters, laboratory and radiology requests and reports, prescriptions and reimbursement claims, which were based on European standards originally developed by CEN TC251. These specifications were piloted, revised and re-tested in fifteen independent locally managed projects. Finally, the experience gained was brought together in voluminous documentation:

> *In such detail and so accurately and precisely that the overwhelming opinion is that MedCom's standards can indeed be used from Gedser to Skagen (from one end of Denmark to the other)* (MedCom 1996) [23].

Even after this preparation, the information sent was not always displayed or was misinterpreted due to ambiguity in data definitions of data elements, local coding schemes and lack clarity about which elements were mandatory or optional. These issues were tackled in a 3-year consolidation project leading to revised standards and compulsory certification. By the end of 2002, 53 software versions had been certified and the error rate was cut by more than 70 % (Johansen 2003) [24].

Today all Danish GPs receive discharge summaries and lab results electronically; most prescriptions and referrals are also sent electronically. One of the lessons is that success requires long-term persistence and political support (MedCom 2008) [25].

Meaningful Use

The term Meaningful Use of health IT was introduced in Obama's HITECH (Health Information Technology for Economic and Clinical Health) Act, 2009, which encapsulates in its name both the financial and care drivers for digital health. The nominal focus is to deliver the promise of electronic health records (EHR), but the real goal was to improve value for money (Blumenthal 2009) [26].

The US healthcare system started from a low base. In 2008, Tom Daschle, Obama's original nominee as Secretary of Health, summarised the problem as follows:

> *Our healthcare system is incredibly primitive when it comes to using the information systems that are common in American workplaces. Only 15 to 20 per cent of doctors have computerized patient records and only a small fraction of the billions of medical transactions that take place each year in the United States are conducted electronically. Studies suggest that this weakness compromises the quality of care, leads to medical errors, and costs as much as $78 billion a year* (Daschle 2008) [27].

By January 2014 93 % of eligible hospitals and 82 % of eligible physicians had registered for the program. By March 2015 more than $30 billion had been paid out.

The government had the good idea that people should be paid for using computers, not just for having them (shelf-ware). To receive incentive payments for being a meaningful user of a certified EHR system, each doctor (or other eligible professional) and hospital has to demonstrate that they are using computers for purposes including e-prescribing with decision support, laboratory results, radiology reports, visit summaries and to exchange coded data and quality reports.

However, physician dissatisfaction has grown. Between 2010 and 2014 satisfaction with EHRs fell from around 61 to 34% [28]. Almost half of respondents (in a self-selected sample of 940) reported that EHRs reduced efficiency, 72% stating it was difficult for EHRs to decrease physician workloads and 54% saying that EHRs increased operating costs. The only positive was that those who have used their system for longer were more satisfied than those who had only recently converted.

The reasons for clinical dissatisfaction are multiple and complex [2]. Many of them have been discussed above; four stand out. First, decisions about what systems to use have usually been made to meet business objectives such as maximizing income rather than to improve clinical quality and patient outcomes which are harder to count. Second, major computer systems are complex to design, build and implement and almost all of the systems in use today were designed in the era before meaningful use. Third, the scheme is seen as overly bureaucratic in its specifications and demands for evidence. Fourth, the regulations failed to incentivize interoperability, which is the subject of this book.

References

1. Topol E. The creative destruction of medicine: how the digital revolution will create better health care. New York: Basic Books; 2012.
2. Wachter R. The digital doctor: hope, hype, and harm at the dawn of medicine's computer age. New York: McGraw-Hill; 2015.
3. Bleich H, Lawrence L. Weed and the problem-oriented medical record. MD Comput. 1993;10(2):70.
4. Collins J. Good to great and the social services. London: Random House; 2006.
5. Wennberg JE. Tracking medicine: a researcher's quest to understand health care. New York: Oxford University Press; 2010.
6. Institute of Medicine. Crossing the quality chasm: a new health system for the 21st century. Washington, DC: National Academy Press; 2001.
7. Smith M, Saunders R, Stuckhardt L, McGinnis JM, editors. Best care at lower cost: the path to continuously learning health care in America. Washington, DC: National Academies Press; 2013.
8. McLuhan M. The gutenberg galaxy: the making of typographic man. Toronto: University of Toronto Press; 1962.
9. Topol E. The patient will see you now: the future of medicine is in your hands. New York: Basic Books; 2015.
10. Ainsworth J, Buchan I. Combining health data uses to ignite health system learning. Methods Inf Med. 2015;54:479–87.
11. Simon H. Designing organizations for an information-rich world. In: Greenberger M, editor. Computers, communication, and the public interest. Baltimore: The Johns Hopkins Press; 1971.
12. Berwick D. Medical associations: guilds or leaders. BMJ. 1997;314:1564.
13. Hodge M. History of the TDS medical information system. In: Blum BI, Duncan K, editors. A history of medical informatics. New York: ACM Press; 1990. p. 328–44.
14. Coiera E. Four rules for the reinvention of healthcare. BMJ. 2004;328:1197–9.
15. Department of Health. The good practice guidelines for GP electronic patient records, vol. 4. London: Department of Health; 2011.

16. Benson T. Why general practitioners use computers and hospital doctors do not – Part 1: incentives. BMJ. 2002;325:1086–9.
17. Westbrook J, Ampt A, Kearney L, Rob MI. All in a day's work: an observational study to quantify how and with whom doctors on hospital wards spend their time. MJA. 2008;188(9):506–9.
18. Brennan S. The NHS IT project: the biggest computer programme in the world ... Ever! Oxford: Radcliffe; 2005.
19. Wanless D. Securing our future health: taking a long-term view. Final report. London: HM Treasury; 2002; Chapter 7 Conclusions and recommendations, p. 121.
20. Department of Health. Delivering 21st century IT support for the NHS: National Strategic Programme. Leeds: Department of Health; 2002.
21. Greenhalgh T, Stramer K, Bratan T, Byrne E, Russell J, Potts H. Adoption and non-adoption of a shared electronic summary record in England: a mixed-method case study. BMJ. 2010;340:c3111.
22. Infoway. http://www.infoway-inforoute.ca/lang-en/standards-collaborative
23. MedCom. A Danish healthcare network in two years. Odense: Danish Centre for Health Telematics; 1996. http://www.medcom.dk/publikationer/publikationer/MedCom1-engelsk.pdf
24. Johansen I, Henriksen G, Demkjær K, Bjerregaard Jensen H, Jørgensen L. Quality assurance and certification of health IT-systems communicating data in primary and secondary health sector. Presentation at MIE 2003, St Malo.
25. MedCom – IT brings the Danish health sector together. November 2008. http://www.medcom.dk/dwn2440
26. Blumenthal D. Stimulating the adoption of health information technology. N Engl J Med. 2009;360:15.
27. Daschle T. Critical: what we can do about the healthcare crisis. New York: Thomas Dunne Books; 2008. p. 35–6.
28. Physicians use of EHR systems 2014. AmericanEHR and AMA 2015. http://www.americanehr.com/research/reports/Physicians-Use-of-EHR-Systems-2014

Chapter 2
Why Interoperability Is Hard

Abstract This chapter explores some of the reasons why healthcare interoperability is hard and why standards are needed. Interoperability can be looked at as layers (technology, data, human and institutional) involving different types of interoperability, technical, semantic, process and clinical. Standards are needed to tame the combinatorial explosion of the number of links required to join up systems, but usually require translation to and from an interchange language. Users and vendors are not always incentivised to interoperate. Apparently simple things such as addresses are more complex than they seem. Clinical information in EHRs is inherently complex, but complexity and ambiguity in specifications creates errors. Any interoperability project involves change management.

Keywords Interoperability definition • Interoperability layers • Technical interoperability • Semantic interoperability • Process interoperability • Clinical interoperability • Interoperability standards • Combinatorial explosion • Electronic health records (EHR) • Translation • Rosetta Stone • Problem-oriented medical records (POMR) • ISO 13606 • Name • Address • Discharge summary • Clinical laboratory reports • GP2GP • Complexity • Errors • Change management

Layers of Interoperability

Few large health IT projects manage to achieve all of their objectives, especially when it comes to interoperability. This chapter looks at some of the reasons why health interoperability is so hard to get right and why standards are essential.

The benefits of joined-up healthcare depend on safe, secure and reliable interoperability to provide the right information when and where it is needed.

We can think of interoperability as having four layers:

- Technology
- Data
- Human
- Institutional [1].

These are not necessarily listed in sequential order. For example, air traffic control is a good example of successful interoperability, where standardisation was achieved first at the human and institutional layers, and the data and technology layers came much later.

In healthcare interoperability each of these four layers is important. Governments, providers and vendors need to work together to achieve good results, especially at the institutional level, where barriers to interoperability are exacerbated by privacy concerns, technology lock-in and lack of appropriate incentives. The art is to enable diversity while ensuring that systems work together in the ways that matter most. We need to aim for optimum interoperability.

> One of the tricks to the creation of interoperable systems is to determine what the optimal level of interoperability is: in what ways should the systems work together, and in what ways should they not [1].

Definitions

The term interoperability means different thing to different people. The *HIMSS Dictionary of Healthcare Information Technology Terms, Acronyms and Organizations* lists 17 definitions from the strictly technical to those that emphasize social, political and organizational factors [2].

The most widely used definition is:

> Interoperability is ability of two or more systems or components to exchange information and to use the information that has been exchanged (IEEE 1990) [3].

This includes both the exchange of information, which is *technical* interoperability and the capability of the recipient to use that information, which is *semantic* interoperability. A third concept, pertaining to the actual use of the information, is *process* interoperability to which we would add *clinical* interoperability (Fig. 2.1) [4].

Technical Interoperability

Technical interoperability moves data from system A to system B, neutralizing the effects of distance. Technical interoperability is domain independent. It does not know or care about the meaning of what is exchanged. Information theory, which shows how it is possible to achieve 100% reliable communication over a noisy channel, is the foundation stone of technical interoperability [5]. Technical interoperability is now taken for granted. This is the technology layer.

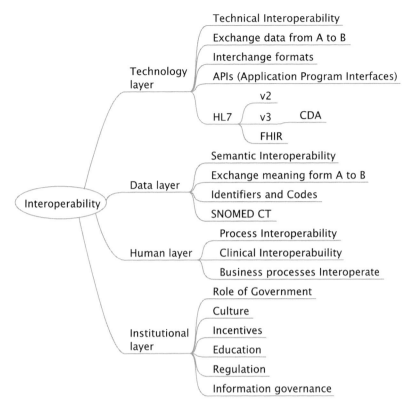

Fig. 2.1 Layers of interoperability

Semantic Interoperability

Dolin and Alschuler define semantic interoperability as *"the ability to import utterances from another computer without prior negotiation and have your decision support, data queries and business rules continue to work reliably against these utterances"* [6]. Both the sender and recipient need to understand the same data in the same way. Semantic interoperability allows computers to share, understand, interpret and use data without ambiguity. Semantic interoperability is specific to domain and context and requires the use of unambiguous codes and identifiers. This is the data layer.

Process Interoperability

Process interoperability is achieved when human beings share a common understanding across a network, business systems interoperate and work processes are coordinated. People obtain benefits when they use information originating elsewhere in their day-to-day work. The importance of re-engineering work processes to take full advantage of electronic systems has long been recognised, but the lessons have not yet been well learnt in healthcare. This is the human layer.

Clinical Interoperability

In healthcare we need to focus on clinical interoperability, which is a subset of process interoperability. Clinical interoperability can be defined as:

> Clinical interoperability is the ability for two or more clinicians in different care teams to transfer patients and provide seamless care to the patient [7].

On its own exchanging data achieves nothing. Only when people use new information in some way that differs from what they would have done without it, can we obtain different results and outcomes. In healthcare clinical interoperability is what matters. This requires changes in workflow and in the way clinicians and clinical microsystems function at a fine level of detail.

The more we understand these different aspects of interoperability, the less likely we are to underestimate the work required to make health systems interoperable. Technical, semantic, process and clinical interoperability are interdependent, and all are needed to deliver significant business benefits.

Interoperability can save an enormous amount of duplication, waste and errors but relatively few of those responsible for commissioning and paying for healthcare know enough about the subject and what is required to achieve the business benefits. This is the institutional layer involving culture, education, regulation and incentives.

Why is interoperability successful in some contexts and not in others? One explanation is to consider the individual and institutional self-interest. It may be in the vendor's financial self-interest to insist on using a proprietary non-standard interface, even though they know well that this will ultimately create an interoperability nightmare. This is technical lock-in. Similarly it can be in a provider's financial self-interest not to share patient information with providers they regard as competitors, thus creating patient lock-in.

In *The Tragedy of the Commons*, it is in each farmer's interest to add an extra cow to the common grazing land, even though that degrades the pasture as a whole [8]. The selfish farmer gains 100% of the benefit from his extra cow, but the downside is shared between everyone.

The traditional solution to this type of problem is for governments to establish an independent regulator, to enforce rules and regulations and impose supervision or oversight for the benefit of the public at large. The regulator would specify what standards should be used within their geographical area, in full and open consultation with all concerned interests, covering interoperability and related security and privacy issues. Many other aspects of healthcare and communications industries have independent regulatory agencies. The case for a regulator to enable healthcare interoperability and related information governance provisions is strong.

Why Standards Are Needed

Part of the problem with standards is not that there are so many to
that we have failed to adequately incentivise the use of those we
problem is that there is no one, such as a regulator, with the power to make deployment happen in an ordered way. Standards that are not deployed are a waste of time
and effort.

An alternative view is that the standards available have been overly complex and
expensive to implement and maintain. This view has led to the development of
FHIR (Fast Healthcare Interoperability Resources), see Chap. 18.

The volume of transactions in healthcare is mind-boggling. For example, in 2007
a single EHR system at one large hospital (the Mayo Clinic in Rochester, Minnesota)
processed more than 660 million HL7 messages a year, about two million messages
a day [9]. This indicates the size of the prize to be won.

Examples of transactions include:

- Requests for tests and investigations.
- Prescriptions for medicines and treatment.
- Orders for nursing care, equipment, meals and transport.
- Test reports.
- Administration notifications for changes in patient details and scheduling.
- Letters from one clinician to another such as referral, clinic and discharge
  letters.
- Transfer and merging of medical records.
- Aggregate information for management, audit and monitoring.
- Commissioning, billing and accountancy.

Combinatorial Explosion

The number of links needed to connect n different systems increases according to
the formula:

$$\text{Number of links} = \frac{n(n-1)}{2} = \binom{n}{2}$$

Linking two nodes needs only a single interface, which can be agreed quite easily
by a couple of people sitting round a table. Linking 6 nodes requires 15 interfaces,
and linking 100 nodes requires 4950 interfaces. This is known as a combinatorial
explosion.

The center of the star at the right of the figure below (Fig. 2.2) indicates a single
specification being used for linking six domains. This replaces the 15 separate links,
each of which could be different, shown on the left hand side.

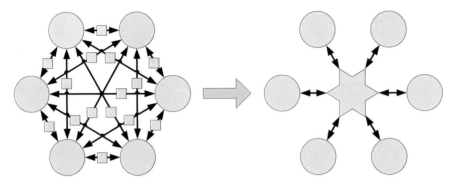

Fig. 2.2 The benefits of one standard

Translations

Every type of computer system stores data internally in a different way. This means that, in order to communicate, data has to be translated from one internal language or format into another. Linkage involves translating to a standard wire format that is understood by each party. This requires two translations, first from the native language of the sender to the wire format and then from the wire format to the native language of the recipient. HL7 provides a common *lingua franca* to do just this.

The Rosetta Stone from ancient Egypt, now in the British Museum, provides an analogy. This contains the same proclamation in three languages, used by the priests (Hieroglyphic), the court (Greek) and the people (Demotic). In our context, the three languages could be those used by a sending system, the receiving system and a common wire format used for information interchange, such as HL7. The meaning of a message is precisely the same in each language but the notation is quite different. The inscribers of the Rosetta Stone only needed to perform their translation once, but in computer interoperability, every message instance has to be translated from one format to another without error.

The choice of interchange language is not sufficient to ensure interoperability. Each transaction must be defined in stringent, unambiguous detail as part of a complete, consistent, coherent and computer-readable set of specifications.

Electronic Health Records

The original vision of an electronic health record (EHR) was a collection of statements, which provide a record of what clinicians have heard, seen, thought, and done [10]. However, the health record is also the key source of information used to support claims for payment, to defend legal actions and for research. This has led to much information being added for bureaucratic financial, legal or research purposes, making it harder to use as a clinical tool.

However, the EHR is not really a collection of facts, but rather a set of observations about a particular patient, made by clinicians, each at a specific time and place and context. It is quite possible for two statements about the same event to disagree with each other. Such disagreements can often be resolved if the context or provenance of each statement is known. This is metadata, typically covering what type of thing it is, who stated it, when and where. As with a work of art, a statement without provenance or context is of doubtful validity.

The ISO 13606 Reference Model for electronic health record communication sets out a hierarchical structure for clinical information [11, 12] (Fig. 2.3).

- **EHR**: The electronic health record for one person.
- **Folder**: High-level organisation of the EHR. Folders may be used as containers for grouping compositions by episode, care team, clinical specialty, condition or time period.
- **Composition**: A composition is a set of information relating to a specific clinical encounter, session or document. Each composition shares common metadata such as the author, subject (patient), date/time and location. Progress notes, laboratory test reports, discharge summaries, clinical assessments and referral letters are all examples of compositions. Once created a composition is immutable (cannot be changed). The EHR is made up of compositions. Compositions may be grouped together into folders, and sub-Folders.
- **Section**: A section is a grouping of related entries within a composition usually under a heading such as history, risk factors, medication, examination findings, diagnoses, investigations and plans, reflecting the workflow and consultation process. Sections may have sub-sections.
- **Entry**: Each entry is a statement about a single observation, evaluation or instruction. Think of it as a single row in a spreadsheet. Examples include the entries about a symptom, test, problem or treatment. Entries may be grouped together in sections. Each composition comprises a number of entries. Entries are also known as clinical statements.
- **Cluster**: Nested multi-part data structures including tables and charts. Related elements may be grouped into clusters. For example, systolic and diastolic blood pressures are separate elements, but are grouped into a cluster (eg 140/90), which represents one item in an entry.
- **Element**: The leaf node of the EHR hierarchy is an Element, which is a single data value, such as systolic blood pressure, a drug name or body weight.
- **Data Value**: Data types for instance values, such as codes, measurements with units etc.

Problem-Oriented Medical Records

Larry Weed's Problem Oriented Medical Record (POMR) was one of the first attempts to structure the patient record [13].

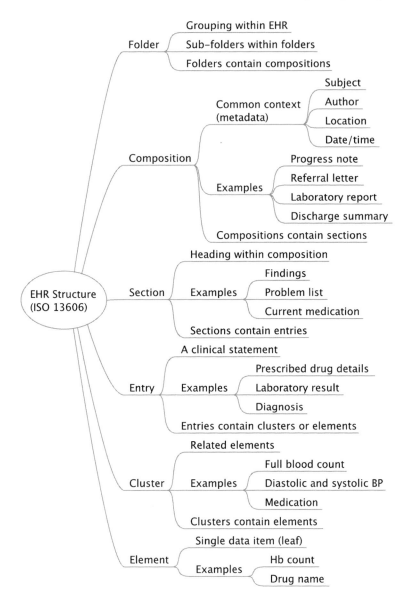

Fig. 2.3 Hierarchical structure of the EHR

The POMR divides the record into two parts, progress notes and database (see Fig. 2.4).

Progress notes are organised under problems. A problem is anything that causes concern, not only a diagnosis. The problem list is a list of all the patient's problems indicating those that are active and those that have been resolved. Each progress note has a problem heading and four sub-headings, using the acronym SOAP:

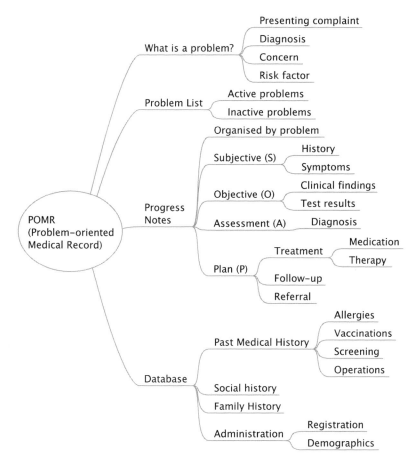

Fig. 2.4 The structure of problem-oriented medical records

S **Subjective**, meaning the information provided about history and symptoms by the patient or relative.

O **Objective**, meaning information obtained by direct examination of the patient or from clinical investigations (laboratory, radiology etc.)

A **Assessment**, meaning the clinician's assessment about what is the matter with the patient (diagnosis), prognosis etc.

P **Plan**, meaning the future plan of action, including investigations and treatment (drug prescriptions, physiotherapy, surgery and so on). Drugs prescribed are also listed in a separate medication list. This section is a problem-specific care plan.

The database covers the patient's social, family and past medical history.

From 1967 to 1982 Weed was funded by the US government to implement a problem-oriented electronic patient record system known as PROMIS, using first-generation touch-screen terminals. This remarkable pioneering project was implemented for many years on medical and gynecological wards at the University of Vermont, but it was withdrawn after federal funding ceased [14]. Although not a long-term success, PROMIS was one of the most influential systems ever.

Weed went on to develop problem-knowledge couplers (PKC), which match detailed patient information with an extensive medical database to provide guidance tailored to individuals [15, 16].

The Devil is in the Detail

The grand vision of joined-up healthcare is predicated on the notion that patient records can be shared electronically between clinicians across specialties. Historically, this has proven difficult, in part because every clinical specialty has its own way of thinking and working.

It is hard to share information between different computer applications even within the same specialty. Each computer application stores data in a different way. Furthermore, even within one specialty, the information is more heterogeneous than may be expected. For example, data collected at a routine outpatient visit differs greatly from that for an elective surgical operation or an emergency admission.

GP2GP

The GP2GP project in England illustrates this point. All NHS patients have a life-long medical record, which follows them when they move from one GP to another. In an ideal world, each patient's records would be sent electronically from their old practice to the new to avoid the need to re-enter information.

The GP2GP project set out to do just that; although the project's leaders recognized that it could a poisoned chalice [17]. The work was as difficult as predicted and each record has to be checked before transmission and on receipt. Early versions met problems, which took several years to overcome. For example there was a technical size limitation of 5 Mb and 99 attachments, which excluded its use on those patients with the most complex records (about 15%); sending practices were required to print copies of all attachments, which was onerous; and it was difficult to quickly integrate records of patients who live in more than one place, such as students, and move back and forth between practices.

After 15 years of effort these problems have been resolved and it is now a contractual requirement for all GPs to use GP2GP. About two million life-long medical records are being exchanged every year in England.

Much of the hard work of health interoperability and digital health in general lies in teasing out the detail of hundreds and thousands of different use cases. Those who pay for IT services like to focus on high volume transactions and pass over the specific needs of smaller specialties. Yet, the common stuff is often not the most important clinically.

It is helpful to distinguish between information that needs to be processed by computer and what needs to be read and understood by human users. Computer processing is required when data has to be matched, retrieved or counted. This type of information should be structured, complete, unambiguous and validated.

Human readers need information in a form they can understand. This should be easy to read and accompanied by supporting contextual data such as who wrote it, when and where, and for what purpose. Humans are good at judging the significance of small discrepancies, while digital computers are unforgiving of a single unexpected bit.

On first thought, names, addresses, clinical laboratory reports and discharge summaries may each appear to be fairly homogeneous, but this is not so.

Names

A person may have several names and several addresses, which they can change at will. A woman may use her maiden name in one context and her married name in another. One person may use several addresses (home, work, previous, holiday etc.) and each address is likely to be associated with different sets of people, such as family members, friends or colleagues. The order in which names and addresses are written varies substantially between countries.

Addresses

An address is a label used to reference a geographical object such as a property through the use of identifiable real-world objects [18]. The postal address is used for the delivery of mail. This is a routing instruction leading to the property. However, most geographic objects have addresses, some of which are postal addresses, but some never receive mail. These include:

- Domestic, commercial and industrial properties.
- Public buildings (schools, hospitals, prisons, halls, leisure facilities, public toilets).
- Churches and monuments.
- Places where events take place (sports fields, parks).

Locations need to be identified and accessed for a range of purposes, which include:

- Uniquely identify people via their place of residence.
- Identify customers and potential customers.
- Identify where people live and work, for planning public services.
- Delivery points for goods or services.

- Levy taxes on people and organizations.
- Property registration and transactions.

Addresses normally have a structure using a nested set of spatial units:

- Sub-unit within a building or property.
- Building or property within a street.
- Street (but some rural areas do not have street names).
- One or more geographic areas (locality, town, county etc.)
- Country.

Part of an address is often abbreviated by a code (eg a postcode or area code).

The definition of each level varies considerably from country to country. Many buildings, both large and small, may have multiple addressable objects within them. Examples include: bed-sits with shared bathrooms and/or kitchen facilities, shared houses, student and worker accommodation, residential care homes for the aged and disabled, flats with third party access to the inside of the property for delivery purposes, flats where there is a single point of delivery for all residents, business premises with residential owners, managers or staff, shared business properties with no particular differentiation (normally associated companies), businesses each with their own private area but shared reception and toilet facilities, and self-contained businesses with one shared entrance. There are no clear rules about how these types of premises should be recorded.

The life cycle of an address is yet another complication. Addresses are often needed before the building itself is built. For example, temporary addresses are often allocated during the planning or construction phase of new developments. Changes to addresses can occur due to merging of two or more properties, extension, subdivision or demolition of a property, change of property number or name, occupancy or use and the names of areas used in the address (for example due to administrative area reorganization).

Discharge Summaries

There is enormous variety in the letters sent from hospitals to GPs, which are sometimes collectively referred to as discharge summaries [19]. Consider the following:

- An elderly patient discharged home after recovering from a fractured femur after a fall.
- Mother and baby following birth.
- A family at the end of a course of counseling by a clinical psychologist.
- Initial consultation report from an ophthalmologist notifying a proposed operation for cataract.
- Notification that a patient has been diagnosed with cancer and outlining the treatment plan.
- Discharge from hospital following hip replacement.

Clinical Laboratory Reports

Clinical laboratory reports differ greatly according to type of laboratory. Histologists examine cells under a microscope; microbiologists cultivate colonies of bacteria; hematologists count blood cells, and clinical chemists measure the intensity of color changes when chemicals are added. The only commonality is that they all work with specimens extracted from patients. But sometimes the requester supplies the sample, sometimes the sample is taken by the laboratory; sometimes the patient is required to be present in person.

Complexity Creates Errors

Building a single link to exchange data between two computers is relatively straightforward. Everyone sits around a table and works out what to do. This works fine for very small projects, where each person is co-located, but does not scale.

The alternative is to provide rigorous implementation guidelines, but these often grow complex and voluminous. For example, the implementation guidelines for the NHS Pathology Message Implementation Project (a successful national project to send clinical chemistry and hematology laboratory test reports to all GPs in England) comprise almost a million words, about ten times the length of this book. The endeavor to be rigorous creates errors caused by the sheer length and complexity of the specifications.

Another problem arises when the domain experts (such as doctors, nurses and managers) do not understand these specifications due to the complexity of language or the time it takes to read them.

Errors multiply according to:

- The probability of misunderstanding any part of the specification. This depends on difficulty of language and terms used as well as the level of domain and technical knowledge of participants. It is rare to find people with adequate technical knowledge and domain knowledge.
- The length of specification. In a long specification, the same idea may be presented in different ways in two places, but each may be understood differently. If large blocks of similar but not identical information are replicated in different sections, key differences can be missed.
- The number of options permitted. Optionality increases the chance of error. The easiest specifications to implement are those that require precisely one instance of each item, without optionality or multiplicity.
- The number of times different implementations need to be made. Each implementation on another system involves mapping or translating the specification into the local implementation language.

All of these issues lead to error [20]. Errors increase costs and reduce quality, create delays and hit profits and reputation. Successful specifications avoid errors by limiting scope, being easy to understand, relatively short and simple, with few if any options.

Many problems could be avoided by more thought and preparation by those responsible for the specification. As deadlines loom, it is all too easy to be vague or offer the implementers a choice of options depending on the local context. This simply increases errors by pushing the problems further down the road.

Users and Vendors

Often, users and vendors genuinely believe that they are in full agreement until the moment when users try to use the final product. Few users fully understand what they want, let alone what other parties can or cannot provide. They seldom commit enough time or effort up front to fully review written requirements specifications. They then won't commit to these, and insist on new features after the schedule and budget have been fixed.

End users are usually technically unsophisticated, do not understand the software development lifecycle and are unable to perform the sort of scrutiny that is often required of them. To do this users would need a much higher level of education in digital health than has been provided in the past.

Vendors are also guilty. Managers often try to shoehorn the users' requirements to fit their existing systems or patterns, believing that it will be quicker, cheaper and lower risk to re-use what already exists, while failing to grasp that the user really needs something else and will never be happy without it.

Vendors often lack specialized domain knowledge and do not understand the user's business processes at the required level of detail to appreciate that their preferred solution will not fit. They focus attention on high volume aspects of digital health, which they understand well, and cover up their lack of knowledge of the idiosyncrasies of every specialty by requiring users to check and sign off on specifications that neither party fully understands.

Shared meaning between computers requires shared understanding between the human participants. As an analogy consider the purchase of a new kitchen. The kitchen designer prepares a plan of the new kitchen. This plan is checked, reviewed and signed off by the customer and this becomes the basis of the contract. This plan uses a precise technical notation, which also provides a means of communicating precisely the user's needs to the implementer (manufacturer), in a form that can be understood by both customer and manufacturer. Manufacture only begins work after the customer has agreed the specification. The challenge in interoperability is similar but even harder; it is to ensure understanding vertically between users and developers and also horizontally across business and clinical processes in different locations (Fig. 2.5).

Fig. 2.5 Reasons why interoperability is hard

Change Management

One of the main lessons that come out of studies on health innovation is that change takes time. Progress often comes from modest incremental changes, which accumulate over time to large gains, over the long, not short, term. Change is driven by a combination of technical, policy, economic, clinical or managerial factors often working together on different actors and at different levels within organisations [21].

Common factors include technology developments, such as Wi-Fi and mobile technologies, the culture of how clinicians and managers work together, and detailed design of patient-centered pathways to give better outcomes at lower cost. Data sharing, including appropriate performance metrics, enables feedback, benchmarking and comparisons of performance and variation. Frontline support enables change, such as quality improvement, education and training and dissemination of best practice. Financial carrots and sticks incentivise and focus on the need for change.

Kotter has identified eight common errors and has proposed remedies for managing innovation and change (Table 2.1) [22].

Table 2.1 Errors and remedies in management of change

	Common errors	Proposed remedy
1	Not establish a great enough sense of urgency	Establish a sense of urgency that something really has to be done (it even helps to have a crisis)
2	Fail to create a sufficiently powerful guiding coalition	Form a powerful guiding coalition of key stakeholders with position power, expertise, credibility and leadership
3	Underestimate the power of vision	Create a vision and strategy that is desirable, feasible, focused, flexible and can be communicated in less than five minutes
4	Under-communicate the vision by a factor of 10 (or 100 or even 1000)	Communicate the change vision simply using examples. This needs to be repeated over again in multiple forums; address apparent inconsistencies, listen and be listened to
5	Permit obstacles to block the new vision	Empower employees to modify structures and systems to bring about the changes required. Changes will be needed in process, workflow and information systems
6	Fail to create short-term wins	Generate short-term wins that are visible, clearly related to the change effort and build momentum
7	Declare victory too soon	Consolidate gains, reduce interdependencies and produce more change
8	Neglect to anchor changes firmly in the corporate culture	Anchor new approaches in the culture. This comes last, not first

References

1. Palfrey J, Gasser U. Interop: the promise and perils of highly interconnected systems. New York: Basic Books; 2012.
2. HIMSS. HIMSS dictionary of healthcare information technology terms acronyms and organizations. Chicago: HIMSS; 2006.
3. IEEE. IEEE standard computer dictionary: a compilation of IEEE standard computer glossaries. New York: Institute of Electrical and Electronics Engineers; 1990.
4. Gibbons P et al. Coming to terms: scoping interoperability in healthcare. Final. HL7 EHR Interoperability Work Group, February 2007.
5. Shannon C. A mathematical theory of communication. The Bell Syst Technical J 1946; 27: 379–423 and 623–56.
6. Dolin R, Alschuler L. Approaching semantic interoperability in health level seven. JAMIA. 2011;18:99–103.
7. Grieve G. Dynamic health IT. Blog 2 Dec 2015 http://dynamichealthit.blogspot.co.uk/2015_12_01_archive.html
8. Hardin G. The tragedy of the commons. Science. 1968;162(3859):1243–8.
9. Anthony J. Personal communication 2008.
10. Rector A, Nowlan W, Kay S. Foundations for an electronic medical record. Methods Inf Med. 1991;30:179–86.
11. Health Informatics – Electronic health record communication – Part 1: Reference Model ISO 13606–1:2008.
12. Kalra D. Electronic health record standards. Year Book Med Inform, IMIA 2006; 45:136–44.
13. Weed L. Medical records that guide and teach. NEJM. 1968; 278: 593–9 and 652–7.
14. Schultz J. A history of the PROMIS technology: an effective human interface. In: Goldberg A, editor. A history of personal workstations. Reading: Addison Wesley; 1988.
15. Weed LL. Knowledge coupling: new premises and new tools for medical care and education. New York: Springer; 1991.

16. Weed LL, Weed L. Medicine in denial. Charleston: Createspace; 2011.
17. Purves I, Fogarty L, Markwell D. The Holy Grail or poisoned chalice: the GP-GP record transfer project. Newcastle: HIRI; 2001.
18. Walker R. A general approach to addressing. ISO Workshop on address standards: considering the issues related to an international address standard. Copenhagen. 2008: 23–7.
19. Benson T. Why industry is not embracing standards. Int J Med Inform. 1998;48:133–6.
20. Benson T. Prevention of errors and user alienation in healthcare IT integration programmes. Inform Prim Care. 2007;15(1):1–7.
21. Alderwick H, Robertson R, Appleby J, Dunn P, Maguire D. Better value in the NHS: the role of changes in clinical practice. London: The Kings Fund; 2015.
22. Kotter J. Leading change. Boston: Harvard Business School Press; 1996.

Chapter 3
Models

Abstract Models provide a way to describe systems as they are now and how we want to change them. Formal methods reduce the probability of misunderstanding. Different types of model are needed at different parts of the project life cycle. Conceptual models differ in detail from implementation-specific models. The early stages use scope statements, storyboards and requirements specification. Examples are provided about Fred and his dog and colorectal cancer referral.

Keywords Models • Graphical models • Model-driven architecture • Modeling maturity levels • CEN TC/251 • Reference model • Scope statement • Storyboard • Lifecycle • Requirements specification • Conceptual design • Esther in Jönköping • Fred and his dog • Technology-specific specification • Colorectal cancer referral

The Importance of Models

Models play a vital part in digital health and interoperability in particular. Models allow us to describe the system as it is now and how we want it to be, before building it, to ensure that everyone understands what is needed and how these needs can be met.

The word *model* has many meanings, but here we are referring to simplified descriptions of either the real world or proposed systems used to help design systems.

Models play a central role in interoperability and a good understanding of modeling is an important foundation skill.

> *Models define the way we learn about the world, interpret what we see, and apply our knowledge to affect change, whether that is through our own actions or through the use of technology like a computer* [1].

People can learn to understand models quite easily, although it is much harder to create good models than to read them. Every model is a simplified representation of aspects of either the real world, such as maps, or the world we wish to create, such as the blueprints used by architects and engineers.

Formal graphical models and diagrams are usually more precise than unstructured narrative. For example, to describe a geographical location, you can:

T. Benson, G. Grieve, *Principles of Health Interoperability*,
Health Information Technology Standards, DOI 10.1007/978-3-319-30370-3_3

- Tell someone face-to-face.
- Write a text narrative.
- Provide a structured report with headings and sections.
- Draw ad hoc diagrams and pictures.
- Provide a map using standard conventions.

These are listed in increasing order of preciseness and their ability to convey meaning reliably. The same ranking applies to specifications, which is why engineers use blueprints not text descriptions to specify how a machine is to be built. Computer engineers and analysts also use graphical models and diagrams.

A model may be thought of as the complete collection of diagrams and supporting documentation that describes a system. Each diagram is one view into the whole model. When designing any complex system, the designer may produce hundreds of different diagrams. Each diagram has a specific purpose within the project. Each diagram shows certain aspects of a situation and everything else is ignored. This simplification provides the power of diagrams, by making the situation understandable, as well as their weakness, because each diagram has a limited scope and things have to be left out.

Computer-based modeling tools help enormously, because each component is recorded once only and is reused unchanged on every diagram that uses it. This ensures that the model and its diagrams are coherent and consistent, while making it much quicker and easier to make changes.

Model Driven Architecture

The Object Management Group (OMG), which is responsible for information modeling standards, defines a framework for the development of object-oriented software, known as Model Driven Architecture (MDA) with four sequential model types [2]:

- Computational-Independent Model (CIM).
- Platform-Independent Model (PIM), which describes the conceptual design of a system.
- Platform-Specific Model (PSM), which specifies the implementable design.
- Code, the actual software code written, also referred to as wire-format.

A feature is formal mapping between each type of model, which provides traceability between each stage in the process.

Six modeling maturity levels (Table 3.1) can be used to classify the role of modeling in any software development project [3].

There is some debate about whether Level 5 is necessarily better than Level 4, but most observers would say that software projects should aim for at least Level 4.

Table 3.1 Modeling maturity levels

0 No Specification	The specification is not written down but kept in the minds of the developers. At this level we find conflicting views between developers and users and it is impossible to understand the code if coders forget what they did or leave (and they always do sooner or later). Sadly this level is all too common even today
1 Text specification	The software is specified using natural language in one or more documents. Such specifications are invariably ambiguous because natural language is ambiguous; it is almost impossible to keep this type of specification up to date when code is changed
2 Text with diagrams	A text specification is enhanced with diagrams to show some of the main structures of the system. This is easier to understand, but still difficult to maintain
3 Model with text	The specification of software is developed in a model with multiple diagrams. In addition to these diagrams, natural language text is used to explain detail, background and motivation of the project, but the core of the specification lies in the model
4 Precise models	The specification of the software is written down in a model. Natural language can still be used to explain the background and motivation of the models but it takes on the same role as comments in source code. At this level, coders do not make business decisions and development may be facilitated by direct transformation from model to code
5 Models only	The model is precise and detailed enough to allow complete code generation. The code generators at this level have become as trustworthy as compilers, and so developers seldom need to look at the generated code

Models in Interoperability Standards

In the early 1990s the European working group responsible for healthcare communications and message standards (CEN TC251) adopted object modeling as a core foundation for defining health interoperability standards.[1] One problem was how to choose an interchange format (syntax) for European interoperability standards. Before the development of XML (extensible markup language, 1996), there were several candidates. It was a time of *syntax wars*, with a range of competitors and people would quip: *the nice thing about standards is that there so many to choose from* [4].

CEN set up a project team to recommend a solution. This created a small set of technology-neutral specifications called General Message Definitions (GMD) using an object model, similar to UML. For each of five target interchange languages (syntaxes), the project team created a set of corresponding Implementable Message Specification (IMS) to test which worked best [5]. The unexpected conclusion was that all of the target interchange languages were adequate for current needs. The selection of an interchange format is mainly a political decision and not a major obstacle for realizing healthcare message exchange.

---

[1] Author TB was convenor of CEN TC251 WG3 (health communication and messaging) from 1991 to 1997 when the work described in this section was done.

While doing this work, the team recognized the value of their GMD models and suggested that syntax-independent specifications should become the core deliverables of future message standardization. GMDs could be implemented in any syntax of choice, such as XML, EDIFACT or HL7 v2.

This approach was picked up by other standards development organizations, including HL7, and evolved into three main types of specification [6, 7]:

- A single domain-wide model, which acts as a reference for all others. This is referred to as a reference model. The HL7 v3/RIM and the ISO 13606 reference model are good examples. FHIR Resource definitions also fall into this category.
- Technology-independent specifications, each of which is a constraint on the domain wide model. These have various levels of granularity from very general to stringent. Examples include HL7 v3/RMIMs, CDA Templates, FHIR Profiles and archetypes.
- Implementable message specifications, which are mappings from the technology-independent message specifications into the selected syntax, such as XML or JSON.

These high-level ideas are summarised in Fig. 3.1.

Lifecycle

The lifecycle of an interoperability project is similar to that of any software development project. Figure 3.2 enumerates the following stages:

1. Scope and objectives, what is to be done, including the business case, objectives, scope and boundaries.
2. Process analysis and design, 'as is' and 'to be'.
3. Conceptual design and specification.
4. Implementable technology-specific specification.
5. Software application coding and development.
6. Testing and certification.
7. Deployment including user education, data migration and installation.
8. Support and maintenance.

This chapter considers the first three stages. Much of the remainder of the book is concerned with the fourth stage – implementable technology-specific specification. The preliminary stages are vital because mistakes made here are expensive to correct later.

The first two stages constitute the preliminary business analysis. The process described is iterative, with continual feedback between work on finalizing the scope statement, business process analysis, storyboards and glossary development. This stage may be fairly short (in comparison with the total project), but it is critical. These early deliverables should not be frozen but need to be reviewed regularly throughout the project and updated as necessary.

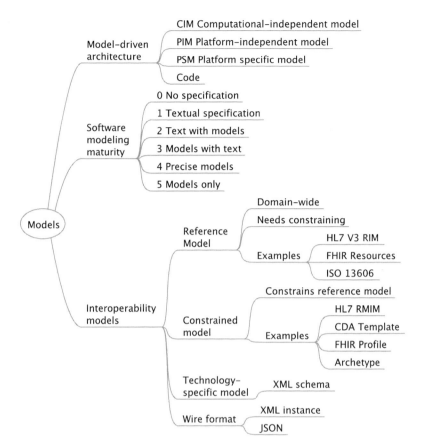

Fig. 3.1 Modeling concepts

Business processes may be described using UML (Unified Modeling Language). More detail of UML is provided in Chap. 4.

Scope

The scope statement provides the big picture of what a project is all about, including the case for action, objectives, scope delineation and reporting structures. This is sometimes called a Project Initiation Document (PID) and it provides a management summary of the project. The PID summarizes why the project is needed, what it should achieve and what is excluded, showing the boundaries and responsibilities of the system of interest and who is responsible for what. Problems can often be traced back to this stage, particularly if developers set the boundaries or responsibilities differently from those expected by the users.

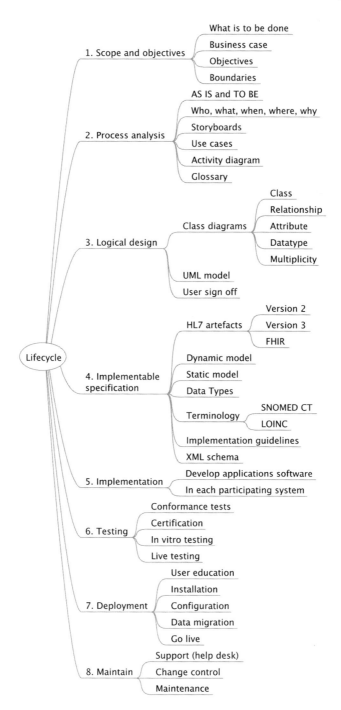

Fig. 3.2 Project lifecycle

Any change in scope must be incorporated into the scope statement, to prevent scope creep. However changes in scope are an inevitable part of most projects as more is understood about the domain and the user needs.

The *case for action* is a concise, comprehensive and compelling statement of why the project is needed. It describes the context, the problem, user needs and the consequences of doing nothing.

The *objectives* provide a clearly stated focus of what the project is about. These goals should be SMART:

- Specific, clearly stating what you want to achieve.
- Measurable, with a means of measuring whether or not you are achieving the objectives.
- Attainable with the available resources.
- Relevant to the organisation's needs.
- Time-bounded, with dated milestones.

Scope *delineation* shows the project boundary – what is in or out of scope. This may also include constraints such as standards and other work that must be used, as well as what is explicitly outside the scope.

Storyboards

Storyboards provide an intimate view of how people use the system to obtain value. They capture domain knowledge and provide specific detail, in contrast to the high level of the scope statement. Storyboards show the actors involved, the information flow and real world situations where the services may be used.

Each storyboard is a story, told in the present tense, describing how a set of named actors use the system to carry out a single instance of a task. Storyboards do not contain options. If there are two ways of doing something then two storyboards are needed, one for each.

Storyboards describe each of the ways the system may be used, including typical and extreme cases. They provide context that everyone can understand and a starting point for developing test data.

Each storyboard should be written by a domain expert, checked by a business analyst, revised and discussed in a group with users to ensure that it captures the process accurately.

Storyboards of the *as is* situation do not change, but *to be* storyboards may need to be updated repeatedly as the system design evolves.

An example of a storyboard for breast cancer triple assessment is:

Jane Sharp attends the One Stop Breast Clinic, having been referred urgently by her GP after noticing a lump in her breast. She sees Dr Lee who takes her history (presenting symptoms, appropriate medical and family history) and performs a physical examination. Jane then proceeds to mammography, where a fine needle aspirate (FNA) is collected. A radiologist reports the mammograph and a pathologist reports the FNA. Jane is asked to return later to hear the result of these tests. On her return she is relieved to hear that the results are negative.

The storyboard development stage provides an opportunity to gather supporting information such as the forms and information sets currently being used, the functionality of existing computer applications including their data dictionaries, relevant national and international standards, regulatory and information governance constraints, future developments and the potential to simplify business processes.

Esther

The Esther Project in Jönköping, Sweden, illustrates the value of storyboarding in healthcare [8]. Using the story of a virtual patient called Esther, they were able to achieve the following improvements:

- Hospital admissions fell 22% from 9300 in 1998 to 7300 in 2003.
- Hospital days for heart failure patients fell 28% from 3500 in 1998 to 2500 in 2000.
- Waiting times for referral appointments with neurologists fell 84% from 85 days in 2000 to 14 days in 2003.
- Waiting times for referral appointments with gastroenterologists fell 71% from 48 days in 2000 to 14 days in 2003.

Fred and His Dog

Dr Mary Hawking, a former GP and sister to Stephen Hawking, has developed a virtual patient, Fred, to help people understand the complexity of coordination across primary, secondary, community, social care and voluntary sectors [9].

Fred is an independent-minded 70 year-old widower who lives alone with his dog. He has multiple problems – insulin-dependent diabetes, rheumatoid arthritis which means he can't give his own insulin, peripheral neuropathy as a complication of his diabetes, and chronic obstructive pulmonary disease (COPD). He still smokes and recently needed to have his left leg amputated due to peripheral vascular disease and gangrene complications of diabetes. He is wheelchair bound, waiting for limb fitting and is understandably depressed.

Fred is on four clinical care pathways (for diabetes, arthritis, COPD and depression). Each pathway requires home visits or transport to hospital or GP surgery because he is housebound.

- His diabetes pathway involves his GP and practice team, community nurses, diabetic specialist nurse, diabetic consultant, diabetic retinopathy screening and diabetic foot services, plus vascular surgery, limb-fitting and wheel chair services for his complications.
- The COPD pathway involves the GP and practice team, respiratory specialist nurse, respiratory consultant, pulmonary rehabilitation and the Stop Smoking Service (who are thinking of giving up).

- The rheumatoid arthritis pathway involves his GP and practice team, the rheumatology specialist nurse, rheumatology consultant, and rheumatology specialist physiotherapy.
- His depression is managed by his GP and practice team with the community mental health team for the elderly (he's over 65), which includes a psychiatrist, community psychiatric nurse and a psychiatric social worker.

Other regular contacts include the ambulance service and paramedics, out of hours service (OOH) and hospital emergency departments. Coordination across care pathways is difficult and becomes even harder where these services are supplied by different provider organisations. His care team is summarised in Fig. 3.3.

Given his combination of conditions, medication interactions are inevitable. For example, it is hard to manage steroids for an acute exacerbation of his COPD against the insulin dose needed for management of his diabetes. A change in one medication may need a dose change in another and an error could lead to a catastrophic consequence such as kidney failure and a medical nephrectomy.

Fortunately, Fred has his friends, who like to take him to the pub on Saturday nights – but with the consequence that he is a frequent visitor to the emergency department on Sunday morning.

His dog is both a comfort and a worry. She needs walking every day and someone has to look after her if Fred has to go into hospital. He also needs the regular care support to administer his insulin, transport for appointments, and help with housework, care and shopping.

He needs a way to manage his numerous medical and social problems safely, and prevent untoward medical events. For example, rheumatology want to put him on NSAIs but need to be warned that he is already on an ACE inhibitor to protect his kidneys. Who decides which is more important?

A couple of years on, Fred's life has improved [10]. He has identified some solutions to some of his problems. He now has an 8 mph pavement scooter and is no longer depressed, although its extra speed is tough on the dog.

A count of the actors in this story shows that he needs the support of more than 20 different healthcare specialists, as well as non-healthcare support.

Who is ultimately in charge? Who ensures the dog is looked after when Fred isn't there? How to keep everyone up to date? Fred wants to be involved – and in charge – as well.

Requirements Specification

The requirements specification describes the important aspects of the system. Clear understanding of the business information flow is vital because mistakes here affect everything else. Failure to nail down and specify information flows now and in the future system is one of the most common causes of systems failure.

It can be useful to prepare two documents: the present system (*as is*) and proposed system (*to be*). The *as is* description is what happens now and can be checked

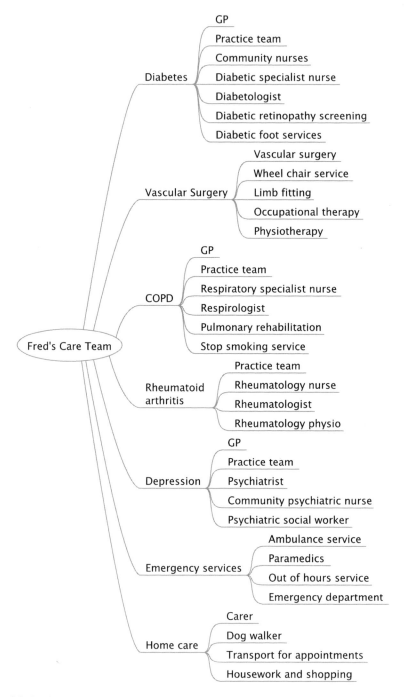

Fig. 3.3 Fred's care team

by current users for being right or wrong. The *to be* system does not yet exist so is harder to check. People often find it useful to test out how the systems should work using storyboards (see above).

The requirements specification is best done as collaboration between one or more business analysts and user domain experts. This should not be frozen early, but should continue to evolve as the work progresses.

There is no one right or wrong way to understand and elucidate business processes. Experienced analysts often use their own approaches. One approach is to capture key aspects of the business processes under the headings of service overview, transactions, actors, locations, identification, evidence, transaction outcome and rules (see Fig. 3.4):

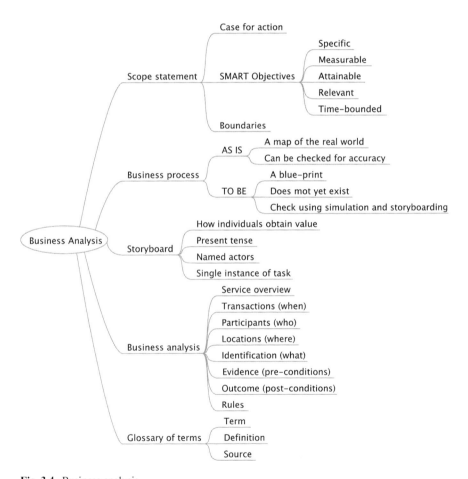

Fig. 3.4 Business analysis

1. Service overview elaborates the domain scope, describing each service provided or expected to be of value. A service typically represents the widest use case of interest. Each service can be decomposed into sub-services and transactions. A sufficiently broad scope is needed to ensure that the whole system is developed in a joined-up way.

2. Actors are the parties, things and systems that are involved. Actors may be physical such as people, things (such as specimens), machines (such as computer systems), or abstract entities such as organizations. Actors are the things about which information is recorded, whether they are actively involved in transactions or third parties referred to.

3. Transactions are communication between actors. Each transaction is described, along with its timing, origin, triggers, pre-conditions, destination, purpose, volume and possible outcomes. In interoperability, transactions are the main use cases of interest. Each transaction is typically an exchange of information between a set of actors requiring evidence and generating some outcome(s) that achieves a useful goal for the primary actor.

4. Locations are physical or virtual places associated with actors and transactions, such as where things take place or the origin and destinations of data. These may be physical or virtual (on a computer network). It is important to specify where each actor needs to be physically when considering differences between the *as is* and *to be* models and between physical (paper-based) and digital systems.

5. Identification of actors, locations and information objects is needed, because computers systems need unique identifiers. We need to specify what identifiers are used, who or what assigns them, and what information may be accessible as a result of knowing this identifier. Many organisations assign their own identifiers (eg patient number), but there may be legal restrictions in using these outside of their main purpose. Furthermore significant numbers of individuals will not know their assigned identifier or may not have one. Soft identifiers such as name, address, date of birth and gender may be needed to match individual people. These are soft because people can change their name or address, give false dates of birth or even change gender. An identifier may refer to an individual instance (such as a person or the serial number on a machine) or to a type of thing (such as the bar codes printed on packets of corn flakes).

6. Evidence is information that needs to be known prior to a transaction being triggered, and is needed to support it. Evidence may be obtained either by direct data input or by querying a database. Evidence is usually associated with one of the actors involved in the transaction.

7. Outcome describes the possible results of each transaction including the post-conditions and responsibilities of each actor. For example, updates to records and letter generation would be described here.

8. Rules include regulations and constraints governing the transactions including non-functional requirements such as security and privacy. Rules may be legislative, policy, logical, procedural or temporal. A rule can be free standing or part of a multi-level hierarchy of decision-making criteria. Rule documentation should include rules, regulations, error handling, reference data and coding

schemes related to each transaction which have not been documented elsewhere. Rules are often important in determining how to handle failures and errors.

A glossary of terms used contains the name of each term, its description/definition and source (if obtained from another reference). Specialised terms should always have their meaning specified in the glossary, which may be populated from forms and information sets currently being exchanged, the functionality of existing computer applications in this domain including their data dictionaries as well as national and international standards.

Business analysis aims to capture the requirements in a form that is fully understood by both users and technical staff. This feeds into the conceptual design, platform-specific specifications and ultimately forms the basis for testing and conformance. UML models provide an effective and flexible way of sharing understanding about the system under consideration. Business processes may be documented using activity diagrams and class diagrams, supported by detailed definitions of every data item.

Conceptual Design

The conceptual design specification is the most detailed description of what is to be provided that users should be able to understand, criticize, review and sign off. It is a model of the *to be* system, although it does not usually cover non-functional requirements in depth. Being technologically neutral, it does not specify what software shall be used. It can be used as the basis of a contract between users and the developers.

The full specification should comprise both the conceptual specification and the technology-specific specification [11]. When there is any doubt as to the meaning of the technology-specific part of the specification, then the conceptual specification, which domain experts can understand and approve, should be regarded as the ultimate authority.

Conceptual design specifications are often based on a conceptual model of the proposed system. The specification should meet the following eight criteria (Fig. 3.5):

- Comprehensive. The conceptual design specification should include all transactions within the scope, being sufficiently expressive to represent and describe each one. It needs to be extensible to incorporate new requirements and local needs.
- Context explicit. The model should describe the business processes surrounding each transaction, specifying trigger events, delays, timing and other constraints, business rules, outcomes and error handling, as well as the static structure of each transaction payload.
- Complete in itself. The model should represent both the data structure (data model) and processes (dynamic behavior). The sequence of activities must be indicated, showing whether the order in which tasks are performed is significant

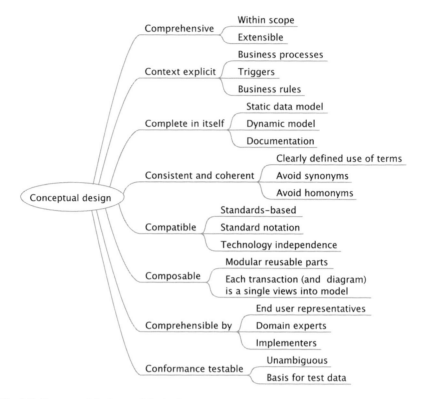

Fig. 3.5 Conceptual design model criteria

or not. The model should be documented internally so it stands alone with no need to reference external documents or manuals.

- Consistent. Each term used should be defined unambiguously. Each concept should only have one name avoiding synonyms, and the same term should not be reused for a different concept to avoid homonyms. A common architecture, notation and terminology should be used to define every element of the model.
- Compatible. Definitions should be compatible with international standards. Use of platform-independent models based on standards facilitates transfer across institutions, helps suppliers to implement compatible solutions using different proprietary technologies, and migrate to new technologies in the future.
- Composable. Modules may be reused and combined freely with each other to produce new ones, possibly for purposes quite different from that for which they were originally designed. There should be minimum dependency between parts so that any change or error in one is not propagated to others.
- Comprehensible. Each part needs to be understood and reviewed independently by clinical end users, domain experts and implementers. Names and definitions should be written in the language of the user. Abstract terms and neologisms (words or phrases with newly coined meaning) should be avoided. Each element needs to be understood by a human reader without any need to consult external reference manuals. A simple graphical notation that is easy to learn and use is ideal.

- Conformance-testable. Implementations based on the model need to be tested against the model to demonstrate conformance. As few alternative methods as possible should be provided for doing any business task. Navigation through the model should be in one direction to generate a hierarchical structure which can be serialized. Recursive structures should be avoided as much as possible. Many-to-many relationships should be avoided and zero-to-many optionalities used as little as possible.

Technology-Specific Specification

The technology-specific specification sets out exactly what is to be built, tested, deployed and supported. For interoperability projects this could be HL7 Version 3 or FHIR, and should include implementation guidelines and XML schema if required.

Technologies change and evolve faster than sound conceptual designs. The same conceptual design can be implemented in many different ways. The mapping from the conceptual design specification to any specific implementation should not involve any changes, either by addition or constraint, in the semantic content.

An Example – Colorectal Cancer Referral

This section illustrates some of the ideas described in this chapter using an example of colorectal cancer referral, covering the process when a GP refers a patient suffering from colorectal symptoms for urgent endoscopy to diagnose or exclude possible cancer.

This example uses a two-stage process, based on NICE Guidelines (2004) [12] and the work of Selvachandran and colleagues (Selvachandran 2002) [13]. Colorectal (CR) cancer is the third most common cancer after breast and lung and the second largest cause of death from cancer in the USA [14]. With about 37,000 cases a year in the UK, each GP is likely to come across about one new case each year. Survival is strongly related to speed of diagnosis. Five-year survival depends on how quickly the cancer is diagnosed:

1. 83% for TNM (Tumor, Nodes, Metastasis) Stage I (localized within the bowel wall)
2. 64% for TNM Stage II (penetrating the bowel wall)
3. 38% for TNM Stage III (cancer in Lymph nodes)
4. 20% for TNM Stage IV (distant metastases, most often in the liver) [15]

Research literature shows evidence of delays, often lasting a year or more, between the onset of symptoms of colorectal cancer and diagnosis. This is due to patient delay in reporting symptoms, and to a lesser extent, delays by the GP and hospital. For example, a national survey of NHS patients in 1999/2000 found that 37% had to wait over 3 months for their first hospital appointment and 13% waited seven or more months, although this is now much improved.

NICE (National Institute for Clinical Excellence) has listed criteria for urgent referral, based on combinations of symptoms and signs, from which a simple questionnaire has been derived. Seven questions relate to presenting history, three to physical examination and one each for age and hemoglobin (iron deficiency anemia). The decision to refer urgently is based primarily on the patient's report of his or her symptoms and medical history, the patient's age, evidence (or lack of it) from physical examination and blood tests (hemoglobin).

This set of criteria is used for initial triage. Although about 85 % of patients with colorectal cancer meet these criteria (sensitivity), the large proportion of all patients who have these complaints do not have cancer (specificity).

The scope is limited to the business process and decision criteria used to make an urgent 'two-week possible colorectal cancer' referral decision by the GP in his or her surgery. All other aspects of the problem are out of scope.

The objectives of the process are:

1. Reduce the number of days between initial reporting of symptoms and final diagnosis.
2. Reduce the number of false negatives (cancer cases missed) and false positives (the number of urgent referrals that are subsequently shown to be free from cancer)
3. Reduce the number of appointments required.

Referral for possible CR cancer involves not only the direct *actors*, such as the patient, the GP and practice staff, but other stakeholders, notably the staff at the units to which the patient may be referred, including doctors, nurses, managers and clerks.

The patient complains of symptoms and may have cancer. The patient is the primary source of information about history and symptoms and must be present for physical and endoscopic examinations and diagnostic imaging as well as for providing samples of blood, feces etc. for laboratory tests.

The second key actor is the GP, who takes the decision of whether or not to refer the patient for endoscopy. Reception and secretarial staff in the practice may also undertake some tasks. Although CR cancer is one of the most common types of cancer, each GP only sees about one new CR cancer patient a year. The incidence of symptoms that warrant detailed assessment is not an every day occurrence, so any tools used to facilitate this need to be unobtrusive. Perhaps a couple of patients a month present with symptoms that warrant further consideration and half a dozen patients a year need to be referred for urgent endoscopy.

Other actors, such as the e-Booking service and the endoscopy unit, receive outputs from the interaction, but are not involved in the decision of whether or not the patient should be referred.

Although the main interaction takes place at the GP surgery, the patient may not know some of the information at the time and may need to consult relatives about details of family history or crosscheck the dates at which they first complained of symptoms. For these reasons, detailed history may best be collected at the patient's home using a web-based questionnaire. Patients who cannot use a web-browser can be given a paper questionnaire, which can be scanned or transcribed. Much of the information used to make this decision is relevant to subsequent care and treatment and may be collected in a form suited for use in a referral letter.

The outcome is a decision of whether or not to refer for urgent (possible cancer) endoscopy. The process can be thought of as two "yes/no" decisions. First, does this patient have any CR symptoms that might be indicative of CR cancer, sufficient to warrant more investigation – this decision is based on the NICE criteria, which include presenting symptoms, physical examination and the patient's age. If this decision is positive, then take detailed history. The second decision – whether to refer the patient for urgent hospital investigation – is based on a detailed structured history covering: symptoms and presenting history, family history and past medical history. If this is also positive, then refer urgently for endoscopy.

A storyboard can provide a brief description of how the GP referral process might work in the future.

> *John Reeves is 64 years old. Over the past couple of months he has noticed that his bowel movements have become loose and more frequent. He makes an appointment to see his GP, Dr Ann Price.*
>
> *Dr Price sees John, takes his history, examines his abdomen and suggests that he complete a detailed colorectal history questionnaire, to be completed at home. John has access to the Internet at home, and the surgery emails John a set of details of the URL for his web-based questionnaire and his instructions.*
>
> *John completes the form on his computer at home with some help from his wife who reminds him about some details of family history. Next morning, the surgery telephones him to say that the data is complete and asks him to come in and see Dr Price the next morning.*
>
> *Next morning, he sees Dr Price, who now has the details of his history on her computer screen. The decision support algorithm indicates that there is some cause for concern. Dr Price notices this and that the symptoms and history warrant urgent endoscopic investigation.*
>
> *She explains the situation to John and makes a referral to the local Endoscopy Unit via an electronic booking service. The information collected by the questionnaire is sufficient to produce a structured referral letter, which Dr Price checks, authorizes and sends.*
>
> *John is naturally anxious and so Dr Price goes into an electronic reference (Map of Medicine), where it lists the main reasons for referral for possible CR cancer as well as other data. She prints out a copy of the relevant page and gives it to John.*
>
> *The next day, John is contacted by the Endoscopy Unit and makes arrangements for the test to be done the next week.*

Here only one storyboard has been provided, but in any real project a number of storyboards should be developed covering each of the main scenarios.

The flow can also be shown as a business process diagram using the BPMN notation shown in Fig. 3.6. This is similar to an activity diagram. The BPMN notation is a formal notation and the diagram can be exported in XML format. BPMN is described further in Chap. 4.

The main locations (GP surgery, patient's home and specialist endoscopy unit) are shown as pools. The GP surgery is subdivided into two lanes (reception and GP consulting room). The rounded rectangles represent separate tasks and the circular icons represent discrete events. The diamond shapes represent decision branches, and the "O" icon inside states that the branches are mutually exclusive (OR). The clock icon represents a time-specific event or delay, while the envelope icon represents a message.

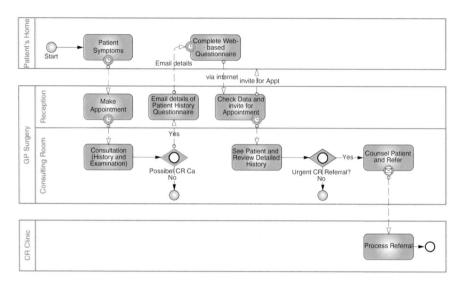

Fig. 3.6 Colorectal cancer referral BPMN diagram

References

1. Coiera E. Guide to health informatics. 3rd ed. London/New York: CRC Press; 2015.
2. Mellor SJ, Scott K, Uhl A, Weise D. MDA distilled: principles of model-driven architecture. Boston: Addison-Wesley; 2004.
3. Warmer J, Kleppe A. The object constraint language: getting your models ready for the MDA. 2nd ed. Boston: Addison Wesley; 2003.
4. Tanenbaum A, Wetherall D. Computer networks. 5th ed. Boston: Pearson; 2010.
5. CEN CR 1350:1993. Investigation of syntaxes for existing interchange formats to be used in healthcare. CEN Report 1993.
6. CEN CR 12587:1996. Medical informatics – methodology for the development of healthcare messages. CEN Report 1996.
7. Beeler W. HL7 Version 3 – an object-oriented methodology for collaborative standards development. Int J Med Inform. 1998;48:151–61.
8. Bodenheimer T, Bojestig M, Henriks G. Making systemwide improvements in health care: lessons from Jönköping County, Sweden. Qual Manag Health Care. 2007;16(1):10–5.
9. Hawking M. Fred and his dog – an update. BCS Primary Health Care Group Blog December 2011. http://primaryhealthinfo.wordpress.com/2011/12/11/fred-and-his-dog-an-update/
10. Hawking M. Fred's saying "You just don't GET IT". BCS Primary Health Care Group Blog November 2013. http://primaryhealthinfo.wordpress.com/2013/11/02/freds-saying-you-just-dont-get-it/
11. Benson T. Prevention of errors and user alienation in healthcare IT integration programmes. Inform Prim Care. 2007;15:1–7.
12. National Institute for Clinical Excellence. Improving outcomes in colorectal cancers: manual update. London: National Institute for Clinical Excellence; 2004.
13. Selvachandran SN, Hodder RJ, Ballal MS, Jones P, Cade D. Prediction of colorectal cancer by a patient consultation questionnaire and scoring system: a prospective study. Lancet. 2002;360:278–83.
14. Corley DA, Jensen CD, Marks AR, Zhao WK, Lee JK, Doubeni CA, et al. Adenoma detection rate and risk of colorectal cancer and death. N Engl J Med. 2014;370(14):1298–306.
15. National Institute for Clinical Excellence. Colorectal cancer: the diagnosis and management of colorectal cancer. Clinical guideline. London: NICE; 2011.

Chapter 4
UML, BPMN, XML and JSON

Abstract UML and BPMN are modelling notations; XML and JSON are simple languages used for structuring information exchanged. They all work best with specialised tools. UML can be used for sketching, detailed design work or even to produce software code directly. UML diagram types include Class, Object, Use-case, Activity, Sequence and State diagrams. Each diagram is one view of a model. BPMN is used for describing workflow. XML is descriptive, extensible markup language for structuring information in documents. The structure of an XML document is specified in schema. JSON is simpler than XML and is widely used on the web.

Keywords UML • BPMN • XML • JSON • Object models • Class diagram • Composition • Aggregation • Navigation • Use case • Activity diagram • Sequence diagram • Pool • Swimlane • Namespace • Schema • XPath • JSON tools

This chapter provides a short introduction to four of the key underlying languages or notations used in healthcare interoperability (UML, BPMN, XML and JSON). A short chapter such as this cannot teach you how to use these, but it is intended to indicate those aspects that you are most likely to need when working in healthcare interoperability.

UML

UML is a standard modeling notation widely used in HL7 and for other health IT purposes. The basic notation is simple and quick to learn, although this hides a good deal of complexity that may be needed for some purposes. It aids the design of software systems before coding and is one of the essential skills of health informatics. A UML model may comprise many diagrams of different types as well as detailed documentation for each element.

UML stands for Unified Modeling Language. The term *Unified* in the name is a clue to its origins. During the early 1990s, a number of competing notations were in widespread use. This was confusing, to say the least. Between 1995 and 1997

Rational Software (now part of IBM) brought together three of the leading methodologists (Booch, Rumbaugh and Jacobson) and, together with the Object Management Group (OMG), developed a notation and language, which they called the Unified Modeling Language, combining many of the best ideas.

UML is a specialised language, not simply a notation for drawing diagrams. It includes a notation used on diagrams and a meta-model, which is of interest primarily to the developers of UML software tools. Simple things are simple, but more complex things require purpose-built tools.

Specialised UML modeling tools, which are widely available, are strongly recommended for serious work involving more than a handful of diagrams. In a modeling tool, all of the information about a model is held in a common repository, which facilitates re-use of work and ensures consistency.

When learning UML it is a convenient simplification to regard a model as the sum of the diagrams. Think of the blueprints for a new building or machine, each diagram showing a small part of the total design. The model is the sum of the specifications, comprising hundreds of diagrams and supporting text.

People use UML in three ways: sketch, blueprint or programming language [1].

Sketches can be made using white-boards, or multi-purpose tools such as Visio or PowerPoint.

We may start off using sketches, but soon need to move on to developing blueprints. The distinction between a sketch and a blueprint is that sketches are incomplete and exploratory, while blueprints are complete and definitive. Serious modeling (blueprints rather than sketches) requires a specialised UML tool. Each tool maintains an internal repository, which facilitates the re-use of common components and avoids the problems produced by describing the same thing in a different way in different places.

The step beyond blueprint is when programs are produced directly from the model. Here UML becomes the source for executable code. Full use of this approach can automate the production of conformant code, schema, documentation and test rigs.

UML is independent of the software used to implement computer applications and is not tied to any development methodology. UML's independence of technology and method is one of the keys to the wide support that it enjoys throughout the IT industry. It fits into any IT organisation.

UML has some weaknesses. Models and diagrams created using different tools are difficult to import and export into and out of different tools reliably. It does not have a neat way of specifying multiple choices, decision tables or other constraints, although it does have a special Object Constraint Language (OCL). However, OCL is opaque to those without formal training in computer science. In many models, unstructured text annotations form an important part of the documentation.

A core premise of UML is that no single diagram (or type of diagram) can provide, on its own, a full representation of what goes on, and so we need to use sets of related diagrams. Each type of diagram only shows certain aspects of a situation – everything else is ignored. This simplification provides both the power (it makes the situation understandable) and the weakness of diagrams (each diagram has a limited scope).

UML diagrams relate to information structure or behaviour. UML has 13 diagram types, although most users of UML make do using three or four types of diagram, depending on what they are using it for.

A danger point is UML's principle of suppressing information, which allows information to be omitted from any diagram in order to make it easier to understand. The corollary is that you must not infer anything from the absence of information in a diagram and that UML diagrams should not be read on their own without access to the rest of the model.

All attributes have default multiplicity (the range of the number of instances that can participate in an association from the perspective of the other end) of one (mandatory one and only one), but if a multiplicity is not shown against an attribute in a diagram, it may either mean that the information is suppressed, or that the default value should be used. For example, no attribute multiplicities are shown in the HL7 RIM (Reference Information Model), because they are suppressed. Structural attributes in the RIM are mandatory while all others are optional (but this is not shown on the diagrams). The only way to check is to look deeper into the model and see what is really there.

Individual organisations often develop their own conventions about what is and is not shown on each diagram.

Diagram layout and style facilitate understanding. Some guidelines apply to all types of diagrams. Diagrams should be laid out so that they can be read left to right and top to bottom. Avoid crossed, diagonal and curved lines. Document diagrams using notes. Use the parts of UML that are widely understood and avoid the esoteric parts. Use colour coding with discretion. Use common naming conventions such as UpperCamelCase for class names and lowerCamelCase for attribute names. Do not put too much on a single diagram. Restrict diagram size to a single sheet of A4. Use consistent legible fonts. Show only what you need to show. It is good practice to suppress unnecessary detail (Fig. 4.1).

Class Diagram

Class Diagrams are the most widely used UML diagrams and show the static structure of classes, their definitions and relationships between classes. A **class** is an abstraction of a thing or concept in a particular application domain. It has properties (attributes), behaviour (operations), and relationships to other objects (associations and aggregations).

On a class diagram, each class is shown as a rectangle with one, two or three compartments. The top compartment shows the class name, the second shows attributes and the third shows operations. Attributes describe the characteristics of the objects, while operations are used to manipulate the attributes and to perform other actions. Attributes and operations need not be shown on a particular diagram.

Figure 4.2 shows a simple class diagram representing a prescription, showing class names only. Each Prescription has a Prescriber (author) and relates to a single Patient. It has one or more PrescriptionLines. Each PrescriptionLine includes details of a Drug and may have any number (zero to many) of DosageInstructions. The

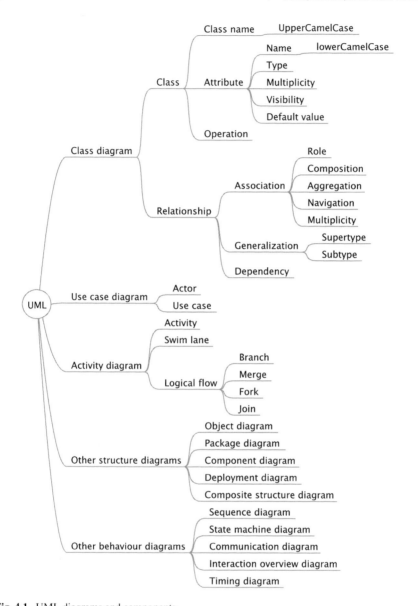

Fig. 4.1 UML diagrams and components

arrowheads on the lines (associations) show navigation. The arrow from Prescription to Patient shows that, in this model, the navigation is from Prescription to Patient but not the other way round.

The notation for **multiplicity**, used in associations and attributes, is:

0..1   optional but no more than one is allowed
* or 0..*   optional but any number of instances is allowed

Fig. 4.2 Simple class
diagram showing
navigation arrows

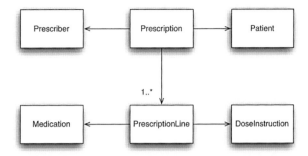

1   required with exactly one (this is the default)
1..*   required with at least one instance

If no multiplicity is shown, the default assumption is multiplicity of 1, meaning that exactly one is required, although caution must always be observed when inferring anything from the absence of data on diagrams.

Figure 4.3 illustrates two notations used for showing containment. The black diamond indicates **composition** between Document and Line (eg an order line). Each Document contains one or more Sections, but Section cannot exist independently of Document. The hollow diamond indicates **aggregation** between Document and Author. Each Document has one Author, but the Author can exist independently of Document.

The multiplicities used in any diagram depend on its purpose. Figure 4.3 is a Document-centric showing that Document has just one Author (a one-to-one relationship), but an Author-centric diagram, Fig. 4.4 would shows that Author has one or more Documents (a one-to-many relationship).

The concept of **inheritance** is illustrated in Fig. 4.5. Patient and Doctor are both specialisations of Person; Person is a **generalisation** of Patient and Doctor. The triangle arrowheads indicate that both Doctor and Patient classes inherit the properties of Person, such as name and address. Patient has attributes: nhsNo, dateOfBirth and gender, and also inherits the attributes name and address from Person. Similarly Doctor has attributes professionalID and organisationID, as well as the properties of Person.

Attribute   Attributes have several properties. For example the notation

        +dateOfBirth:Date[0..1]

indicates:

- Visibility (+) is public, meaning that it is fully accessible.
- Attribute name is dateOfBirth. Attribute names are usually written in lowerCamelCase
- Attribute type is Date.
- Multiplicity is [0..1] meaning that this attribute is optional with a maximum number of occurrences of one.

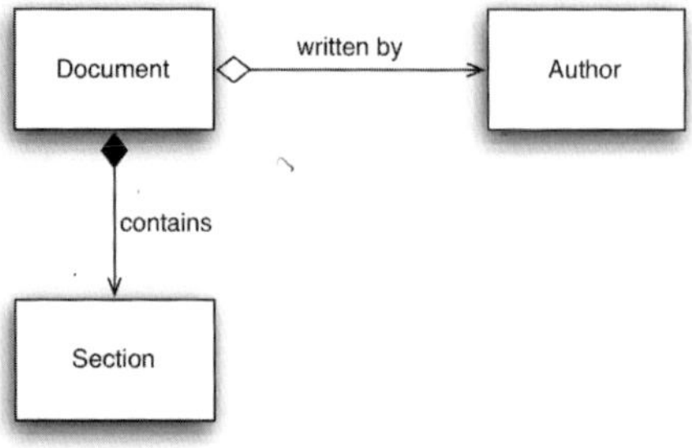

Fig. 4.3 Aggregation and
composition – document
has one author

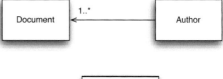

Fig. 4.4 Author has
associations with many
documents

Fig. 4.5 Person
specialisation – patient and
prescriber inherit attributes
of person

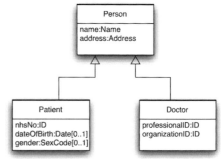

Initial values and defaults may also be specified.

Operation   Operations implement the functionality of a software object. They are the
actions that an object knows how to carry out. The syntax for operations includes: visibility, operation name, parameter list (in parenthesis), return type and property string.

An **object** is a unique instance of a class. An object diagram, such as Fig. 4.6,
shows the relationships between objects. Each object may have an identity (name),
state (attributes) and behaviour (methods). The object name is underlined (to distinguish it from a class), and comprises the object's name, which is optional, followed
by a colon and the class name (eg TimBenson:Author).

Package   Packages are used to divide up a model in a hierarchical way. Each package may be thought of as a separate name space. Each UML element may be allocated to a single package. Packages provide a useful means of organising the model.
Classes that are closely related by inheritance or composition should usually be
placed in the same package.

In Fig. 4.7, the Participants package might include all classes related to people
and organisations, including patients, doctors and nurses. The Interactions package

Fig. 4.6 Simple object diagram

Fig. 4.7 Package diagram

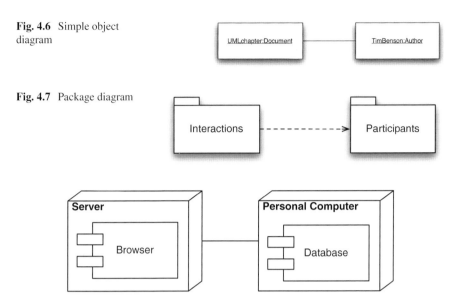

Fig. 4.8 Deployment diagram

might include messages and entries in clinical records. The dashed arrow from Interactions to Participants indicates that Interactions have a dependency on Participants.

Deployment Deployment is the physical organisation of computer systems and is shown in deployment diagrams (Fig. 4.8). Each piece of the system is referred to as a node. The location of software can be shown as components.

Modeling Behaviour

Use case Use cases capture the behavioural requirements of business processes and provide a common linkage across all aspects of a project from initial analysis of requirements through development, testing and customer acceptance. They show how people will ultimately use the system being designed. Each use case describes a specific way of using the system. Any real system has many use cases. Each use case constitutes a complete course of events, initiated by an actor (or trigger). A use case is essentially a sequence of related transactions performed by an actor and the system in a dialogue.

An actor is an external party such as a person, a computer or a device, which interacts with the system. Each actor performs one or more use cases in the system. By going through all of the actors and defining everything they are be able to do with the system the complete functionality of the system is defined.

Each use case is a description of how a system can be used from an external actor's point of view; it shows the functionality of the system, yielding an observable result of value to a particular actor. A use case does something for an actor and represents a significant piece of functionality that is complete from beginning to end.

The collected use cases specify all the ways the system can be used. Non-technical personnel can understand use cases. Thus they can form a basis for communication and definition of the functional requirements of the system in collaboration with potential users.

A simple use case diagram is shown in Fig. 4.9. Stick-men represent actors; ellipses represent use cases.

Use case descriptions should be documented using simple templates, to include:

- Metadata, such as use case name, unique ID, author, date, version and status
- Scope and context
- Primary and other actors
- Pre-conditions and trigger event
- Main success scenario describing the normal flow of events using numbered steps from trigger through to post-conditions
- Post-conditions
- Alternative flows, e,g, when errors occur.
- Importance and priority
- Open issues

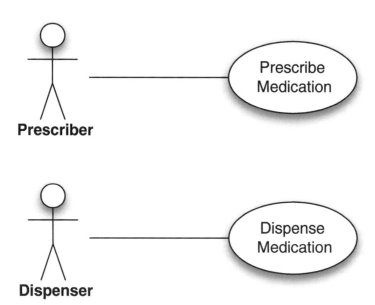

Fig. 4.9 Use case diagram

A **scenario** is an instance of a use-case. It is one path through the flow of events for the use case and can be documented using an activity diagram or a storyboard free text description (Fig. 4.10).

Activity diagram Activity diagrams describe business processes undertaken by each actor or role in diagrammatic form. Each role may be shown in a separate **swim lane**. Interoperability transactions are usually communications that cross swim lanes.

Activity diagrams display a sequence of actions (including alternative execution paths) and the objects involved in performing the work. They are useful for describing workflow and behaviour that has branches and forks. Figure 4.11 shows a sim-

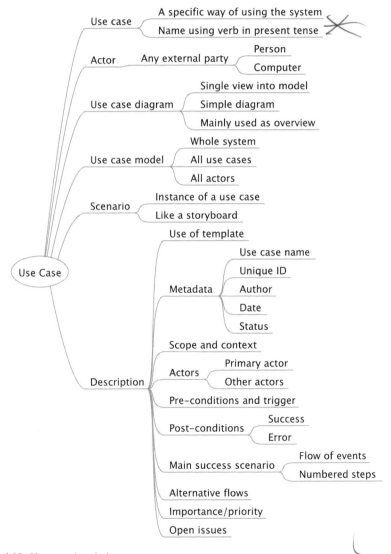

Fig. 4.10 Use case descriptions

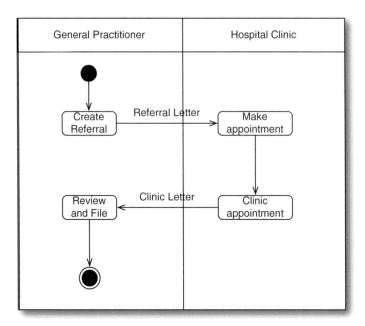

Fig. 4.11 Activity diagram

plified activity diagram for the exchange of a referral and clinic letter between GP and hospital. It is organised in swim lanes to show who or what is responsible for each activity.

Activity diagrams can be used to show logical data flows. A branch has a single entry point, but a choice of exits depending on some condition. Only one route can be taken. Branches end at a merge. A fork has one entry and multiple exits, which can be undertaken in parallel, and the order of activities is not important. A fork ends at a join.

Sequence diagram Sequence diagrams (eg Fig. 4.12) show how objects interact with each other. Sequence diagrams show when messages are sent and received. A sequence diagram depicts object interactions arranged in time sequence, where the direction of time is down the page. The objects, which exchange information, are shown at the top of a vertical line or bar, known as the object's lifeline. An arrow between the lifelines of two objects represents each message.

A **statechart** diagram (Fig. 4.13) shows an object life cycle, and can be used to illustrate how events (messages, time, errors and state changes) affect object states over time. State transitions are shown as arrows between states.

An object state is determined by its attribute values and links to other objects. A state is the result of previous activities of the object. A state is shown as rectangle with rounded corners. It may optionally have three compartments (like classes) for name, state variables and activities.

Fig. 4.12 Sequence
diagram

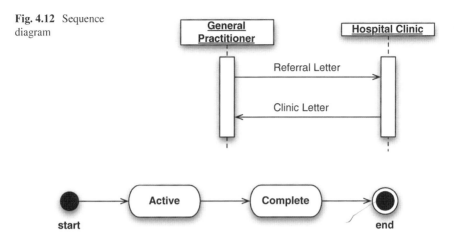

Fig. 4.13 Statechart diagram

Business Process Modeling Notation (BPMN)

Clinical processes are difficult to model with existing tools, in part because they are inherently complex, but more importantly because each clinician may adopt a variety of different paths, depending on the specific clinical situation of the individual patient. This requires business process specifications that take account of the full range of choices that clinicians have open at any one point in time.

BPMN is a standard for business process modeling [2] with a notation that is similar to that used in UML Activity Diagrams. BPMN shows who does what, where and in what sequence using the pool and swim-lane notation. It distinguishes between messages which flow between actors from the flow of activities of a single actor. BPMN shows trigger events, delays and messages that precede or follow on from each activity. It allows drill down of sub-processes into greater detail of activities and tasks and provides additional structured and/or free text documentation for any element. The output is executable, using Business Process Execution Language (BPEL), an XML-based language.

BPMN defines business processes, processes, sub-processes and tasks.

Business process Business process is the top of the activity hierarchy in BPMN. It is defined as a set of activities that are performed within an organisation or across organisations, shown on a Business Process Diagram (BPD). Process is limited to the activities undertaken by one participant (organisation or role). Each business process may contain one or more processes.

A **Process** is an activity performed within an organisation, and is depicted as a set of activities (sub-processes and tasks) contained within a single pool.

Each **Sub-Process** may be expanded as a separate, linked diagram, showing its component sub-processes or tasks. The facility to expand or consolidate sub-processes is a useful feature of BPMN.

A **Task** is an atomic activity, showing that the work is not broken down to a finer level of detail. Sub-processes and tasks are shown as rounded rectangles. Sub-processes, which can be expanded, are shown with a "plus sign" at the bottom centre of the icon.

Participant Participants are each represented by a **Pool**, which may contain **Lanes**. Each pool contains a single process. A pool may be subdivided into lanes (like swim lanes in UML activity diagrams). lanes may represent different roles within an organisation. If a diagram contains a single pool, the pool boundaries need not be shown. A pool is a container separating each Process from others and showing the sequence flow between activities.

Event Events are things that happens during the course of a business process that affects the flow, such as triggers for activities to begin or their outcomes. Events are shown as a small circular icons. Start, intermediate and end events are indicated by the thickness of the circle perimeter. An additional icon inside the circle shows the type of trigger or result such as a message, timer, error or cancelation.

Gateway A **Gateway**, shown as a square diamond, is used to control branching, forking, merging and joining of paths. An icon inside the diamond shows the type of control (exclusive XOR, inclusive OR, parallel AND or complex).

Connector Connectors link the flow objects (activities, events and gateways). There are three types of connector. Sequence flow (a solid line with arrow head) shows the order that activities are performed within a process. Message flow (a dotted line with arrow head) shows connections between processes, crossing pool boundaries. Association (dotted line, no arrow head) is used to associate information, such as data objects and annotations, with flow objects (Fig. 4.14).

A complete business process from start to finish is shown in Fig. 4.15, which illustrates the traditional OP referral pattern for a patient suffering from a bowel problem (which was described in Chap. 3). The pools and lanes show who does what in what order. The dotted lines represent movement of information, such as messages, or of information sources (eg the patient). Each of the tasks shown could be represented as sub-processes and analysed further in subsequent diagrams. Clinical care is essentially fractal and can usually be decomposed into smaller and more detailed sub-processes and tasks. Trigger events are shown as circles, with an icon indicating the type of trigger – an envelope indicates a message and a clock indicates a time trigger, such as an appointment slot.

XML

XML (eXtensible Markup Language) is a universal format for encoding documents and structured data, which is used in interoperability between different applications. XML was developed from SGML by the World Wide Web Consortium (W3C),

Fig. 4.14 Business
process modeling notation

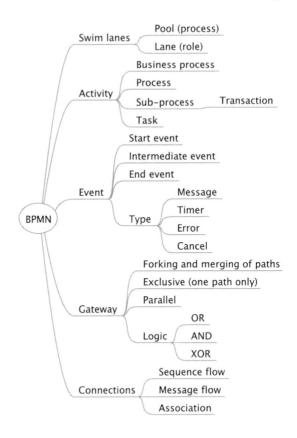

which is the main international standards organization for the World Wide Web [3]. XML documents are independent of the applications that create or use them. This is one of the reasons why XML has become the preferred mechanism for sharing data between systems.

XML is a markup language. Markup is a term that covers any means of making an interpretation of a text explicit. In electronic documents, the system does this by inserting special coded instructions into the text, which are not normally seen on a printed copy. A markup language specifies:

- what markup is allowed
- what is required
- how markup is to be distinguished from text
- what the markup means

Word-processors include markup instructions embedded within their text. However XML differs in one vital respect from this sort of markup. XML is a descriptive markup scheme, while most schemes traditionally used in word processors are procedural. A descriptive scheme simply says what something is (for instance, a heading), while a procedural scheme says what to do (for instance, print in 18 point Ariel font, bold, left-hand justified).

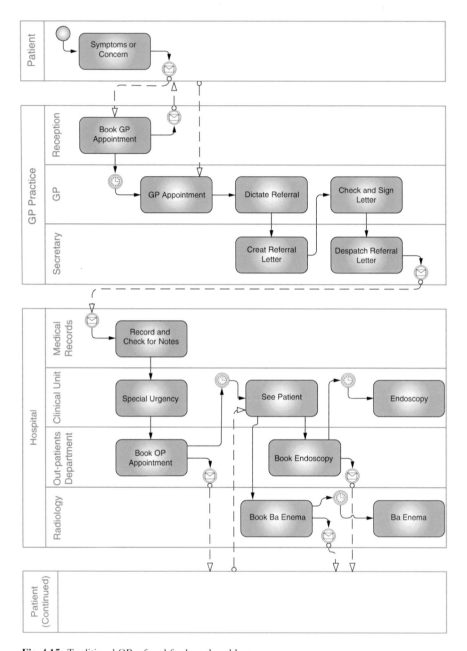

Fig. 4.15 Traditional OP referral for bowel problem

XML uses style sheets to specify how to render information. Style sheets are coded using specialised languages such as CSS (Cascading Style Sheets) and XSL (Extensible Stylesheet Language).

The separation of description (providing names for parts of a document) from procedure is key to platform independence and universality. A text marked up using a descriptive scheme can be processed in different ways by different pieces of software. Procedural markup can only be used in one way.

XML documents are text files designed for computer processing. People should not have to read these, but they are human-readable when the need arises, for example, when debugging applications. XML is verbose by design and XML files are nearly always larger than comparable binary formats. The rules for XML files are strict, and forbid applications from trying to second-guess an error. These were conscious decisions by the designers of XML.

XML documents can include images and multimedia objects such as video and sound. The data may also include metadata – information about itself – that does not appear on the printed page.

XML and HTML are closely related. Indeed, XHTML is a version of HTML which is fully XML-compliant. Like HTML, XML makes use of tags (words bracketed by '<' and '>'), attributes (of the form name="value") and entities (of the form & xx). The key difference is that while HTML specifies in advance what each tag and attribute means and how the text between them will look on a browser, XML uses the tags only to delimit pieces of data, and leaves the interpretation of the data to the application that reads it.

All XML documents comprise nodes (elements, attributes, text content etc.) organised in a tree structure, with a parent-child relationship with other nodes. All nodes descend from a single root node, which is called the document element, corresponding to the document itself. The root node has two child nodes, the XML declaration and the root element.

Every XML document begins with an XML declaration, such as:

```
<?xml version ="1.0" encoding="UTF-8"?>
```

XML documents are made up of elements. All elements are delimited by both a start tag (eg<tag>and a matching end tag </tag>. A start-tag takes the form<name>while the end-tag takes an identical form except that the opening angle bracket is followed by a slash character </name>. An empty element, which does not contain either text or child elements (but may contain attributes) can be abbreviated to a single tag closed with /> eg<name/>. Element names are case sensitive.

A diagnosis element in a text might be tagged as follows:

```
<diagnosis>Diabetes mellitus</diagnosis>
```

Elements may be nested (embedded) within elements of a different type. For example, the line of a poem may be embedded within a stanza, which is embedded within the poem, which is embedded within an anthology.

The following XML fragment contains the name and age of a person. The<name>and<age>elements are nested within<person>.

```
<person>
        <name>Mary Jones</name>
        <age unit="years">30</age>
</person>
```

Element nodes can contain other elements, comments, attributes and a single text node.

Comments can be placed anywhere in an XML document except inside an element. Comments start with a less than symbol, an exclamation mark and a dash (<!–) and end with two dashes and a greater than symbol (––>), for example:

```
<!-This is a comment-->
```

XML Attribute

Attributes are used in XML to add information to the start-tag of an element to describe some aspect of a specific element occurrence. Attribute values are written in quotes and are separated from the attribute name by an equals sign. For example a hypertext link in HTML is shown as <a href="url">, where url is the address of the uniform resource location (URL). Any number of attribute value pairs may be defined for any element.

XML Entity

Entities are named bodies of data, referenced by an entity reference. Entity references begin with "&" and end with ";". A small number of entities are used to represent single characters that have special meanings in XML, such as:

```
<       &lt;
>       &gt;
&       &amp;
```

Numeric character references can be used to represent Unicode characters. For example "©" is used to represent the © symbol. Other entities can be defined.

An XML document is *well-formed* if it complies with a small set of formal rules, which include the following. There is only one root element in an XML document. Each opening tag has a corresponding closing tag although for empty elements the abbreviated form can be used. XML is case sensitive, so opening and closing tags must be identical. Elements must be nested without overlapping. The value of each attribute is enclosed by either single or double quotation marks. Attributes within the same element must have unique names (no repeats).

An XML document is *valid* if it conforms to specified Schema or DTD (document type definition). Validation ensures that an XML document has the correct elements and attributes, the correct relationships between elements and attributes, that child elements have the correct sequence and quantity and that the correct data types are used.

XML Schema

The structure of an XML document is specified in a schema, which is also written in XML. The schema defines the structure of a type of document that is common to all documents of that type. It identifies the tags (elements) and the relationship among the data elements. This means that any document of a known type can be processed in a uniform way, including checks that all of the elements required are present and in the correct order.

The development of schemas is the central analysis and design task of working with XML. The schema makes the rules explicit that need to be specified where uniformity of document structure is required. A large part of each schema comprises XML element definitions for the form and content of each XML element and attribute.

Schema processing tools are used to validate XML documents using one or more schemas. Schema validation is applied to elements within a well-formed XML document. Two schema languages in widespread use in HL7 are W3C's Schema Definition Language (XSD) and Schematron. XML schema are usually defined as a separate file with extension.xsd and linked to the document using a namespace declaration. In the example below, the prefix xs is used to declare that the URL indicates a schema against which the document must be validated.

Schema definitions can be verbose. The following example shows a schema definition for a simple Name element:

```xml
<?xml version="1.0" encoding="UTF-8"?>
<xs:schema xmlns:xs="http://www.w3.org/2001/XMLSchema"
    elementFormDefault="qualified" attributeFormDefault="unqualified">
  <xs:element name="Name">
    <xs:complexType>
      <xs:sequence>
        <xs:element name="Forename" maxOccurs="unbounded">
          <xs:simpleType>
            <xs:restriction base="xs:string">
              <xs:minLength value="1"/>
              <xs:maxLength value="35"/>
            </xs:restriction>
          </xs:simpleType>
        </xs:element>
        <xs:element name="Surname">
          <xs:simpleType>
            <xs:restriction base="xs:string">
              <xs:minLength value="1"/>
              <xs:maxLength value="35"/>
            </xs:restriction>
          </xs:simpleType>
```

```
            </xs:element>
          </xs:sequence>
        </xs:complexType>
      </xs:element>
    </xs:schema>
```

Namespace

Namespaces are provided to eliminate confusion when combining formats. This is used in XML schema to combine two schemas, to produce a third which covers a merged document structure.

   Namespace definitions include an abbreviation that is used as an element prefix. This becomes the local name for the namespace within the XML document. Some namespaces are predefined.

Stylesheet

Stylesheets are used to format XML documents into other formats such as HTML format for use in web browsers. Two commonly used methods are CSS (Cascading Style Sheets) and XSLT (eXtensible Stylesheet Language Transforms). XSLT can also be used to transform an XML document into another form. XSLT stylesheets consist of a set of rules that are applied to different elements and attributes of an XML document. XSLT uses pattern syntax to select specific elements or attributes and provides a broad set of instructions to program the transformations required. Only one XSL stylesheet can be applied to a document at a time.

XPath

XPath is a language, defined by the W3C, which is used to find information in XML documents. Each XML document is treated an inverted tree of nodes, with the root element at the top and branches below.

   XPath gets its name from the path notation used to navigate the hierarchical structure of an XML document. These path expressions look much like the expressions you see when you work with a traditional computer file system. XPath uses path expressions to select nodes. Nested classes are separated by / and XML attributes are prefixed by @.

   XPath includes over 100 built-in functions. There are functions for string values, numeric values, date and time comparison, node, qualified name (QName) and sequence manipulation, boolean values, and more. XPath expressions, often referred to simply as XPaths, return either a set of nodes or may be used to compute values such as strings, numbers or boolean values.

The most important nodes are elements, attributes and text. Atomic values are nodes with no children or parent. Each element and attribute has one parent. Siblings are nodes that have the same parent. Element nodes can have any number of children.

XPath is a major element in XSLT and other XML parsing software. You need XPath knowledge to create XSLT documents. XQuery and XPointer are both built on XPath expressions.

XPath notation is also used in HL7 Implementation Guides to represent classes and attributes (Fig. 4.16).

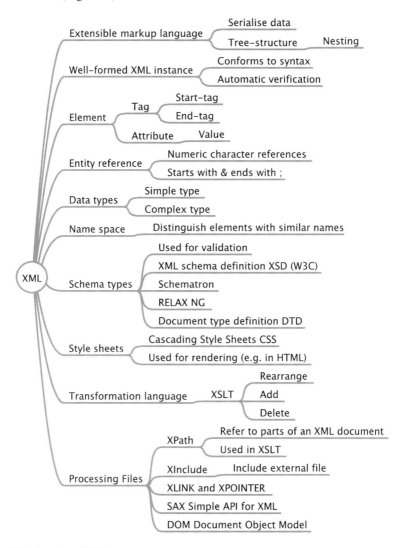

Fig. 4.16 Aspects of XML

JavaScript Object Notation (JSON)

Origin of JSON

Through the late 1990s, the steady increase in both the number of people connected to the web, and the functionality provide by the standard browsers led to an explosion in web-based applications. Initially, web-based applications were based on the http/html request response paradigm:

- the server would construct an HTML page for the browser, and send it
- The browser would render the page, and wait for a user to take action
- The browser would convert the page to an HTTP operation,
- The server would construct an HTML page…

However it became clear that this simple paradigm does not scale to deliver really workable applications – sending whole HTML pages hammers both the client and server networks and it was impossible to maintain the user's sense of state as the content grew richer.

New applications started to be written based on sending fragments of pages (IFRAMES) instead, but these still were an imperfect solution. Finally, the internal scripting language that the browsers contained became solid and standardized enough to develop meaningful applications. These were called AJAX (Asynchronous JavaScript and XML) applications. Instead of the server constructing HTML pages to send to the client, the server sent the data the client needed as raw XML and the JavaScript code would determine how the data should be presented to the user.

XML is a document description language and has many subtle and marvellous features that are unnecessary for the purpose of transferring data between client and server. However, application developers soon found a simpler approach.

The JavaScript Object Notation (JSON) is based on features that are inherent in the JavaScript language [4] – namely, the way that constant values are declared. By reusing the native language format for the data, there was no need to write any parser – the language interpreter itself was able to decode the data, and the programmer could just address the data as native JavaScript objects directly with no need to parse it.

JSON has rapidly become the format of choice for rich client web-based applications. Its relative simplicity compared to XML (discussed below) and ubiquity have meant that it has increasingly become the preferred format for exchange data between systems in other contexts.

The JSON Syntax

JSON is described at http://json.org/. JSON is a text format that is easy to use, and programming libraries exist for most (if not all) languages. This is an example JSON instance:

```
{
        "Image": {
            "Width":   800,
            "Height": 600,
            "Title":   "View from 15th Floor",
            "Thumbnail": {
                "Url":"http://www.example.com/image/481989943",
                "Height": 125,
                "Width":   100
            },
            "Animated" : false,
            "IDs": [116, 943, 234, 38793]

        }

}
```

This is a simple object (surrounded by "{"and "}"), with a set of named properties, each of which has a value. Some of the properties are simple values (Width, Height) while Thumbnail is itself an object and IDs is an Array of values.

Note that a JSON instance contains a single JSON value. There are 5 types of value.

String	A sequence of characters with surrounded by "". See below for escaping rules
Number	A number, written in decimal (23.4) or exponential notation (2.34e1)
Array	A sequence of values, surrounded by [] and separated by ",". Note that the values do not need to be the same type, though they usually are
Object	A sequence of properties, surrounded by {}. Each property has a string name and a value separated by a ":", and properties are separated by a ","
Literal Name	One of the three special values null, true, or false.

Note: a JSON instance is allowed to be any kind of value, but many implementations assume that the value is an object (eg starts with a {), so it is a good idea to start with an object.

Other than some rules around text representation in strings (see below), that is all that needs to be said about the JSON notation.

The current formal specification for JSON can be found at http://rfc7159.net/rfc7159. Due to some legacy standards development issues, there is a slight difference between the current JavaScript object notation and the formal JSON standard. This means that it is not always possible to use the inbuilt JSON interpreter to read

ıt is not appropriate for security reasons anyway, and there are multiple
ers for JavaScript to use instead.

Comparing JSON to XML

Table 4.1 summarizes the important differences between JSON and XML.

XML has many capabilities compared to JSON, but JSON is a much simpler
format.

Whether JSON or XML is easier to use for a particular problem depends partly
on whether the representation needs to use the XML features that JSON does not
have (mixed content, namespacing). If these features are not needed (eg a hierarchical set of data values) then JSON is easier in principle.

However the really important factor for whether to use JSON or XML involves
what tooling stack you do or can use. XML has a deep, well standardised and widely
used tool chain, including:

- DOM/SAX for reading documents
- XPath/XQuery for extract data from documents
- XSLT for transforming XML documents to other forms
- W3C Schema or Relax NG for describing XML formats
- Many code generation and system integration tools

If you use these kinds of tools for working with the data, then XML is a natural
fit. JSON, on the other hand, has a different tool chain with a different focus.

Table 4.1 Differences between XML and JSON

XML	JSON
Supports many character sets and encoding	Supports only Unicode, using a UTF encoding (usually UTF-8)
Has both elements and attributes	Only has properties
Elements can have a mix of text and child elements	Properties can only have a single value
Element names can be namespaced	No namespacing system for property names
It can be ambiguous whether whitespace matters or not	The meaning of whitespace is never ambiguous (it matters in string values, and does not matter elsewhere)
Elements can repeat	Repeating values are represented explicitly with an Array

JSON Tools

Browser + JavaScript

The easiest way to handle JSON is with a JavaScript application in a browser. The JavaScript language is loaded with useful features for working with JSON – it is the native form, after all – and there are many powerful libraries for extracting data out of the objects and populating an HTML-based user interface. Browsers are now powerful IDEs (Integrated Development Environments) with script debuggers and page inspectors that support this process.

Mobile applications are often built around the same architecture internally, but the language and platform libraries they are developed in include explicit support for retrieving and working with JSON.

JSON Query

Because so much of the focus of working with JSON is in the context of a browser using JavaScript, and because JavaScript provides solid support for extracting content from the objects, there has not been enough interest in a standard language independent query. Thus there is no widely adopted standard language for query. There is "JsonPath" (https://code.google.com/p/json-path/) but this is not widely adopted.

This means that it is not possible to write widely understood declarative data extraction statements or invariant rules on JSON objects.

JSON Schema

JSON schema is a method for using a JSON instance to describe what are the valid contents for another JSON value. At the time this was written, JSON schema was still a draft specification undergoing further development (http://json-schema.org/).

Functionally, JSON schema is equivalent to XML schema, in that it is intended to be used for validation of instances, and to support with code generation. However the JSON schema specification works slightly differently than XML schema and it includes additional features while omitting some basic XML schema features. The result is that JSON schema is roughly equivalent when it comes to validation, but presents a tougher challenge when it comes to code generation.

JSON Implementation Issues

Although JSON is a simple specification, there are still several subtle issues that arise when using it.

Text Escaping

A JSON string is a set of Unicode characters surrounded by a double quotes (""), such as "string". A few characters inside the string must be escaped, by prefixing with a backslash (\):

> "a \"quote\" needs to be escaped properly".

Obviously this means that a backslash also needs to be escaped:

> "c:\\temp".

Any character with a unicode code point < x20 (space) also must be escaped and any other Unicode character can be escaped. So, for instance 20 °C, which includes the Unicode character °, can be represented like this:

> "23\u00B0C".

where the escape is \u and then the 4 digit hexadecimal value of the character.

Note: this supports Unicode characters up to the value of xFFFF (16bits). Unicode characters actually have a 32bit identifier, so there are many characters with a value greater than xFFFF. The characters with a value below xFFFF called the Basic Multilingual Plane, and are widely supported. Characters above this value are not so well supported and implementers should avoid them. However, it is possible to represent them in JSON if required – see RFC 7159 section 7.

Numbers

There is no inherent precision limits specified in the JSON syntax – numbers can have any size and/or precision. However, implementers should not assume that values outside a standard 8 byte floating point representation will be supported, though values like this rarely arise in healthcare anyway.

There is also no inherent syntactical representation for non-real numbers such as NaN (not a number) and these concepts do arise in healthcare, particularly with measured values. An implementation can choose to provide the literal JSON value null, or a string value of "NaN", but would need to ensure that the other systems with which data is exchanged are capable of supporting this.

Dates

There is no standard date/time representation built into JSON. Several different forms have been proposed and are in use in some circumstances, but no single form is widely supported. Given this, it is best to represent dates in a string using some agreed format by everyone you are exchanging data with. The XML date format YYYY-MM-DDThh:mm:ssZ is widely used (and supported by JSON schema) and implementers should always consider this format as their first option.

Property Names

There is no restriction on a property name; it can be any valid string, including the empty string. Property names are not required to be valid object property names in JSON or any other language. However most JSON libraries assume that the property names will be valid JavaScript property names (eg a single token containing a mix of letters, numbers and the underscore character). Implementers should not use any other characters in their property names.

Property Uniqueness

The JSON specification does not say that the property names in a single object must be unique. For example, this is not specifically disallowed by the syntax:

```
{
    "Image": {
        "Width": 800,
        "Width": 600
    }
}
```

However the actual result of reading this JSON fragment varies between libraries – most will either report the first or the last, or throw some kind of error when reading the instance.

Implementers should avoid using duplicate property names.

Property Order

The JSON syntax makes no comment on the order in which the property names appear, as shown in the following two instances:

```
{
"Image": {
    "Width":   800,
    "Height": 600
  }
}
{
"Image": {
    "Height": 600,
    "Width":   800
  }
}
```

Most libraries read the JSON into a series of objects that represent the JSON instance, and provide named access to the properties. For these libraries, the order of the properties does not matter – the application using them cannot even find out what the order was, and the two examples above are indistinguishable (for the XML centric reader, this is the DOM approach).

However, some libraries use an approach where the JSON instance is read straight into a higher level set of objects (this corresponds to the way SAX is used for XML). If the value of one of the attributes controls what kind of higher level object is created to match the JSON object, then this property needs to be read first so the right kind of object can be created. Hence these approaches can be highly dependent on the order, since the contents of the object cannot be evaluated properly until the key property value is known.

However, many JSON generator libraries are unable to control the order in which property names are written to a JSON instance, so relying on the order of property names may make different implementations unable to exchange data. Note that some prominent JSON libraries do depend on the order of properties.

Converting Between XML and JSON

A common question for integrators is whether it is possible to automatically convert from XML to JSON, and vice versa. The general answer is that it is possible to convert from XML to JSON or vice versa, but to do it reliably, the conversion process must know about the definitions of the XML and the JSON. In practice, this means that a custom converter is required.

There are a number of decisions that need to be made when converting from JSON to XML or vice versa:

- What to do about namespaces
- What to do about attributes in the XML
- What to do about element text, and mixed content

All these could be resolved, though everybody makes different decisions about them. However, the most difficult problem is related to the JSON use of Arrays. Consider converting this XML fragment to JSON:

```
<Concept>
  <ID>116</ID>
  <ID>943</ID>
  <ID>234</ID>
  <ID>38793</ID>
</Concept>
```

Assume that the correct java script representation is

```
{
  Concept: {
    "IDs": [116, 943, 234, 38793]
  }
}
```

An automatic converter can reasonably be imagined that would be able to perform this conversion. However, if it encounters this XML:

```
<Concept>
  <ID>116</ID>
</Concept>
```

then the XML ->JSON converter would not know that ID was a repeating element and would not produce a JSON array.

In order to get reliable interconversion between XML and JSON, the conversion process must have specific knowledge of the XML and JSON formats, or the underlying logical model, so that the conversion can be performed correctly.

References

1. Fowler M. UML distilled: a brief guide to the standard object modeling language. 3rd ed. Boston: Addison Wesley; 2004.
2. White SA. Introduction to BPMN. IBM Cooperation 2.0. 2004.
3. Bray T, Paoli J, Sperberg-McQueen CM, Maler E, Yergeau F. Extensible markup language (XML). World Wide Web Consortium 1998 Recommendation REC-xml-19980210. http://www.w3.org/TR/1998/REC-xml-19980210
4. Bray T. The JavaScript Object Notation (JSON) Data interchange format. Internet Engineering Task Force (IETF) 2014 RFC 7157.

Chapter 5
Information Governance

Abstract Information governance and security is a large topic, which has at its heart the ethical issue when it is right to share information. Data protection is built around some core principles, which are incorporated in HIPAA and other legislation. Healthcare staff are usually required to sign a confidentiality code of conduct. Computer systems use the concepts of consent, authentication (including OAuth) and authorization to implement access control policies. Cryptography is used to protect data from unauthorized reading. Individuals and organizations have rights and responsibilities, which may include anonymization or pseudonymization of data. These are usually set out in legal contracts.

Keywords Information governance • Privacy • Security • Data protection • HIPAA • Confidentiality • Consent • Authentication • Authorization • OAuth • Access control • Cryptography • Public key infrastructure (PKI) • Digital signature • Encryption • Rights • Responsibility • Anonymization • Pseudonymization • Data controller

To Share or Not to Share

At the heart of healthcare interoperability is a conflict about when it is right to share information and when not to share [1].

People using health and social care services are entitled to expect that their personal information will remain confidential. They must feel able to discuss sensitive matters with a doctor, nurse or social worker without fear that the information may be improperly disclosed, whether by malice, poor practice or carelessness. These services cannot work without trust and trust depends on confidentiality. On the other hand, people also expect professionals to share information with other members of the care team, who need to co-operate to provide a seamless, integrated service. Sharing of information, when sharing is appropriate, is as important as maintaining confidentiality.

All organisations providing health or social care services must succeed in both respects if they are not to fail the people that they are there to serve. People also need to be able to see their own personal confidential data by gaining access to their files, allowing them to make choices and participate actively in their own care.

© Springer-Verlag London 2016   83
T. Benson, G. Grieve, *Principles of Health Interoperability*,
Health Information Technology Standards, DOI 10.1007/978-3-319-30370-3_5

Additionally, anonymised patient data needs to be shared to enable the health and social care systems to plan, develop, innovate, conduct research and be publicly accountable for the services they deliver to the people they serve.

Computers bring risks to individuals' privacy. They make healthcare information more easily accessible, re-usable and more easily manipulated than ever before. Unlike paper records, electronic records can be read, copied and even amended from remote locations. The user is unseen, file access may pass unnoticed, changes may be hard to detect and it is all done very fast. Security threats to electronic data are of a different level of magnitude than to paper records.

Access to electronic data must be regulated to avoid risks from criminal access, social control, discrimination and surveillance. Health information has to be protected from unauthorized access, use, disclosure, disruption, modification, perusal, inspection, recording or destruction. Risks derive from intentional and unintentional human activity including unauthorized access and disclosure, theft, virus and denial of service attacks. There are physical risks such as fire, floods and earthquakes, power loss, equipment failure, and software crashes, not directly caused by users.

Privacy and security needs to be addressed holistically, accounting for policy, risk assessment, procedure, training, and technology. Many healthcare organizations still take a traditional *perimeter* approach to privacy and security, with over-reliance on perimeter controls such as firewalls in the logical sense and buildings in the physical sense.

Cloud technology moves personal confidential data out of this perimeter and into the cloud provider's data center, but problems, including cybercrime malware, routinely occur inside the security perimeters. Healthcare data should be protected directly using encryption and other security services wherever it is, at rest or in transit, on EHR clients, servers, databases and backup systems.

Information systems security and governance is a broad subject and one short chapter in a book like this cannot do it justice. The ISO 27001 standard on information security management systems (ISMS) specifies the requirements for establishing, implementing, operating, monitoring, reviewing, maintaining and improving one within any organization [2]. It covers:

- Information security policies
- The organisation of information security
- Human resource (HR) security before during and after employment
- Asset management
- Access control
- Cryptography
- Physical and environmental security
- Operations security
- Communications security
- System acquisition, development and maintenance
- Supplier relationships
- Information security incident management
- Business continuity management security aspects
- Compliance with policies, regulations and laws

The HIPAA security rule (2003) [3] sets out detailed requirements for administrative, physical and technical safeguards, policies, procedures and documentation.

Healthcare organizations need to plan for every aspect of information governance, business continuity and disaster recovery. Most of these subjects are well beyond the scope of this chapter. Here we focus is on a few technical aspects of privacy management relevant to interoperability, including data protection, consent management and cryptography.

Data Protection

The term **personal confidential data** describes personal information about identified or identifiable individuals, which should be kept private or secret. This includes dead as well as living people and both information given in confidence and that which is owed a duty of confidence.

The NHS Information Governance Review, 2013, set out the following revised list of Caldicott principles.

1. Justify the purpose(s)
   Every proposed use or transfer of personal confidential data within or from an organisation should be clearly defined, scrutinized and documented, with continuing uses regularly reviewed, by an appropriate guardian.
2. Don't use personal confidential data unless it is absolutely necessary
   Personal confidential data items should not be included unless it is essential for the specified purpose(s) of that flow. The need for patients to be identified should be considered at each stage of satisfying the purpose(s).
3. Use the minimum necessary personal confidential data
   Where use of personal confidential data is considered to be essential, the inclusion of each individual item of data should be considered and justified so that the minimum amount of personal confidential data is transferred or accessible as is necessary for a given function to be carried out.
4. Access to personal confidential data should be on a strict need-to-know basis
   Only those individuals who need access to personal confidential data should have access to it, and they should only have access to the data items that they need to see. This may mean introducing access controls or splitting data flows where one data flow is used for several purposes.
5. Everyone with access to personal confidential data should be aware of their responsibilities
   Action should be taken to ensure that those handling personal confidential data—both clinical and non-clinical staff—are made fully aware of their responsibilities and obligations to respect patient confidentiality.

6. Comply with the law
   Every use of personal confidential data must be lawful. Someone in each organisation handling personal confidential data should be responsible for ensuring
   that the organisation complies with legal requirements. In the NHS such people
   are referred to as Caldicott guardians.
7. The duty to share information can be as important as the duty to protect patient
   confidentiality
   Health and social care professionals should have the confidence to share information in the best interests of their patients within the framework set out by these
   principles. They should be supported by the policies of their employers, regulators and professional bodies.

The seventh principle (the duty to share) is new.

These principles build on a great deal of work, going back to the OECD Privacy
Guidelines (1980) [4], issued more than 30 years ago. The OECD Privacy Guidelines
formed the basis for legislation throughout the world to protect personal data and
enable trans-border flows of personal data.

Privacy

Privacy is a critical non-functional requirement for the exchange of personal
confidential information, which is also referred to as individually identifiable health
information (IIHI).

Trust is fundamental in health and social care. All health information is sensitive,
but what each person deems sensitive depends on his or her individual circumstances. Some types of information (eg psychotherapy notes, terminations of pregnancy, substance abuse treatment) are more sensitive than others. People must be
confident that health services will protect the confidentiality of their personal health
information and that providers will be held responsible for any breach in the privacy
and security of records.

High-level data protection requirements include:

Confidentiality – ensuring that sensitive and/or business critical information is
appropriately protected from unauthorized 3rd parties and can only be accessed by
those with an approved need to access that information. Only fully identified and
authenticated entities, equipped with access control credentials, are able to use
services.

Integrity – ensuring that information has not been corrupted, falsely altered or
otherwise changed such that it can no longer be relied upon.

Availability – ensuring that information is available at point of need to those
authorized to access that information. The data and keys associated with encryption
for the purposes of confidentiality are recoverable.

Accountability – users are accountable for and unable to repudiate their actions.
A system's accountability features show who performed any action and what actions
have taken place in a specified time period.

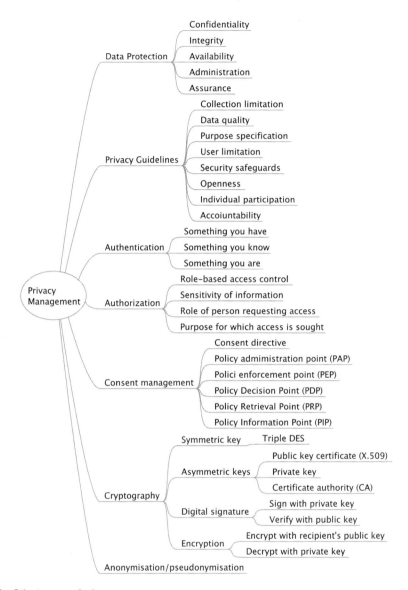

Fig. 5.1 Aspects of privacy management

Administration – those responsible for security policy have secure, usable interfaces for defining, maintaining, monitoring, and modifying security policy information.

Assurance - the claimed level of protection can be demonstrated with periodic validation that the protection is still effective (Fig. 5.1).

Rule

~~87~~

...rance Portability and Accountability Act of 1996 (HIPAA) set
. privacy and security protections for health information. The
...us Rule extends patients protection and increases the penalties
to... ...pliance up to $1.5 million per violation [5].

The HIPAA Privacy Rule [6] sets out how organizations need to
de-identify protected health information by removing all of the following
elements that could be used to identify the individual or the individual's
relatives, employers, or household members. See http://privacyruleandre-
search.nih.gov/pr_08.asp

1. Names
2. All geographic subdivisions smaller than a state, including street address,
   city, county, precinct, ZIP Code (postal code), and their equivalent geographical codes, except for the initial three digits of a postcode if according to the currently publicly available data the geographic unit formed by
   combining all ZIP codes with the same three initial digits contains more
   than 20,000 people, or the initial three digits of a ZIP Code for all such
   geographic units containing 20,000 or fewer people are changed to 000.
3. All elements of dates (except year) for dates directly related to an individual, including birth date, admission date, discharge date, date of death;
   and all ages over 89 and all elements of dates (including year) indicative
   of such age, except that such ages and elements may be aggregated into a
   single category of age 90 or older
4. Telephone numbers
5. Facsimile numbers
6. Electronic mail addresses
7. Social security numbers
8. Medical record numbers
9. Health plan beneficiary numbers
10. Account numbers
11. Certificate/licence numbers
12. Vehicle identifiers and serial numbers, including licence plate numbers
13. Device identifiers and serial numbers
14. Web universal resource locators (URLs)
15. Internet protocol (IP) address numbers
16. Biometric identifiers, including fingerprints and voiceprints
17. Full-face photographic images and any comparable images
18. Any other unique identifying number, characteristic, or code, unless oth-
    erwise permitted.

Information Governance Policy

Each organisation requires a formal information governance policy, which outlines the principles that underpin the policy, detailed procedures and sets out what is expected of staff, including action to be taken if the policy is breached. This policy needs approval at a high level and all staff made aware of it. Leaders are accountable for information security.

There needs to be an information governance lead, responsible for managing and monitoring the implementation of information governance policy.

All contracts with staff, including temporary and contract staff, who may have access to sensitive data, should contain clauses that clearly set out their responsibilities for ensuring confidentiality, information security and data protection. Staff need to be provided with appropriate training and made aware of their information governance responsibilities.

Components of a confidentiality code of conduct typically include:

- Responsibility for compliance with the law
- Definition of material considered to be confidential
- Guidelines on passwords, smartcards and security
- Systems and processes for protecting personal information (secure storage, safe havens etc.)
- Use of email and web-based services
- Circumstances under which confidential information can be disclosed
- Subject access
- Abuse of privilege
- Off-site/home working arrangements
- Who to approach in case of difficulty
- Sanctions for breach of confidentiality.

A key aspect of information governance assurance is to identify and map all routine flows of personal health information in and out of the organisation, in order to identify risks associated with data transfer and to mitigate such risks. Safe haven procedures for the receipt of confidential information need to be documented.

Patients need to be informed about how patient information is used and stored, who is able to access patient information, how they can gain access to information about themselves and who they should talk to for more information.

Authentication

Authentication verifies that people are who they say they are.

The mechanisms required to ensure identification and authentication are continuing to evolve, but generally involve:

- Something only they have, such as a smart card or other token

- Something only they know, such as a personal identification number or PIN, which they may choose
- Something they are, such as a finger print.

Strong authentication requires providing at least two of these three types of information. All unsuccessful attempts at authentication are logged.

Authorization

Authorization is the next stage after authentication. Authorization determines what people are able to do – what programs they can run and what they can view, add, change or delete.

One of the most commonly used forms of authorization is role-based access control (RBAC), which authorizes users according to their job functions. Permissions to do certain tasks are assigned to roles. Users are assigned functional roles. Users acquire permission to perform tasks through role assignments. Users are not assigned permissions directly, but acquire them through their roles.

Any user may be assigned to many roles. Similarly any role may undertake many different task permissions, and any permission may be authorized to many roles.

For management review, it is important to know all of a user's roles and all of a role's users. There is a complementary need to review all of a role's permissions and all of a permission's roles.

One benefit of RBAC is that roles are fairly stable and this helps reduce the cost, complexity and potential for error in assigning user permissions. More complex schemes can include role hierarchies, where one role can include all the permissions of another role.

People are demanding more control over who may view their records and to restrict access to certain types of data. An international standard (ISO/TS 13606-4) provides a structure to classify access to healthcare data according to its sensitivity, the role of the person seeking access and purpose [7].

Information sensitivity depends on purpose:

- Personal care – information for which the person decides to whom they will reveal that information, such as a past history of abortion.
- Privileged care – information, which would be made available only on a strict need-to-know basis, such as psychiatric illness, drug and alcohol abuse and sexually transmitted disease.
- Clinical care – information used for direct and indirect clinical care.
- Clinical management – information needed by staff in clinical laboratories, pharmacy and other support services.
- Care management – administrative and financial information about patients.

Role played by the person wishing access to the data:

- Subject of care – people or patients.

- Subject of care's agent, such as the parent of a child.
- Personal health professional explicitly named and authorized by the patient.
- Privileged health professional, by virtue of his or her role and relationship with the patient.
- Health-related professional such as a laboratory technician.
- Administrative and clerical staff.

Purpose for which access is sought:

- At the request of the subject – no reason need be given.
- For clinical care.
- Payment.
- Medico-legal purposes.
- Teaching.
- Research.

Consent

Consent is the approval or agreement for something to happen. For consent to be legally valid, the individual giving consent must be informed, must have the capacity to make the decision in question and must give it voluntarily. This means individuals should know and understand how their information is to be used and shared (no surprises) and understand the implications of their decision, particularly where refusing to allow information to be shared may affect the care they receive. This applies to both explicit and implied consent.

Explicit consent is clear and can be given in writing or verbally, or conveyed through another form of communication such as signing. Explicit consent is required when sharing information with staff who are not part of the team caring for the individual. It may also be required for a use other than that for which the information was originally collected, or when sharing is not related to an individual's direct health and social care.

Implied consent applies only within the context of direct care of individuals and consent of the individual patient can be implied without having to make any positive action. Examples of implied consent include doctors and nurses sharing personal confidential data during handovers without asking for the patient's consent. Alternatively, a physiotherapist may access the record of a patient who has already accepted a referral before a face-to-face consultation on the basis of implied consent.

Implied consent may be used as the legal basis for sharing relevant personal confidential data in communications such as letters and discharge summaries. However, there is less consensus on the legal basis for sharing whole records. When whole records are shared, patients do not have the ability to block access to individual information items, which does not align with the principle of sharing only relevant information.

For example, a patient may tell a GP she is pregnant, but not by her husband, and she does not consent to this information being shared with any other doctor. Or a professional in a particular field, such as a physiotherapist treating a patient's knee, may have no need to know about his erectile dysfunction.

Third party data may create other problems. Third party data includes data from or about third parties. Data from a third party would be Mrs. X reporting her husband's headaches, personality change and refusal to visit the doctor. Data about a third party includes a family history of premature stroke in the patient's siblings listed in the patient record. With increasing patient access to records, information from a third party should not be added to a patient's record unless the provider of the information understands that the patient may become aware of this information and its source.

Consent is one way in which personal confidential data can be legally shared. Consent decisions should be recorded and be available to be shared, so people's wishes can be respected, bearing in mind that people can change their consent at any time.

In an opt-in consent system, the patient has to provide explicit consent for their data to be used. In an opt-out system, consent is implied if a patient does not opt out. However, both approaches can be implemented in ways that fail to allow the patient any meaningful choice. For example, if a registration clerk just gives the patient a form that broadly describes all potential uses and disclosures of personal health information and asks the patient to sign and consent to all of it, that is not meaningful explicit consent. Similarly, if patients have a right to opt out, they must be provided with the time, information and means to exercise that right.

Some people would like to have control over who can access what aspects of their information, when and where. For example, a patient might wish to obtain a second opinion before deciding to have an operation, but this may be of limited value without access to medical history and test results. Other use cases include the need to share health information with the wider circle of care, such as family members and other health and social care organizations. People should be in control over the use of their personal confidential information outside the information governance perimeter of the organisation that creates it.

Healthcare services are moving from the era of delegated consent to shared decision making. In delegated consent patients hand control of their healthcare to one or more trusted doctors and let the experts do as they think best, while in shared decision-making, patients are active informed participants in their own care ("no decision about me without me").

The same considerations and customary practice that apply to paper and fax exchange apply to electronic exchange. Organizations need operating procedures to ensure that people are asked to give consent to their personal confidential information being used for any purpose not directly related to the service for which it was collected. Such operating procedures need to cover:

• How and when to obtain consent.
• How to inform people about who may have access to personal health information.

- Patients' right to choose.
- Patients' right to change their mind.
- How to record consent and dissent.
- Exemptions, such as when the patient is unconscious.

Consent is documented in a consent directive, which is an agreement (a document) between the person and a care provider (individual or organization), granting or withholding authorization to access their personal confidential information.

HL7 has produced a draft standard for trial use for implementing consent directives using CDA (Clinical Document Architecture) [8]. Consent directives may be encoded using OASIS eXtensible Access Control Markup Language (XACML) [9].

Access Control

Access control enforcement has several logical functions, described in the XACML documentation. These include:

- Policy administration point (PAP)
- Policy information point (PIP)
- Policy retrieval point (PRP)
- Policy decision point (PDP)
- Policy enforcement point (PEP)

The Policy Administration Point (PAP) creates and manages policies and consent directives.

A Policy Information Point (PIP) may also be required, which is the source of information about the participants that may be needed to make decisions. For example the PIP may be used to look up the current role of an individual requesting access.

Policies relating to access control, including consent directives, are stored at a Policy Retrieval Point (PRP).

They are acted upon by a Policy Decision Point (PDP), which evaluates the applicable policy and consent directives, and issues an authorization decision.

Policies are enforced by Policy Enforcement Points (PEP), which send authorization requests to the PDP and implement the decisions returned. For example, when the PEP receives an access request, it queries the PDP, which decides whether to allow access for that request.

Access control includes procedures to add, edit and remove user accounts. Each member of staff, including temporary staff, should have individual logins, with an appropriate level of access. Access management procedures include setting user ID, access levels, rights and passwords.

OAuth

OAuth is an open protocol to allow secure authorization in a simple and standard way from web, mobile and desktop applications. The OAuth framework enables third-party applications to obtain limited access to HTTP services.

For app developers OAuth provides a simple way for clients to interact with personal confidential information, referred to in OAuth as a protected resource, held on another data store (resource server), while protecting end users' (resource owner) passwords.

OAuth eliminates the need for end-users to reveal their passwords to apps. It can also restrict the level of data available to an app and allow end-users to revoke access to their data, when it is no longer needed.

For app developers, it helps keep end-users safer. It minimises how many passwords users need, discourages password reuse and simplifies the process of signing on new users.

The key to OAuth is that the resource owner (typically a person or end-user) provides explicit consent to generate an access token that verifies end-user identity and specifies what data can be accessed.

OAuth introduces an authorization layer to separate the client (app) from the end user (person). In OAuth, the client requests access to resources controlled by the end user and hosted by the resource server. However, instead of using the end user's username and password to access protected resources, the client obtains a special access token, which when presented grants access. An authorization server issues the access tokens to the client after successfully authenticating the resource owner and obtaining authorization.

An access token is a string denoting specific scope, lifetime, and other access attributes. An authorization server with the consent of the end user issues the access token to the client. The client uses the access token to access the protected resources hosted by the resource server.

For example, an end-user can grant a printing service (client) access to her protected photos stored at a photo-sharing service (resource server), without sharing her username and password with the printing service. Instead, she authenticates directly with a server trusted by the photo-sharing service (authorization server), which issues the printing service delegation-specific credentials (access token).

OAuth is designed to work with HTTPS. HTTPS provides bidirectional encryption of communications between a client and server. It provides a reasonable guarantee that one is communicating with the website that one intended to communicate with (as opposed to an impostor), as well as ensuring that the contents of communications between the user and site cannot be read or forged by a third party.

Cryptography

The principles of cryptography are relatively simple, but the processes involved are quite complex.

Symmetric key cryptography uses the same key for both encryption and decryption. The most commonly used algorithm is the Triple DES (Data Encryption Standard). This is fast and efficient, but has the obvious weakness that both the encryptor and the decryptor need to know the same key, providing an obvious route of attack. One way to share a symmetric key is to share it using asymmetric encryption.

Asymmetric encryption uses public key infrastructure (PKI), which requires public key certificates.

Public key certificates are also known as digital certificates and as X.509 certificates [10]. They are used for a variety of purposes such as sharing a secret key used to exchange encrypted information, to digitally sign documents and to authenticate the identity of a person or entity using a challenge-response mechanism.

A certificate is simply an electronic document, which typically contains:

- Serial number of the certificate.
- Subject's public key.
- Subject's name.
- Validity date range.
- Name of the certification authority (CA) that issued the digital certificate.
- Signature of the CA that issued the certificate.
- Thumbprint – hash of certificate to ensure that it has not been tampered with.
- Details of algorithms used.
- Optional extensions such as the purposes for which the certificate may be used such as SMTP and S/MIME.

The current recommended standard is the X.509 v3 certificate profile defined in RFC 5280.

Public key certificates are generated by a *Certificate Authority* (CA), which needs to be a trusted third party, who generates random private keys and corresponding public keys bound to a particular user. The CA undertakes a detailed process of identity verification, which involves checking to ensure that the person requesting a certificate is who they claim to be. This identity verification may require attestation by a public notary and exchange of information using hard copy. Different levels of public key certificates are available according to the rigour of the process used to verify the owner's identity. Standard validation certificates verify the owner's email address, while extended validation involves additional verification of the owner's personal identity. Typically, certificates need to be renewed every year.

The validity of a public key, contained in a certificate, is signed by the CA's private key and can be checked using the CA's own public key. The trustworthiness of a digital certificate is dependent on how much you trust the CA, and indeed the CA that issued the private key to the CA that issued the certificate, back to the root CA. The practices used by CAs in issuing and managing certificates are described in their certification practice statements (CPS).

The user's private key may be hidden on something that belongs to the user, such as within their PC or server or on a smart card. When stored on a computer system, either within a browser or on a secure server, the private key is protected by a local password, but this has two disadvantages. The user can only sign documents on that particular computer and the security of the private key depends on the security of the computer. A better alternative is to store the private key on a smart card, which is protected by a personal identification number (PIN). This is two-factor authentication (something you have – the smart card – and something you know – the PIN). The private key never leaves the smart card. If the smart card is stolen, the public key can be revoked immediately. The CA issues a certificate revocation list (CRL).

There are a number of practical difficulties in using certificates. For example, in some schemes, the owner's name on a public key certificate has to be exactly the same as the user's logon name for the computer being used and the certificate can only be used within the logon under which it was installed.

Digital Signature

Digital signatures are used for authentication, integrity and non-repudiation. Authentication of the source of a document is based on the assumption that the secret key, which is known to have been used in creating the digital signature, is bound to a specific user. The integrity of a document signed with a digital signature is assured because the document itself is used in creating the signature. Any change to the document after it is signed will invalidate the signature. Non-repudiation of document origin is assured because the signer of a digital signature cannot later deny that he or she signed the document.

The names Alice and Bob are commonly used placeholder names for archetypal characters in security-related discussions. It is easier to understand "Alice sends a message to Bob" than "Party A sends a message to Party B". Other characters can be introduced in alphabetical order, such as Carol and Dave. Eve is usually an eavesdropper.

In PKI, the Alice uses Bob's public key to send him an encrypted message. Bob reads the document using his own private key.

The digital signature process involves signature by the sender and verification by the receiver. The process of signing a specific document has three steps:

- Alice uses a standard hash function, such as SHA-1 (Secure Hash Algorithm) on the document to produce a short hash string.

- Alice encrypts the hash string with her own private key to produce the digital signature.
- She attaches the digital signature and her own digital certificate, which includes her public key, to the document.

The verification process also has three stages, to accept or reject the authenticity of a specific document, the originator's public key and the digital signature:

- Bob uses the hash function (SHA-1) on the document to recreate the hash string.
- Bob decrypts the digital signature using the Alice's public key, from her digital certificate to produce a test string.
- Bob checks that the test string is exactly the same as the hash string. If the two are equal, the signature is valid.

Encryption

Encryption using PKI works in a similar manner to digital signatures but with some important differences:

Secure MIME (S/MIME) [11] is a specification for secure electronic messaging to prevent the interception and forgery of e-mail and other point-to-point messaging. The process of encrypting a document using S/MIME has the following stages:

- First, Alice generates a one-off key, known as a one-off session key; think of this as a random number. This session key is used for symmetric encryption of the document.
- The session key is encrypted first using sender Alice's public key, second using recipient Bob's public key to create two different strings.
- The encrypted document is then sent to Bob.
- Both Alice and Bob can decrypt the session key using their respective private keys and hence both can decrypt the original message.

It is best practice for encryption and digital signature functions to use different sets of keys.

Rights and Obligations

The 2013 NHS Information Governance Review [1] set out that as a service user:

1. You have the right of access to your own personal records within the health and social care system.
2. You have the right to privacy and confidentiality and to expect the health and social care system to keep your confidential information safe and secure.

3. You have the right to be informed about how your information is used.
4. You have the right to request that your confidential data is not used beyond your own care and treatment and to have your objections considered, and where your wishes cannot be followed, to be told the reasons including the legal basis.

It also sets out the responsibilities of health and care providers to:

1. Ensure those involved in providing care and treatment have access to patients' health and social care data so they can care safely and effectively.
2. Anonymise the data collected during the course of care and treatment and use it to support research and improve care for others.
3. Give patients the opportunity to object to the use of identifiable data wherever possible.
4. Inform patients of research studies in which they may be eligible to participate.
5. Share with patients any correspondence sent between staff about your care.

Other Security Services

Anonymization is the removal of identifiable personal elements from the data, making it less sensitive and less subject to stringent regulations governing privacy of personal data – while retaining its value for legitimate secondary uses like research and reporting. One approach is to strip out person identifiers to create a data set in which person identifiers are not present.

Pseudonymization replaces person identifiers with other values (pseudonyms) from which the identities of individuals cannot be inferred, such as replacing a patient ID with a random number. In some situations it may be possible to reverse the process.

Audit Trails maintain a record of actions related to electronic health information. The date, time, patient identification, and user identification are recorded when electronic health information is created, modified, accessed, or deleted, with an indication of which actions occurred and by whom. Every transaction is tracked and logged. The log file can be subject to routine surveillance to detect abnormal activity patterns.

Contract Requirements Relating to Use and Sharing of Data

Appropriate, legally enforceable contractual arrangements are needed to provide protection for processing personal confidential data or de-identified data for limited access in the public interest. The 2013 NHS Information Governance Review (2013) offers the following checklist for contracts [1].

1. Status and relationship of the parties as data controller or data processors. This includes clarity about sole, joint or in common data controllership, and where an organisation's relationships with data may fall across these categories in different circumstances, to clarify the circumstances in which the different relationships with the data will apply.
2. Scope and term of the contract.
3. Whether the contract will be supported by service level/data sharing agreements. Where applicable define data set and disclosures for specific purposes. This should include any variation to the data controller relationships set out in the contract.
4. Terminology used.
5. Legal, professional and contractual requirements. Definition of the governing law, requirement to adhere to legal and professional requirements, and the provisions of this contract in particular in relation to data protection, human rights and common law obligations such as the duties of care and confidentiality. These include but are not limited to:

   - When personal confidential data may lawfully be disclosed.
   - For de-identified data for limited disclosure or access the requirement for this data to be held separately from personal confidential data within a safe haven (to ensure it does not become identifiable, and therefore personal data requiring a legal basis to process).
   - Having mechanisms to prevent re-identification where de-identified data may be linked together in a safe haven.
   - A requirement not to disclose data to other parties other than in anonymised form, or as authorized by the data controller, or where required by law.
   - for data processors the requirement only to process data as instructed by the data controller.

6. Duty to co-operate with other parties.
7. In relation to personal confidential data, a definition of the purposes and the legal basis for processing for each specified purpose, with a restriction to confine processing to these purposes. Where there is a need to re-identify individuals, this must be in the purposes and authorized. It is helpful to include this within the contract so all parties are assured of the legal basis for processing and the boundaries of that legal basis. Privacy impact assessments are helpful in clarifying whether there is a secure

(continued)

basis in law and the nature of that basis as part of the pre-contract checks and ongoing management of the contract. In relation to de-identified data for limited disclosure or access, clarity of the purposes and assurance that the purposes of processing are in the public interest.

8. Confidentiality and protection of commercially sensitive information and intellectual property.
9. Fair processing information responsibilities, including service user involvement in its development.
10. Policies and procedures on: consent both for treatment and for the use of data; conflicts of interest management; and agreement more broadly about whose policies are used. This may be specific to the policy in question.
11. Timely communication of transfer or discharge information to other care professionals.
12. Online access to records and communication of care plans to the service user.
13. Conformance with requisite Information and data standards.
14. Staff recruitment checks, education and training, and terms and conditions of employment—this also needs to address honorary and seconded staffing arrangements to ensure the failure to adhere to policies and procedures are addressed through disciplinary action via the substantive contract of employment.
15. Maintenance of Information asset registers, data flow mapping and data sets for extraction and reporting requirements.
16. Data extraction processes.
17. Responsibility for FOI (freedom of information), EIR (environmental information regulations) and subject access requests—in particular attention needs to be given to who will undertake the clinical review of records for Subject Access Requests to ensure that seriously harmful information, or information provided by third parties is not disclosed.
18. Housekeeping measures covering:

    • Business continuity
    • Disaster recovery
    • Monitoring and auditing of access controls and reporting
    • Transfer, retention, archiving, and disposal of records at end of data lifecycle in line with record retention schedules or termination of contract.

19. Security requirements (ISO 27001 and 2) in information security management systems (ISMS) to include:

    • Network security
    • Device security (including encryption)

(continued)

- Software security including protection against malware
- Data and system back-up
- Secure transfer of data
- Physical security
- Access control functionality, logging, alerts, auditing and reporting
- Software control of printing and USB devices
- Use of security and privacy enhancing technologies
- Risk assessment, audit and reporting (including penetration testing)
- Review and updating
- Incident reporting.

20. Registration Authority (RA)—Legitimate Relationship (LR) and Role Based Access Control (RBAC) authorisation and implementation.
21. Change control, authorized officers and approvals processes.
22. Sub-contracting notification to data controller of intent to sub-contract, identity of sub-contractor(s), contracting and oversight arrangements of sub-contractor and authorization by data controller requirements.
23. Location of data storage and arrangements ie within EEA, outside EEA, or cloud. Need for binding corporate rules or other means of satisfying DP principle 8.
24. Serious incidents/data breaches (duty of candour): monitoring, reporting, investigating, publishing with outcomes.
25. DC contract performance management including right of access to visit site(s) and audit procedures/use of data including any sub-contractors. Additionally, mandatory independent audit of the IG Toolkit submission or equivalent statements of compliance should also be considered, with the scope set annually by the data controller.
26. Process for agreeing variations to the contract including novation to new bodies.
27. Dispute resolution process.
28. Exit from contract:

- Natural end of contract considerations such as record management
- Premature end of contract from failures of any party eg bankruptcy, serious data breach
- Continuing obligations, eg not using data subsequently for own purposes and maintaining confidentiality of personal data indefinitely.

29. Charges, liability and indemnity, remedies and penalties for breach of contract—care needs to be taken to ensure that this clause includes unlimited recovery of costs arising from a breach by data processor and data processors need to maintain insurance supporting liability in the contract.

(continued)

30. Definition of roles and responsibilities—senior responsible officers for implementation and oversight of different elements of the contract for each party to the contract.
31. Signatures of senior responsible officers of all parties.
32. An appendix to the contract, with the day-to-day contact details for the senior responsible officers and other key staff.

References

1. Caldicott F. Information: to share or not to share: the information governance review. London: Department of Health; 2013.
2. ISO/ICE 27001:2013 – information technology – security techniques – information security management systems – requirements. International Organization for Standardization. 2013.
3. Health Insurance Reform: security standards; Final Rule. Department of Health and Human Services. Federal Register Vol. 68, No. 34. February 20, 2003.
4. OECD. Guidelines on the protection of privacy and transborder flows of personal data. Paris: OECD; 1980.
5. Rothstein MA. HIPAA privacy rule 2.0. J Law Med Ethics. 2013;41(2):525–8.
6. Gunn PP, Fremont AM, Bottrell M, Shugarman LR, Galegher J, Bikson T. The health insurance portability and accountability act privacy rule: a practical guide for researchers. Med Care. 2004;42(4):321–7.
7. ISO/TS 13606-4:2009 Health informatics – electronic health record communication – part 4: security.
8. HL7 implementation guide for clinical document architecture, release 2: consent directives, release 1. HL7 draft standard for trial use, CDAR2_IG_CONSENTDIR_R1_DSTU_2011JAN. January 2011.
9. Extensible Access Control Markup Language (XACML) Version 2.0. OASIS Standard 2005. oasis-access_control-xacml-2.0-core-spec-os.
10. Cooper D et al. Internet X.509 public key infrastructure certificate and Certificate Revocation List (CRL) Profile. IETF Network Working Group RFC 5280. May 2008. http://www.ietf.org/rfc/rfc5280.txt
11. Ramsdell B (ed). Secure/Multipurpose Internet Mail Extensions (S/MIME) Version 3.1 message specification. IETF Network Working Group RFC 3851. July 2004. http://www.ietf.org/rfc/rfc3851

Chapter 6
Standards Development Organizations

Abstract Standards are documents, established by consensus and approved by a recognised body. ISO is the international standards organization with membership from national standards bodies. US-based standards developers (SDOs) such as HL7 and DICOM are represented by ANSI. HL7 International is the leading SDO for healthcare interoperability and is described in more detail than the others. DICOM leads in medical imaging; IHTSDO for SNOMED CT. IHE develops standards profiles for specific purposes, Continua works on consumer medical devices and CDISC on clinical trials data. OpenEHR develops clinical models.

Keywords Standard • Consensus • ISO • CEN • ANSI • HL7 International • IHTSDO • Recognised body • Standards development organizations • Joint Initiative Council (JIC) • OSI model • Ballot process • Membership • TSC • DICOM • IHE • CDISC • Continua alliance • OpenEHR

Healthcare interoperability is based on the application of standards. This chapter introduces the major international Standards Development Organizations in digital health.

When new requirements emerge, such as in response to the development of personal apps on mobile phones and wearable devices, the SDOs need to respond.

What is a Standard?

ISO defines a **standard** as a document, established by consensus and approved by a recognized body, that provides, for common and repeated use, rules, guidelines or characteristics for activities or their results, aimed at the achievement of the optimum degree of order in a given context [1].

Two of the key terms are *consensus* and *recognized body*. **Consensus** is general agreement, characterized by the absence of sustained opposition to substantial issues by any important part of the concerned interests and by a process that involves seeking to take into account the views of all parties concerned and to reconcile any conflicting arguments. Consensus need not imply unanimity.

© Springer-Verlag London 2016
T. Benson, G. Grieve, *Principles of Health Interoperability*,
Health Information Technology Standards, DOI 10.1007/978-3-319-30370-3_6

A **recognized body** is understood to be an internationally recognized standards development organization such as ISO, CEN, BSI, ANSI and its accredited SDOs including HL7.

The activity of **standardization** consists of the processes of formulating, issuing and implementing standards. Important benefits of standardization are improvement of the suitability of products, processes and services for their intended purposes, prevention of barriers to trade and facilitation of technological cooperation.

There are two main types of standard:

- Exact specifications, which enable interworking of nuts and bolts, paint colors and computers
- Minimum thresholds to ensure the safety and quality of processes, materials and the environment.

For healthcare interoperability we mainly need stringent specifications at the technical and data layers, although minimum thresholds may be needed to ensure safety and security at the human and institutional levels.

Interoperability has been a limiting factor in market growth for health information systems, in part due to lack of suitable standards. Standards have a multiplier effect, the more people can interoperate, the more cost effective is every new application and the larger the IT market becomes.

Interoperability standards are the foundation of whole industries [2]. This is well illustrated by the explosive growth of the airline, Internet and mobile telephone markets, and should also be true of healthcare computing. In healthcare digital imaging has been relatively successful because all suppliers adopted DICOM (see below).

One reason is that health-care standards development organisations have failed to provide sufficiently stringent standard specifications to enable plug and play, leaving this to local implementers.

A report produced in 2008 for the European Union concluded

> Despite a generally large number of conflicting e-health standards, versions and implementations, there may be a lack of the "right" standards. For particular applications and for concrete processes there may be no well-developed standards. In an expert survey, 80% of the respondents stated that there is a lack of sufficiently developed standards, and 64% said that there is a lack of standards for electronic health records (EHRs) [3].

The benefits of using standards increase exponentially with the number of different systems that need to be linked.

Purchasers of computer systems should insist on open interoperability standards to avoid supplier lock-in and give them choice and flexibility in procurement, allowing them to shop around for whatever meets their needs most closely. Open standards offer a guarantee for future migration, growth and evolution, foster competition between suppliers, drive down costs and push up cost-effectiveness.

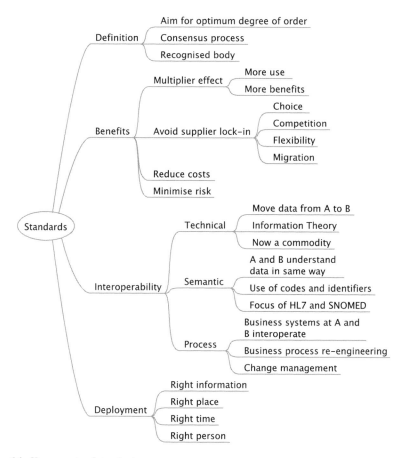

Fig. 6.1 Key aspects of standards

Suppliers also benefit; their criterion for success is return on investment. The actual return is often outside each supplier's direct control, so their priority is always to minimise investment costs and risk (Fig. 6.1).

How Standards Bodies Work

Most of the work in developing standards is performed by volunteers, often working over many years in small committee meetings. Their proposals are then presented to a much larger group to achieve a consensus.

For example, HL7 volunteers meet together three times a year in weeklong working group meetings at which more than 30 specialised committees meet face to face. Work continues throughout the rest of the year coordinated by regular telephone conferences.

The writing of the draft standard is usually the work of a few dedicated individuals – typically people who represent the vendors in the field. Other people then review that draft; controversial points are discussed in detail and solutions are proposed and finally accepted. Writing and refining the standard is further complicated by the introduction of people new to the process who have not been privy to the original discussions and want to revisit points which have been resolved earlier. The balance between moving forward and being open is a delicate one. Most standards-writing groups have adopted an open standards development policy; anyone can join the process and can be heard [4].

Ten commandments for effective standards (originally created in the arcane world of electronic design automation of computer chips) apply equally well in health information and interoperability standards [5]:

1. Cooperate on standards, compete on products. This is the Golden Rule of technical standards. The essence of standardization is to provide interfaces for multiple products to work together well, while encouraging suppliers to develop the best products possible.
2. Use caution when mixing patents and standards. Perhaps the biggest challenge faced in creating technical standards is making them available for everyone to use without restrictions while protecting the intellectual property of invention.
3. Know when to stop. Not every standards project should be completed. Not every standards project should be started. Not everyone wants to join. Timing is important, as is having the right participants.
4. Be truly open. The word "open" has many definitions. When it comes to standards, open means available to everyone, without discrimination or conditions.
5. Realize that there is no neutral party. Everyone participating in a standards project has a reason for being there, whether it is obvious or not. Technical standards can be political.
6. Leverage existing organizations and proven processes. Reinventing the wheel isn't necessary. It is more effective to work within experienced standards development organizations.
7. Think relevance. Technical standards can be expensive to produce, so it's important that they address a real need or solve a real problem.
8. Recognize that there is more than one way to create a standard. Formal standards committees are just one way to create technical standards for an industry. Different methods have pros and cons.
9. Start with contributions, not from scratch. Producing standards from technology that has already been developed can speed up the standardization process and increase the quality of the resulting standard.
10. Know that standards have technical and business aspects. Getting the technical details right for a standard is necessary, and so is understanding the commercial implications.

SDOs

The organization of health informatics standards development internationally is complex, changes frequently, and has created a fog of acronyms. The intent here is to introduce the most important players and to provide information that does not readily fit into other chapters.

The *International Standardization Organization (ISO)* was established in 1947 to provide a focal point for all international standards. ISO is a membership organisation, with one member in each country. In the USA the member is the *American National Standards Institute (ANSI)* and in the UK the member is the *British Standards Institute (BSI)*. The *Vienna Agreement* specifies how conflicts between different standards should be handled. In particular, work done at the International level takes precedence over national standards.

ISO has established a committee for Health Informatics (ISO TC215). The main task of this committee is to ratify existing standards as full international standards.

The *European Committee for Standardisation (CEN)* was founded in 1961, along the same lines as ISO as a national member organization. In 1990, CEN set up the first formal international standards organization in health informatics, CEN/TC 251. Its scope is:

> *Standardization in the field of Health Information and Communications Technology (ICT) to achieve compatibility and interoperability between independent systems and to enable modularity. This includes requirements on health information structure to support clinical and administrative procedures, technical methods to support interoperable systems as well as requirements regarding safety, security and quality.*

CEN/TC 251 is organised into four working groups: (WG 1) information models, (WG 2) terminology and knowledge representation, (WG 3) security safety and quality, (WG 4) technology for interoperability. Each European country established its own mirror committee; for example in the UK, the mirror committee is BSI IST/35.

In the USA, ANSI was established in 1918 to represent existing SDOs. ANSI now accredits 220 standards developers. ANSI accreditation dictates that any standard submitted to ANSI for approval be developed and ratified by a process that adheres to ANSI's procedures for open consensus. It must meet a balance of interest requirement by attaining near equal participation in the voting process by the various constituencies that are materially affected by the standard (eg, vendors, providers, government agencies, consultants, non-profit organizations). This balance of interest goal ensures that a particular constituency is neither refused participation nor is it allowed to dominate the development and ratification of a proposed standard.

SDOs and related organisations operating in the healthcare IT domain include:

- ISO/TC 215
- CEN/TC 251
- IHTSDO The International Health Terminology Standards Development Organization, responsible for SNOMED CT

- *HL7 International for clinical and administrative data.
- DICOM (ACR/*NEMA) – Digital Imaging and Communication in Medicine
- *ASTM International – Continuity of care record (CCR)
- *IEEE – Bedside devices
- *ASC X.12 – Claims processing.
- *NCPDC – National Council for Prescription Drug Programs
- *ADA – American Dental Association
- GS1/*ICC for bar code standards
- *OASIS – XML schema
- IHE Integrating the Healthcare Enterprise, which develops profiles for specific use cases leveraging existing standards.
- Continua Health Alliance focuses on home Telehealth devices.
- CDISC (Clinical Data Interchange Standards Consortium) responsible for coordinating data capture for clinical trials
- OpenEHR focuses on elements of EHR architecture.

Organizations marked with a star (*) are ANSI accredited standards developers.

Joint Initiative Council

Harmonization between international health informatics SDOs is performed by the Joint Initiative Council (JIC), which includes eight SDOs (CEN/TC 251, CDISC, DICOM, GS1, HL7, IHE, IHTSDO, and ISO/TC 215). In the San Francisco Declaration (April 2015) the executives of the Joint Initiative Council declared that their objective is to *"contribute to better global patient health outcomes by providing strategic leadership in the specification of sets of implementable standards for health information sharing"*.

The Council operates at the strategic level to identify emerging requirements for standardization and to resolve gaps, overlaps, and counterproductive health informatics standardization efforts. In the spirit of openness, transparency and flexibility, it seeks to promote common policies across participating SDOs, including full sharing of deliberations with all stakeholders from the health informatics standards community in support of standards harmonization.

HL7 International

HL7 International is an international standards development organisation (SDO), with Affiliates in 31 countries.[1] HL7 produces the world's most widely used standards for healthcare interoperability. Most of the leading suppliers use and support the development of HL7 standards across six continents.

---

[1] See www.hl7.org

HL7's vision statement is

A world in which everyone can securely access and use the right health data when and where they need it.

Its mission reads:

HL7 empowers global health data interoperability by developing standards and enabling their adoption and implementation.

HL7 creates standards for the exchange, management and integration of electronic healthcare information for clinical and administrative purposes. HL7 does not develop software, but simply provides healthcare organisations with specifications for making their systems interoperable. It develops coherent extensible standards using a formal methodology. It collaborates with and provides a meeting place for healthcare information experts from the healthcare IT industry and healthcare providers to work together and with other standards development organizations. And, it promotes its own standards and provides education for the healthcare industry and policy makers.

The name Health Level 7 is derived from the 7th level of the ISO's Open Systems Interconnect (OSI) model: the application layer, which provides a framework for communication between disparate computer systems. The OSI model has seven layers; the top three layers are concerned with applications (interworking); the lower four layers are concerned with the transmission of data (interconnection):

Layer 7 – Application: addresses definition of the data to be exchanged, the timing of the interchange, and the communication of certain errors to the application.
Layer 6 – Presentation: concerned with the syntax of information transfer between end systems.
Layer 5 – Session: provides mapping between physical and logical sessions, including checkpoint recovery and restart.
Layer 4 – Transport: provides end-to-end transmission of data to the required quality of service (eg error-free).
Layer 3 – Network: concerned with routing and relaying between multiple sub-networks
Layer 2 – Data-link: transmit a stream of bits from one network node to another with indication of errors and limited error correction.
Layer 1 – Physical: provide the interface to the physical communications medium

Enveloping is a key concept in the OSI model. Data from a source system enters the OSI stack at layer 7 (application) and is encapsulated by another envelope at each layer, so that by the time it reaches the communication medium (the wire) at Layer 1, it has collected seven envelopes. At the destination, each envelope is checked and removed, one by one, so that the data exiting from layer 7 at the destination is exactly what the source system sent.

Layers 1–6 of the OSI model deal with various aspects of technical interoperability. The only domain-specific aspect is the application layer – Layer 7, which deals with the semantics or meaning of what is exchanged. This is why the founders of HL7 chose the name Health Level Seven.

HL7 Products

Since 2013, HL7 has licensed its standards and other intellectual property free of charge. The change was made so that there would be no financial or political barriers to the adoption of HL7 standards worldwide.

HL7 produces four types of document: documents can be:

- Normative Standard: content is balloted by the general membership and is considered a structural component of the HL7 Standard. Negative ballots must be resolved.
- Draft Standard for Trial Use (DSTU): Content is balloted by the general membership as the draft of a future standard which will, following a pre-specified period of evaluation and comment (usually 2 years), be expeditiously incorporated into normative standard. Because many DSTUs are used for operational purposes, these may be renamed as Working Standards with different levels of maturity.
- Reference: content is harmonized during HL7 meetings or approved by the HL7 Board. It is not subject to ballot acceptance.
- Informative: content is balloted by the general membership. However, it is not considered to be a structural part of the Standard but only supporting information.

HL7 Balloted Standards are introduced first as a DSTU and must show some successful implementation before being advanced as a Normative Standard.

Ballot Process

Ballots normally progress through two or more cycles of ballots. The ballot pool is limited to declared interested members. Negative votes must be accompanied with a specific reason justifying the negative vote. Work Groups must resolve negative votes either by accepting the voters comment and recommended solution, negotiating with the voter and get them to agree to withdraw their negative or declare the vote non-persuasive.

Voters may appeal to the TSC and Board. They can also re-vote their same negative vote on the next round of balloting. Substantive changes to a ballot (either to fix a negative or add new material) merit another ballot round. When 75 % (for normative documents) of the responses are registered as affirmatives and (hopefully) all negatives withdrawn, a document is ready for publication as an HL7 Standard.

HL7 has produced a "Version 3 Publishing Facilitator's Guide", which is a style guide for v3 documentation.

The stringency of conformance statements is specified by use of SHALL, SHOULD and other modal verbs. For example the word SHALL conveys the sense of being mandatory or required; SHOULD implies best practice or a recommendation, and MAY implies acceptable or permitted.

Membership

HL7 offers two main types of membership: individual membership is for those with a personal interest in the standards, while organizational membership includes benefits of importance to those who rely on the standard as part of their business. Organizational benefits are summarized as:

- Influence the technical and policy environment of the future by voting on standards. The ability to vote on HL7 standards is one of the most important benefits of membership. HL7 standards, including HL7 Fast Healthcare Interoperability Resources (FHIR®) and Consolidated Clinical Document Architecture (C-CDA®), have featured prominently in recent Meaningful Use legislation and discussions. Voting on HL7 standards is the best way to influence the use and implementation of standards at the national and international level.
- Show the industry that you are a leader who is helping to make interoperability a reality. Interoperability leaders create and shape industry standards. Being an active member of HL7 lets your business partners, clients, and the industry know that you are a leader who is helping to make interoperability a reality.
- All organizational members are encouraged to use the HL7 organizational partner logo in their marketing materials. Aligning yourself and your marketing with HL7 gives your partners and customers the confidence that your products and services are being developed using the world's most-widely adopted healthcare standards.
- Manage your implementation costs and speed time to market. In addition to training, members get access to the HL7 Help Desk, staffed by HL7 professionals who can help answer questions about the problems you have that slow down your interoperability projects and drive up costs of implementation.
- Access industry information to help make more informed business decisions. HL7 membership offers exclusive access to thought leaders and market intelligence that can help give your organization a competitive advantage in the industry and support effective strategic decision-making.
- Learn best practices from industry leaders. The ability to network at HL7 meetings and user groups (free for members) gives you valuable opportunities to collaborate with and learn from industry leaders and international thought leaders.
- Reduce your training budget. Members get access to free and significantly discounted training opportunities that will help alleviate implementation roadblocks and keep your team on the cutting edge of healthcare IT standards.

Members of HL7 who meet together electronically or in person are collectively known as the Working Group and are self-organized into a number of different technical committees. There are usually 3 weeklong working group meetings each year (Fig. 6.2).

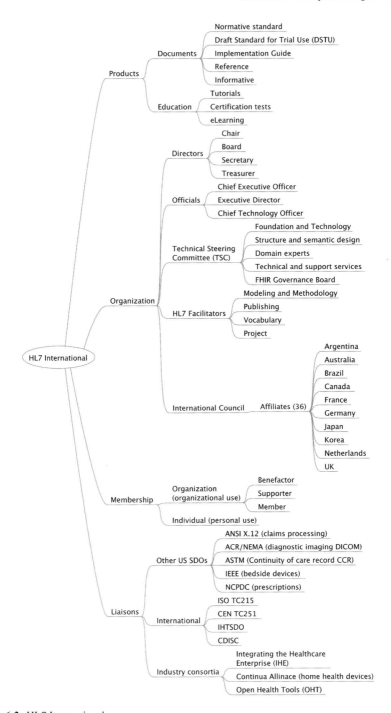

Fig. 6.2 HL7 International

Technical Steering Committee

The HL7 Technical Steering Committee oversees and coordinates the technical effort contributed by the HL7 volunteers who make up the HL7 Working Group. Its mission is to assure that the efforts of the Working Group are focused on the overall HL7 mission. There are four steering divisions:

Foundation and Technology work groups provide the fundamental tools and building blocks for all HL7 activities.

Structure and semantic design focuses on creation of basic patterns and common messages that could exist on their own, but are mostly used by others.

Domain Experts committees and projects in this space focus on creation of messages, services, documents using many of the common structures in place, yet expanding it in key areas as well.

IHTSDO

The International Health Terminology Standards Development Organization (IHTSDO, pronounced 'itzi-doo') is the custodian for SNOMED CT. IHTSDO is an international not-for-profit organisation, based in London (from 1 January 2016).[2] It was established in 2007, when it acquired the IP of SNOMED CT from the College of American Pathologists (CAP).

The IHTSDO vision is to enhance the health of human-kind by facilitating better health information management; to contribute to improved delivery of care by clinical and social care professions and to facilitate the accurate sharing of clinical and related health information, and the semantic interoperability of health records.

The achievement of this vision for broad, demonstrable and successful use of SNOMED CT requires a globally coordinated effort to gain agreement on a core terminology for recording and sharing health information, pooling resources to share costs and benefits relating to the development of terminology products and consistent promotion of the uptake and correct use of the terminology.

An important strand is active harmonization activity with other SDOs, including HL7 and the Open Health Tools consortium.

The 28 current (2016) members include Australia, Belgium, Brunei, Canada, Chile, Czech Republic, Denmark, Estonia, Hong Kong (China), Iceland, India, Israel, Lithuania, Malaysia, Malta, Netherlands, New Zealand, Poland, Portugal, Singapore, Slovak Republic, Slovenia, Spain, Sweden, Switzerland, United Kingdom, United States and Uruguay.

IHTSDO is responsible for the core content of SNOMED CT, while each member country has a National Release Centre, which distributes SNOMED CT and has responsibility within its territory for liaison with IHTSDO, licensing and distribution

---

[2] See www.ihtsdo.org

of SNOMED CT, quality assurance and conformance with IHTSDO standards, issues tracking, change control and monitoring IP (products, trademarks, etc.).

DICOM

DICOM (Digital Imaging and Communications in Medicine) is an international standard for medical images and related information (ISO 12052:2006).[3] It defines the formats for medical images that can be exchanged with the data and quality necessary for clinical use. DICOM is used for most imaging modalities including radiography, magnetic resonance imaging (MRI), nuclear medicine, ultrasound, tomography, echocardiography, X-ray, CT, MRI, ultrasound and other modalities used in radiology, cardiology, radiotherapy, ophthalmology and dentistry. Mainly implemented in medical equipment it is used by all of the main manufacturers.

DICOM was originally developed under the auspices of the National Electrical Manufacturers Association (NEMA) in collaboration with the American College of Radiologists (ACR). The first DICOM standard was released in 1985, pre-dating the founding of HL7, and the present version, DICOM 3.0 was released in 1993. The standard has not stood still since 1993, but is fully backward compatible with it.

DICOM is the universal format for PACS (picture archiving and communication systems) image storage and transfer. PACS comprise digital image acquisition devices (modalities), image archives and workstations. All of the main manufacturers use DICOM, although each unit only implements a subset of DICOM, as described in a *DICOM Conformance Statement*. DICOM is a large standard and devices will only interoperate with other equipment if

DICOM has been credited with revolutionizing the practice of radiology, allowing the replacement of X-ray film with a fully digital workflow [6]. Much as the Internet has become the platform for new consumer information applications, DICOM has enabled advanced medical imaging applications that have "changed the face of clinical medicine". From the emergency department, to cardiac stress testing, to breast cancer detection, DICOM is the standard that makes medical imaging work — for doctors and for patients.

IHE

IHE (Integrating the Healthcare Enterprise) was established in 1999 by the Healthcare Information Systems and Management Society (HIMSS) and the Radiological Society of North America (RSNA) to help improve the way healthcare computer systems share information.[4] The initial focus of IHE was in radiology,

---

[3] http://dicom.nema.org/

[4] See www.ihe.net

where it developed profiles that specify how to use DICOM and HL7 together, but it has moved on to cardiology, clinical laboratories and other specialties.

A second dimension to IHE's work has been the development of IT infrastructure standards for use across departmental and institutional boundaries. The XDS (Cross-enterprise Document Sharing) profile (see Chap. 17) is one example.

IHE has established a four-stage approach:

- Identify interoperability problems. Clinicians and IT experts work to identify common interoperability problems with information access, clinical workflow, administration and the underlying infrastructure.
- Specify integration profiles. Experienced healthcare IT professionals identify relevant standards, define how to apply these to address the problems and document these profiles in the form of IHE integration profiles. For example, IHE XDS is a profile of the OASIS ebXML Registry standard.
- Test systems at a Connectathon. Vendors implement IHE integration profiles in their products and test their systems for interoperability at an annual IHE Connectathon. This allows them to assess the maturity of their implementation and resolve issues of interoperability in a supervised testing environment.
- Publish integration statements for use in requests for proposals (RFPs). Vendors publish IHE integration statements to document the IHE integration profiles their products support. Users can reference the IHE integration profiles in requests for proposals, simplifying the systems acquisition process.

Continua Alliance

The Continua Health Alliance is a non-profit, open industry coalition of healthcare and technology companies working to establish a system of interoperable personal health solutions.[5] The main driver is that use of Telehealth solutions in the home can foster independence, empower individuals and provide the opportunity for personalized health and wellness management.

Continua has set out to develop an ecosystem of connected technologies, devices and services that will enable the more efficient exchange of fitness, health and wellness information. The foundation of this ecosystem is a set of interoperability guidelines that specify how systems and devices made by different companies can work together. Such products are expected to become common over the next few years.

The first set of Continua standards includes specifications for using existing standards such as Bluetooth, USB, medical devices (IEEE 1173) and HL7 to enable people to use home-based devices to monitor their weight, blood pressure, glucose and blood oxygen levels and share this with their healthcare professionals.

---

[5] See www.continuaalliance.org

Four groups of interfaces have been designed covering personal, local area, wide area and health record networks.

Personal Area Network The PAN (Personal Area Network) Interface uses IEEE 11073 Personal Health Device (PHD) standards, the wireless Bluetooth Health Device Profile (HDP) or the wired USB Personal Healthcare Device Class (PHDC) to link portable medical devices, such as a pulse oximeter, blood pressure monitor, thermometer, weighing scale or blood glucose meter, to a local application hosting device, which could be a smart phone or console.

Local Area Network The LAN (Local Area Network) interface links hard-wired devices, such as laboratory or fitness equipment to a local application-hosting device

Wide Area Network The WAN (Wide Area Network) interface links local application hosting devices to a central WAN device, such as a Remote Patient Monitoring (RPM) server. This could use the Internet or a mobile telephone network.

Health Record Network The HRN (Health Record Network) interface links WAN devices (HRN Senders) to an Electronic health record device (HRN Receiver), which could be a hospital Enterprise health Record (EHR), a physician's Electronic Medical Record (EMR) or a Personal Health Record (PHR) service used by the patient. This uses HL7 CDA R2 PHM (Personal Health Monitoring) message profile and the IHE XDR (Cross-Enterprise Document Reliable) interchange profile. PHM re-uses HL7 CCD templates. XDR is one of the IHE XDS family of profiles that use common services.

Continua has developed a product certification program with a recognizable logo signifying interoperability with other certified products, intended to build trust and confidence among customers.

CDISC

The Clinical Data Interchange Standards Consortium (CDISC) has been founded by the pharmaceutical industry to develop worldwide industry standards to support electronic acquisition, exchange, submission and archiving of clinical trials data and metadata for medical and biopharmaceutical product development.[6] The CDISC mission is to lead the development of global, vendor-neutral, platform-independent standards to improve data quality and accelerate product development.

---

[6] See www.cdisc.org

CDISC standards include:

- Study Data Tabulation Model (SDTM) for the regulatory submi. Report Tabulations, including the Standard for the Exchange of Nonc (SEND).
- Analysis Data Model (ADaM) for the regulatory submission of ...alysis datasets.
- Operational Data Model (ODM) for the transfer of case report form data.
- Laboratory Model (LAB) for the transfer of clinical laboratory data, including pharmacogenomics.
- Biomedical Integrated Research Domain Group (BRIDG) model.
- Case Report Tabulation – Data Definition Specification (define.xml).
- Clinical Data Acquisition Standards Harmonization (CDASH)
- Terminology standard containing terminology that supports all CDISC standards.
- Glossary standard providing common meanings for terms used within clinical research.

OpenEHR

OpenEHR is a not for profit foundation, which has developed a technology independent architecture, including a Reference Model, Archetypes and Templates, for health computing platforms.[7] OpenEHR is based around the OpenEHR reference model, which has a close resemblance to that of ISO 13606–1 Electronic Health Record Communication - Part 1: Reference Model, although there are some significant differences.

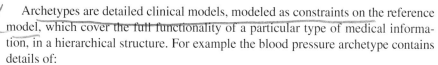

Archetypes are detailed clinical models, modeled as constraints on the reference model, which cover the full functionality of a particular type of medical information, in a hierarchical structure. For example the blood pressure archetype contains details of:

- BP Data: systolic, diastolic, mean arterial pressure, pulse pressure and comment.
- Protocol: cuff size, instrument, location of measurement, Korotkoff sounds, device etc.
- State: position, exertion level, exercise, tilt.
- Events: baseline reading, 5 min reading, 10 min reading, postural change, paradox.

For use in clinical records, Templates are specified which constrain the archetypes to just the data required for this situation, such as systolic and diastolic pressures.

The main activities are to promote the uptake of openEHR technologies globally; to maintain the openEHR specifications and control the change management

---

[7] See www.openehr.org

process for the openEHR model; to protect the copyright of open source software components based on openEHR; and to act as a forum for discussion and contribution on openEHR and related technologies.

References

1. ISO/IEC Guide 2: standardization and related activities – general vocabulary. 2004; definition 3.2.
2. ISO. ISO strategic plan 2005–2010: standards for a sustainable world. Geneva: ISO; 2004.
3. EU. ICT standards in the health sector: current situation and prospects. A Sectoral e-Business Watch study by Empirica. Special Study No. 1. 2008. http://www.epractice.eu/en/library/281850
4. Hammond WE, Cimino JJ. Standards in biomedical informatics. In: Shortliffe EH, Cimino JJ, editors. Biomedical informatics: computer applications in health care and biomedicine. 4th ed. New York: Springer; 2014. p. 211–54.
5. Bartleson K. The ten commandments for effective standards. Mountain View: Synopsys Press; 2010.
6. Pianykh OS. Digital imaging and communications in medicine (DICOM): a practical introduction and survival guide. 2nd ed. Berlin: Springer; 2012.

Part II
Terminologies and SNOMED CT

Chapter 7
Clinical Terminology

Abstract Unlike most sciences, medical terminology is poorly structured. This creates major problems for semantic interoperability, where terms need to be used in a precise and unambiguous way. This chapter introduces the core concepts of clinical terminology, sets out a list of requirements (desiderata) and illustrates these with the story of the Chocolate Teapot.

Keywords Terminology • Synonym • Homonym • Code • Classification • Hierarchy • Concept • Coding scheme • Display term • Relationship • Value set • Identifier • Reference terminology • Interface terminology • Ontology • Expression • Vocabulary • Polyhierarchy • NEC (not elsewhere classified) • NOS (not otherwise specified) • Redundancy • Chocolate teapot

Why Clinical Terminology is Important

When, in the fifteenth century, Gutenberg's invention of the movable type led to the mass production and dissemination of books and written information, language was still relatively unformalised. It took until the eighteenth century before the great dictionaries and nomenclatures such as Dr Johnson's English Dictionary and Linnaeus' biological taxonomy were produced.

Sciences such as biology and chemistry have an internationally agreed formal structure for their terminology. Every living organism has a generic and specific Latin name expressed within a comprehensive biological taxonomy, which in many ways anticipated the full understanding of the evolution of life. All chemical structures are expressed in internationally standardised ways.

Medical terminology escaped formalization, leading to problems of ambiguity that are now recognised as a significant risk to patient safety. The lack of agreed medical terminology has been recognised as an issue for at least 250 years. There is even an old word, "nosology", to describe the development of medical terminology, but the need has increased with the use of computers. Modern nosologists call themselves clinical terminologists.

The historical, eclectic and ad hoc origins of medical terminology have encumbered anyone interested in health-care with the need to learn a whole new language,

© Springer-Verlag London 2016
T. Benson, G. Grieve, *Principles of Health Interoperability*,
Health Information Technology Standards, DOI 10.1007/978-3-319-30370-3_7

replete with homonyms (where the same term means different things depending on context), synonyms (where there is more than one term for exactly the same concept), eponyms named after people, three letter acronyms and abbreviations. Nobody, who has not learnt the eponym, can guess the meaning of Hodgkin's (lymph node cancer), Bright's (kidney disease) and von Recklinghausen's disease (hereditary neurofibromatosis).

Information scientists classify knowledge in a series of levels. For example the Dewey Decimal Classification, used in libraries, attempts to organize all knowledge into ten main classes:

000 – Computer science, information and general works
100 – Philosophy and psychology
200 – Religion
300 – Social sciences
400 – Language
500 – Science (including mathematics)
600 – Technology and applied Science
700 – Arts and recreation
800 – Literature
900 – History, geography, and biography

Blois in his seminal book *Information and Medicine* showed how healthcare is unique amongst scientific endeavors in that day-to-day medical information relates to so many different levels [1]. The breadth of healthcare covers an exceptionally broad scope, ranging from radiation and subatomic structures, complex molecules including DNA and proteins, cells including hematology and cancers, microorganisms such as bacteria and viruses, anatomical structures including the different body systems, mental activity, the whole person, groups, societies and populations.

Each aspect of healthcare mixes multiple overlapping theories, each with their own sub-terminology. Any classification system is inevitably just one way of slicing up a very complex reality, made even more difficult because key medical concepts such as diseases are abstractions, defined using information from a variety of information levels; diseases are not objects which can be seen or touched.

People use terms in the way that they and their immediate colleagues understand. Each user of a term assumes that everyone else understands precisely what he or she intends it to mean; over time groups develop their own local dialect. Medical records staff can often identify not only a doctor's specialty but also the institution where he or she was trained from the way they use certain terms.

Lewis Carrol expressed the same problem in an exchange between Alice and Humpty Dumpty in *Through the Looking Glass*: [2]

> *'I don't know what you mean by "glory"'* Alice said.
>
> *Humpty Dumpty smiled contemptuously. 'Of course you don't – till I tell you. I meant "there's a nice knock-down argument for you!"'*
>
> *'But "glory" doesn't mean "a nice knock-down argument"'* Alice objected.
>
> *'When I use a word,' Humpty Dumpty said in a rather scornful tone, 'it means just what I chose it to mean – neither more nor less.'*

> *'The question is,'* said Alice, *'whether you can make words mean so many different things.'*
>
> *'The question is,'* said Humpty Dumpty, *'which is to be master – that's all.'*

The representation of written information has become more and more specific over the centuries. The first way of representing information was by a picture or drawing, such as in Stone Age cave paintings. The earliest writing was based on pictograms, such as Egyptian hieroglyphics and Chinese characters, but the need for cheap and quick writing materials led to the development of cruciform characters on wet clay blocks in Mesopotamia and the development of phonetic alphabets such as those of Greece and Rome. Modern English does everything using just 26 letters, 10 digits and a few punctuation marks.

Computers hold information as sequences of binary bits (0s and 1s) and work by matching strings; they need precisely coded data. A computer can instantly check if two strings are the same but, if a difference is detected, it cannot judge whether that difference is important. In spite of decades of effort we do not yet have computers that cope well with the ambiguity inherent in natural language.

Coding and Classification

People often confuse the terms coding and classification and use them almost synonymously. This may be because the process of classification involves recording the codes used to name specific classes. However, coding schemes and classifications do different jobs. Classification allocates things into groups or classes, while coding is the allocation of identifiers, which can apply to anything (including classes in classifications).

A **code** is a sequence of symbols, usually digits or letters, which designate an object or concept for identification or selection purposes. It is simply an alternative name for something, an identifier, designed for computer processing. Coding systems are an indispensible part of healthcare computer applications and interoperability specifications for exchanging data between computers.

The primary challenge for the designers of coding schemes is to produce something that will be widely and willingly adopted and endorsed by clinicians and managers. However, clinicians and managers have no more interest in codes than a retail customer has in the bar code on a packet of corn flakes. Codes are needed and used by computers, not humans.

Clinicians need to record information in the form, language and detail that is of most benefit to them when treating individual patients. Clinical records require precise and comprehensive detail about each individual patient, creating a tension with statistical analysis, which requires patients to be classified into a manageable number of discrete and mutually exclusive groups.

Clinicians and managers should to be interested in classification, because it is the basis for most statistical analysis, quantitative management, accountancy and research.

Classification is the systematic placement of things or concepts into categories or classes, which share some common attribute, quality or property. There is no limit to the number of ways that any set of objects can be classified and so no possibility of a perfect classification that is good for everything.

The choice of what classification system to use is often determined by payment agencies, insurance companies and national governments that control whether or not a doctor or institution gets paid. Such bodies usually specify the precise classification system that they require, often in collaboration with representatives from the professional and trade associations, medical colleges and educational bodies. Once chosen it has to be accepted by users and implemented in computer software.

In *The Endangered Medical Record*, Slee argues that the choice of scheme used for electronic patient records represents a serious real threat to the truthfulness and completeness of medical record content. By using of broad categories, such as those specified by the International Classification of Diseases (ICD), rather than precise diagnoses, we throw away detail that should be preserved permanently. His plea is for detailed, permanent and unambiguous codes [3].

For example, a trauma surgeon might describe a typical skiing accident as: a closed spiral fracture of the shaft of the right tibia with fractured fibula. In ICD-10, the code for fracture of shaft of tibia has the following logical structure:

```
Chapter XIX: Injury, poisoning and certain other consequences of exter-
nal cause (S00-T98)
    Block: Injuries to the knee and lower leg (S80-S98)
      S82: Fracture of lower leg, including ankle
              S82.2 Fracture of shaft of tibia (with or without mention
              of fracture of fibula)
                      S82.2.1 Closed fracture of shaft of tibia
```

The selected ICD-10 code S82.2.1 does not specify whether the leg is left or right, whether the tibia fracture is simple, spiral or compound or how the fibula is affected.

A **hierarchy** is an ordered organization of concepts. General concepts are at the top of the hierarchy; at each level down the hierarchy, concepts become increasingly specialized. This can be thought of as an inverted tree with its trunk or root at the top. For example, biological classification places animals and plants into a hierarchical classification (a taxonomy) according to similarities in structure, origin etc., which indicate a common relationship. The main levels in the biological taxonomy are Kingdom, Phylum (animals) or Division (plants), Class, Order, Family, Genus and Species.

Healthcare computer systems use nationally prescribed coding systems. Many of these, such as the ICD-10, CPT-4 and Read Codes use a *position-dependent hierarchical coding structure*. The internal structure of the code specifies its meaning relative to other codes. The structure of the code increases in detail from left to right, with the first character of the code specifying the chapter, the second the main subdivision and so on until down the branches of the tree until the final leaf codes are reached.

One of the technical problems of position-dependent hierarchical coding systems is that they cannot be modified easily without changing the meaning of codes in different versions, creating major problems when, as inevitably happens, one version needs to be replaced by another.

Coding Systems

Any coding system has various components.

Concept: The fundamental idea is that of a concept, which is a medical idea. Each concept is identified by a concept code.

Coding Scheme: Each concept code originates from a coding scheme. A coding scheme defines a set of concept codes, which are unique within the namespace of the coding scheme, and are globally unique when coupled with the name of the coding scheme itself.

Display Term: This is a human readable term. In some cases more than one display term may be provided for the same concept, to cover true synonyms, such as translations into different languages. One display term is usually designated as the preferred term.

Relationship: Concepts may be related to other concept via a relationship, which allows the generation of hierarchical structures. One concept may be part of more than one hierarchical structure. Often these relationships will be defined as part of original coding schemes, but other relationships are also possible.

Value Set: A set of values that are allowed for a particular data item. Message specifications refer to value sets as the allowed values for a field. Codes from a single coding scheme may be referenced using a value set table, which has a heading and includes metadata such as: value set name, unique identifier, coding scheme, author, time validity, version and other notes. Each entry in the table contains concept code value, display term and notes about applicability.

Identifiers: Computer systems need unique identifiers for people, things and places, which have similar properties to codes. One way of achieving uniqueness is to treat each identifier as a pair, comprising a unique name for the assigner plus a value for the identification number, which is unique within assigner. It is the responsibility of the assign or to ensure that all such values are unique.

Terminologies

Terminology: a set of concepts designated by terms belonging to a special domain of knowledge, or subject field.

Reference terminology: a terminology in which every concept designation has a formal, machine-usable definition supporting data aggregation and retrieval. Reference terminologies are designed to provide exact and complete representations of a given domain's knowledge, including its entities and ideas, and their interrelationships, and are typically optimized to support the storage, retrieval, and classification of clinical data.

Interface Terminology: Systematic collections of clinically oriented phrases or terms aggregated to support clinicians' entry of patient information directly into computer programs, such as clinical documentation systems or decision support tools. They may mediate between a user's colloquial conceptualizations of concept descriptions and an underlying reference terminology.

Ontology: hierarchical structuring of knowledge about things by sub-categorizing them as a set of concepts within a domain according to their essential qualities and relationships between those concepts.

Expression: A collection of references to one or more concepts used to express an instance of a clinical idea. An expression containing a single concept identifier is referred to as a pre-coordinated expression. An expression that contains two or more concept identifiers is a post-coordinated expression.

The scope and some of the terms used in clinical terminology are summarized in Fig. 7.1.

User Requirements of Terminologies

A key design requirement for any coding and classification system is to satisfy the needs of the different stakeholders. Roger Côté, the father of SNOMED, views this as a pyramid with three levels of use:

1. At the tip, case-mix classifications such as DRGs, used for payment.
2. In the middle, classifications of diagnoses and procedures used to monitor and audit clinical activities.
3. At the base, clinical terminology used for individual patient care.

Healthcare managers and researchers need classified data, which enable comparisons and data exchange with existing data sources. Links between classifications must be explicit with one-to-one or many-to-one links. A many-to-one link involves loss of information, the extent of which is determined by how closely one classification is based on the other.

A multilevel classification with both coarse and fine granularity may allow two-way mapping from another classification. High levels of compatibility can usually

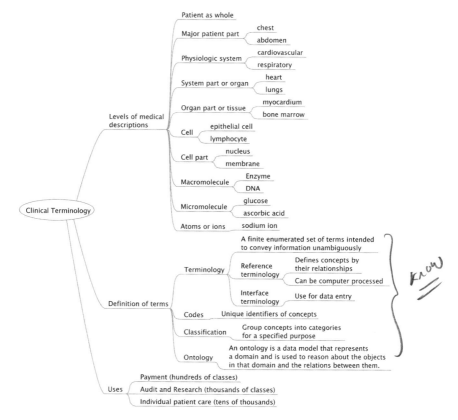

Fig. 7.1 Scope and terms used in clinical terminology

be obtained only by basing a new classification directly on the target, using the same class boundaries. This requirement for cross-mapping with existing classifications inevitably drives the developers of clinical classifications to build on existing schemes, even if they are not suitable for the need in hand. For example the ICD is organised around body systems, which is helpful in some circumstances, but not in others. Early versions of SNOMED reflected its origins in the College of American Pathologists as an extension of the Systematized Nomenclature of Pathology (SNOP), which gave a pathological slant.

Doctors and nurses will not take the trouble to learn how to use any system unless it is quick and easy to use and provides information in the form and language that best helps them treat individual patients. Automatic or semi-automatic encoding software is needed. Clinical records need to be as specific as possible. Hence clinicians require a comprehensive nomenclature of medical terms covering everything that could occur within any patient's medical record. That is, all of clinical medicine and health service administration, but not the whole of bio-medical science.

In 1984, the IMIA working conference on clinical terminology concluded that

In future healthcare information systems, the user interface should be based upon natural language. The generation of numerical or alphanumeric codes should occur within the computer. Automatic encoding of natural language should be used. The morbidity and mortality statistical classification requirements of national and international groups should be the by-product of medically based healthcare information systems.

It was not anticipated that almost 30 years later most clinical coding in hospitals would continue to be done by coding clerks.

Desiderata

Desiderata for Controlled Medical Vocabularies in the Twenty-First Century [4] brought together together a number of common requirements for clinical terminologies, which had been developed in leading terminology projects such as GALEN [5], UMLS (Unified Medical Language System) [6], SNOMED RT (Reference Terminology) [7] and the NHS Clinical Terms Project [8]. This paper was highly influential in the design of SNOMED CT. The desiderata are:

1. **Vocabulary Content:** In terms of scope and quality, content is paramount. Any practical clinical terminology needs to be comprehensive in terms of both domain coverage (concepts) and human readable terms (descriptions and synonyms). A methodology is required to allow the content to be expanded as and when required, including translation into other languages and dialects, while maintaining quality.
2. **Concept Orientation:** This means that each concept term has one meaning (non-vagueness), and only one meaning (non-ambiguity). However, each concept can be described by several terms (synonyms) in the same language plus different terms in other languages and dialects. Note also that the same term can have different meanings (homonyms), each relating to a different concept.
3. **Concept Permanence** Once a concept is created its meaning persists. It must not be changed or deleted by updates. However, a concept may be marked as *retired* where its meaning is found to be ambiguous, redundant or otherwise incorrect.
4. **Non-semantic Concept Identifiers** Each concept should have a unique identifier, which should be meaningless. All semantic information (relating to meaning) is an attribute of the concept, and should not be part of its identifier. Further examples of the problems with position-dependent hierarchical coding schemes are discussed with examples in Chap. 8.
5. **Polyhierarchy** While it is useful to organize medical concepts in a hierarchical way, many clinical concepts are naturally multi-dimensional, with more than one parent (super-type) concept. For example, a *fractured tibia* is both a type of *fracture* and a type of *leg injury*.
6. **Formal Definitions** The means of classifying a concept is independent of the means of identifying it. The development of formal, descriptive logic to define and classify clinical concepts is a major development away from the traditional position-dependent coding schemes and dictionary forms of definition. For

example *pneumococcal pneumonia* may be defined using a hierarchical (*is a*) link to the concept *pneumonia* and a *caused by* link to the concept *streptococcus pneumoniae*.

7. **Rejection of "Not Elsewhere Classified" Terms** Many existing classifications include one or more catchall categories for concepts not covered. The problem with such *not elsewhere classified* or *NEC* categories is that they change their meaning, as and when a new category is added that covers some of the NEC scope. The meaning is not permanent, which was a previous criterion.

8. **Multiple Granularities** Different users require different levels of granularity. Different levels of granularity are needed for defining concepts, navigation, decision support and reporting. For example, a manager may only need to know that a patient has a *broken leg*; the finance department that it is a *fractured tibia*, but the clinician needs to know that it is a *closed spiral fracture of the shaft of the right tibia*. In principle, there should be no limitations on the number of levels in the display tree hierarchy.

9. **Multiple Consistent Views** When a concept has multiple parents in a hierarchy, the view of that concept should not depend on whether it was reached by following the hierarchy from a particular parent. The complete structure of a terminology, including all hierarchies and relationships can be complex and difficult to use. Each end user needs one or more views that reflects his or her own needs and understanding, but in a way that is consistent with the underlying model.

10. **Context Representation** Information is recorded within a particular context and cannot be interpreted without that understanding. The context needs to be computer-processable. One approach is to provide a means of recording context explicitly within the terminology.

11. **Graceful Evolution** Terminologies change over time. It creates problems for users if the meanings of aggregated time series data change in an uncontrolled manner. Care is needed to design the whole structure to support graceful evolution of concepts, terms and relationships.

12. **Recognize Redundancy** When terminologies change, some components will become redundant and so it important to recognise explicitly that this has happened.

The Chocolate Teapot

The apocryphal story of the Chocolate Teapot, developed by Dr Malcolm Duncan [9] illustrates a number of the issues involved in classification and terminology. I am grateful for his permission to reproduce this in an edited form.

Because not all readers will be confident they know what, for example, asthma really is (even doctors disagree), this discussion is organised around the classification of teapots. Most people think they know what a teapot is, or do until they read this.

Consider a fragment of a simple but functional crockery classification that exists solely to document the appearance of table settings as described in Victorian literature.

Classification of Tableware Version 9 (CTV-9, 1875)

```
Crockery
---Teapot
------Brown teapot
------White teapot
------Blue teapot
------Teapot, color not elsewhere classified (NEC)
------Teapot, color not otherwise specified (NOS)
```

Users are obliged to *code to the leaf*, which is how ICD-9 and ICD-10 are used. Only brown, white and blue crockery were fashionable in 1875. The uncouth might deploy other colors (NEC) or worse, might not care (NOS). However it is axiomatic that your teapot is made from earthenware, ideally from quality porcelain.

Elsewhere there is the frequently updated *Systematized Nomenclature of Kitchen Terminology (SNoKitch)* intended to support all catering applications. A fragment of this terminology is followed through a number of iterations:

SNoKitch Release n

```
Crockery
---Teapot
```

Teapot has no children and could be equivalent to any of the five classification leaves of CTV-9. People then ask for further concepts to be added, leading to the next release.

SNoKitch Release n + 1

```
Crockery
---Teapot
----- Brown teapot
------White teapot
```

Anyone wanting to specify a brown or white teapot using *SNoKitch* can now do so. Coding to leaf is not mandatory and so NOS and NEC are absent. The Teapot concept in *SNoKitch* is the equivalent of both Teapot, color (NEC) and Teapot, color (NOS) categories in CTV-9, although in effect the meaning of Teapot in this version of SNoKitch is skewed to mean a teapot that is neither brown nor white.

SNoKitch Release n + 2

```
Crockery
---Teapot
```

```
----- Brown teapot
------White teapot
------Blue teapot
------China teapot
```

In release n + 2 Blue teapot (forgotten in the previous version) is added as a child. This further alters the meaning of the concept Teapot in its role as Teapot, color not elsewhere classified.

Also a China teapot child has been added, but this is not a discrete sibling: it could be any color. Teapot now has non-disjoint subclasses. China teapot has perhaps acquired part of the meaning of Teapot, color (NEC) from Teapot (ie China teapots which are not brown, white or blue). However, when users interested in whether a teapot is china receive a White teapot code they are out of luck. All they can infer is that it represents a teapot.

As both the material and color are important, new concepts White china teapot and Blue china teapot are added in the next release. Unfortunately we cannot rely on everyone using them because they don't have to code to leaf. Some people may only choose to specify whether their teapots are china and not capture color at all.

However we still have a common understanding of the meaning of the parent Teapot concept as superclass of its children. Concepts added in the next release will overturn this.

SNoKitch Release n + 3

```
Crockery
---Teapot
----- Brown teapot
------White teapot
----------White china teapot
------Blue teapot
----------Blue china teapot
------China teapot
----------White china teapot
----------Blue china teapot
------Chocolate teapot
------Ornamental teapot
------Industrial teapot
```

The addition of chocolate, ornamental and industrial teapots means that ontological continuity with previous versions is deeply in question. Industrial teapot indicates where it is used but nothing else except presumably that it is not ornamental or made of chocolate. Teapots made from metal or chocolate are not earthenware and hence not crockery. From release to release, there is little consistency in what we can infer about color, material or use.

Worse, the arrival of certain children has even altered the meaning we can infer from the unadorned `Teapot` parent concept. For example, the *Kansas Tea Company (KTC)* operate decision support software designed against release n + 2:

```
If Concept = (teapot or child of teapot)
Safe to add tea leaves plus boiling water
End If
```

KTC had not anticipated confectionary in this hierarchy. As for `ornamental teapot`, who knows if it can be used to make tea? KTC must now add multiple additional nodes to their decision support rule and be able to exclude subtypes eg it would not be wise to add boiling water to `Chocolate teapot`.

The validity of these relationships depends on your interest in teapots. There is no longer a universal understanding of the concept. There is merely a shared assumption about the *term* now exposed as not applying in all contexts of use. If we had started with a clear definition of teapot (as a 'free text' scope note) this might have been avoided. Most people can express the teapot that is in their head most of the time, but this may be at the expense of loss of predictable machine readability.

The *Comestibles Supply Consortium* recognises the incoherent use of the *SNoKitch* terminology across their systems and mandates use of a small subset.

```
Teapot
------Brown teapot
------White teapot
------Blue teapot
```

All within the Consortium will now understand what was recorded and satisfy their leading concern which is to make billable returns using *Classification of Tableware Version 9*. They still lack the distinction between NOS and NEC and may need to employ professional coders to abstract and map records manually.

Classifications such as the ICD family include items like `Chronic airway obstruction, not elsewhere classified` and `Other specified excision of adrenal gland`. A problem with such constructs is that they are not stable in meaning across versions of the classification ie what is classified elsewhere may change with addition or removal of other content. In contrast to classifications such as ICD-9, modern medical terminologies such as SNOMED CT typically do not mandate coding to leaf and do not permit *not otherwise specified* or self-referential entities such as *not elsewhere classified*.

The widespread assumption is that this provides immunity from semantic discontinuity across releases. It does not. In the absence of NOS and NEC, these static 'known unknowns' become 'unknown unknowns' mobile between releases. To quote Donald Rumsfeld:

> ...there are known knowns; there are things we know we know. We also know there are known unknowns; that is to say we know there are some things we do not know. But there are also unknown unknowns — the ones we don't know we don't know.

The situation is further exacerbated if a terminology is neither exhaustive at any one *level* of siblings nor disjoint (two sets are said to be disjoint if they have no element in common). As terminologies evolve, applications involving data reuse (such as messaging, billing, clinical audit and active decision support) need to recognise what version of the terminology a code is drawn from. Successive versions may improve content but these alterations create circumstances where interpretation of recorded data requires reference to the ontology, as it was when the concept was chosen, not as it is now.

A concept within a terminology is not entire unto itself. Addition, retirement and movement of other concepts alter its use and interpretation. Such changes are common. As well as additions and retirements there may be many hierarchy changes in each release of a terminology such as SNOMED CT.

References

1. Blois MS. Information and medicine: the nature of medical descriptions. Berkeley: University of California Press; 1984.
2. Carrol L. Through the looking glass and what alice found there. London: Macmillan; 1871.
3. Slee V, Slee D, Schmidt HJ. The endangered medical record: ensuring its integrity in the age of informatics. St. Paul: Tringa; 2000.
4. Cimino JJ. Desiderata for controlled medical vocabularies in the twenty-first century. Methods Inf Med. 1998;37:394–403.
5. Rector A, Solomon W, Nowlan A, Rush T, Claassen A, Zanstra P. A terminology server for medical language and medical information systems. Methods Inf Med. 1994;34:147–57.
6. Lindberg DA, Humphreys BL, McCray AT. The unified medical language system. Methods Inf Med. 1993;32(4):281–91.
7. Spackman K, Campbell K, Cote R. SNOMED RT: a reference terminology for healthcare. Proc AMIA Symp. 1997: 640–4.
8. O'Neil MJ, Payne C, Read JD. Read codes version 3: a user led terminology. Methods Inf Med. 1995;34:187–92.
9. Duncan M. Medical terminology version control discussion paper: the chocolate teapot. Version 2.3. Medical Object Oriented Software. 2006. www.mrtablet.demon.co.uk

Chapter 8
Coding and Classification Schemes

Abstract This chapter describes a number of important coding and classification systems that have been and remain influential in healthcare. We briefly discuss the International Classification of Diseases (ICD), Diagnosis Related Groups (DRGs), the Read Codes, SNOP and SNOMED, LOINC and the Unified Medical Language System (UMLS).

Keywords International Classification of Diseases (ICD) • Diagnosis related groups (DRG) • The Read codes • Systematized Nomenclature of Pathology (SNOP) • Systematized Nomenclature of Medicine (SNOMED) • Logical observation identifiers names and codes (LOINC) • Unified Medical Language System (UMLS)

International Classification of Diseases

The ICD (International Classification of Diseases) provides a common language for reporting and monitoring diseases, used throughout the world to compare and share data in a consistent standard way between hospitals, regions and countries and over periods of time [1]. It is used to classify diseases and other problems for payment, management and research, as recorded on many types of health records including medical records and death certificates. It enables international comparisons of mortality and morbidity by WHO member states.

The origins of the ICD have been traced back to John Graunt's London Bills of Mortality in the seventeenth century and the work of John Farr and Jacques Bertillon in the late nineteenth century to produce the International Classification of Causes of Death.

WHO published ICD-9 in 1977 and ICD-9-CM (clinical modification) was used until 2015 in the USA for payment purposes with annual updates, extending the original ICD-9 with more morbidity and procedure codes.

ICD-10 was published in 1992 and is now used by 117 countries to report mortality data. The full title is *The International Statistical Classification of Diseases and Related Health Problems*, and it is published in three volumes. Volume 1 is a tabular list. Volume 2 is the Instruction Manual and Volume 3 is an Alphabetical List.

ICD-10 has an alphanumeric coding scheme with one letter followed by three numbers at the four-character level.

ICD-10 has 21 chapters, corresponding roughly to body systems. Medical conditions have been grouped in a way that was felt to be most suitable for general epidemiological purposes and the evaluation of healthcare. Within each chapter the various diseases are listed with 3 digit codes with an optional fourth or fifth digits for additional detail. The following tables illustrate the hierarchical structure of ICD-10 showing the chapters (Table 8.1), then the main blocks or sections in one chapter (Table 8.2) and then the third level codes within one block (Table 8.3).

At the next level of detail, Chap. 10 (diseases of the respiratory system) is subdivided into blocks:

At the third level, acute upper respiratory infections block has the following categories:

Each category is specified with inclusion and exclusion criteria. Most groups have a further level of detail. For example, J04 *acute laryngitis and tracheitis* is divided into *acute laryngitis* J04.0, acute *tracheitis* J04.1 and *acute laryngotracheitis* J04.2. An additional code from the section on *bacterial, viral and other infectious agents* may be used to identify the infectious agent.

Table 8.1 ICD-10 Chapters and code ranges

ICD-10 Chapter		Code range
I	Certain infectious and parasitic diseases	A00-B99
II	Neoplasms	C00-D48
III	Diseases of the blood and blood-forming organs and certain disorders involving the immune system	D50-D89
IV	Endocrine, nutritional and metabolic diseases	E00-E90
V	Mental and behavioral disorders	F00-F99
VI	Diseases of the nervous system	G00-G32
VII	Diseases of the eye and adnexia	H00-H59
VIII	Diseases of the ear and mastoid process	H60-H95
IX	Diseases of the circulatory system	I00-I99
X	Diseases of the respiratory system	J00-J99
XI	Diseases of the digestive system	K00-K93
XII	Diseases of the skin and subcutaneous tissue	L00-L99
XIII	Diseases of the musculoskeletal system and connective tissue	M00-M99
XIV	Diseases of the genitourinary system	N00-N99
XV	Pregnancy, childbirth and the puerperium	O00-O99
XVI	Certain conditions originating in the perinatal period	P00-P96
XVII	Congenital malformations, deformations and chromosomal abnormalities	Q00-Q99
XVIII	Symptoms, signs and abnormal clinical and laboratory findings not elsewhere classified	R00-R99
XIX	Injury, poisoning and certain other consequences of external causes	S00-T98
XX	External causes of morbidity and mortality	V01-Y98
XXI	Factors influencing health status and contact with health services	Z00-Z99

Table 8.2 Blocks of codes within ICD-10 Chap. 10

Section	Code range
Acute upper respiratory infections	J00-J06
Influenza and pneumonia	J10-J18
Other acute lower respiratory infections	J20-J22
Other diseases of the upper respiratory tract	J30-J39
Chronic lower respiratory diseases	J40-J47
Lung diseases due to external agents	J60-J70
Other respiratory diseases principally affecting the interstitium	J80-J84
Suppurative and necrotic conditions of the lower respiratory tract	J85-J86
Other diseases of the pleura	J90-J94
Other diseases of the respiratory system	J95-J99

Table 8.3 Three-character codes within ICD-10 Acute upper respiratory infections

Rubric	Code
Acute nasopharyngitis [common cold]	J00
Acute sinusitis	J01
Acute pharyngitis	J02
Acute tonsillitis	J03
Acute laryngitis and tracheitis	J04
Acute obstructive laryngitis [croup] and epiglottitis	J05
Acute upper respiratory infections of multiple and unspecified sites	J06

ICD-10 is used primarily by professional coders working with clinical records to provide data to summarize and compare hospital caseloads. The terminology used is not intended or suitable for use directly by clinicians at the point of care and is not detailed enough to meet the needs of hospital specialists. For example, the most detailed description of a fractured tibia is that it is S82.2.0 *closed fracture of shaft of tibia*. This says nothing about the type of fracture (simple, spiral, or compound), laterality, or whether the fibula is also fractured.

The beta draft of the new ICD-11 has been published and the final version is due to be released in 2018. ICD-11 builds on and has substantial backward comparability with ICD-10. It differs from ICD-10 where justified by new knowledge, changes in mortality or morbidity and changed user requirements and in order to remedy technical deficiencies in ICD-10. ICD-11 is developed in a way that takes advantage of developments in information technology so as to facilitate its use and maintenance. It is designed to be a suitable basis for clinical modifications that may be developed by some states and is expected to be accompanied in future by specialist classifications to serve special purposes.

Diagnosis Related Groups

Diagnosis related groups (DRG) were originally created as a research tool to help answer questions such as *why do some patients stay in hospital longer than others?* [2] The product of any hospital may be defined as the sum of the set of goods and services provided to individual patients. A hospital produces patient care, which involves treatment, tests and other services. However, no two patients are ever exactly the same and so a method is needed to classify similar patients such that the costs, process and outcome of care can be studied systematically.

Robert Fetter and others at Yale University approached the problem in the late 1970s by classifying patients into groups based primarily on diagnosis and the amount of resources usually required, in order to identify those patients who had an unusually long length of stay for their condition. They did this using retrospective statistical analysis of hospital in-patient returns to identify clinically relevant groups of patients with similar expected lengths of stay (iso-resource groups). The original diagnosis-related groups (DRGs) were based on analysis of what had actually happened, not on what people thought should happen.

DRGs can be used to define the products of hospitals, in order to compare activities, length of stay, services used and so on. Using the DRG paradigm, each hospital can be thought of as a producer of DRGs. In many cases individual DRGs may be the appropriate units of attention while in other cases it is useful to work with strategic product line groups (SPGs), which are clusters of similar DRGs performed by the same specialists.

DRGs might have remained a useful research tool had it not been for the introduction of the Medicaid prospective payment scheme (PPS) in 1983 [3]. At that time Medicaid stopped paying hospitals on the basis of cost incurred, which provided incentives for hospitals to keep patients in longer, but paid a fixed price per patient based on their DRG category.

DRGs have a hierarchical structure. The original DRGs had 23 major diagnostic categories, based on principal diagnosis, each of which is partitioned into medical and surgical groups according to whether the patient had an operation during their stay. The medical groups are further divided according to primary diagnosis, age and the presence of complications and co-morbidities (CC), which have an impact on length of stay of more than 1 day. Similarly surgical groups are divided by type of operation, age, complications, co-morbidities and the presence of malignancy. After a number of refinements this led to a classification of 467 groups such that patients in any one group might be expected to use broadly the same range of hospital resources.

Over the years the basic DRG scheme has evolved into a family of systems, including Medicare DRGs, Refined DRGs, All Patient DRGs, Severity DRGs, International-Refined DRGs and Health Related Groups (HRGs) used in the NHS.

Use of DRGs depends on high quality coded data, with direct symmetry between coded data and the real world health events to which they relate. Ultimately this depends on the quality, clarity and organization of the medical record.

The Read Codes

The Read codes are used in primary care and are one of the two direct predecessors of SNOMED CT. Without the Read Codes there would be no SNOMED CT as we now know it. The Read Codes have been used by all GPs in the UK and New Zealand since the mid 1990s. New Zealand still uses the original 4-byte codes, described here.

Table 8.4 Read Codes version properties

Concept	4 Byte	Version 2	Version 3 (ctv3)
Hierarchy representation	Code-dependent	Code-dependent	Link-based
Multiple parents	No	No	Yes
Hierarchy depth	4 levels	5 levels	Unlimited
Hierarchy relationships	Mixed	Mixed	Subtype
Meaningless identifiers	No	No	Yes
Compositionality	No	No	Constrained
Cross maps	ATC	OPCS4, ICD9, ICD10, ATC	OPCS4, ICD9, ICD10, ATC
Flexibility	No	No	Yes
Simplicity	Yes	Yes	No
Term identifiers	No	Yes	Yes
Semantic definitions	No	No	Yes
Number of concepts*	40,927	88,995	187,598
Number of terms*	57,128	125,914	220,840

* 1997 figure includes pharmacy

In England the NHS has decided to withdraw support of the Read Codes from 2016 and require all systems that use them to migrate to SNOMED CT by 2018. There have been three versions of the Read Codes, known as 4-byte, Version 2 and Version 3 (also referred to as ctv3). The main differences are summarised in Table 8.4 [4].

Health management relies on comparable coded data. It is hard to imagine that the government in England would entrust 80% of the healthcare budget to GPs to manage, without having coded data to monitor what is going on. This happened as a result of the Health and Social Care Act 2011. The Read codes are a good example of a successful clinical coding system, which was fit for purpose, at least in general medical practice.

The development of the Read Codes began in 1983, when, with colleagues Dr James Read and Dr David Markwell, the author (TB) helped design a new computer system for use in general practice. An early design decision was to use a development tool that used fixed-length fields, requiring all codes, terms and look-up keys to have a fixed pre-defined length.

The original design used alphanumeric codes with four characters (later extended to 5 characters) and terms up to 30 characters long (later extended). A key requirement was that the coding scheme should be comprehensive, covering everything that might be entered into a patient's computerized record. No existing coding scheme could be found which met all these criteria so we chose to write one from scratch (as did several other suppliers during that period).

The motivation was commercial, a point of view shared by other GP suppliers. GPs do not want to do any extra work and were mildly computer-phobic. GPs had little interest in developing their own local coding schemes, and wanted a system that worked out of the box. We wanted a coding scheme that would allow one-finger typists to enter data in the consulting room, by typing in a few letters and the computer doing the rest. They also wanted a system that could generate reports almost instantly.

Table 8.5 Read Codes (4-Byte), 1986

	Coded terms	Synonyms	Total terms	%
Diseases	2598	2575	5173	22 %
Procedures	6023	2483	8506	36 %
Occupations	1749	936	2685	11 %
History	1299	890	2189	9 %
Examination	1480	890	2370	10 %
Prevention	1279	460	1739	7 %
Administration	696	416	1112	5 %
Total	15,124	8650	23,774	

The idea was to take existing classifications and convert these into the appropriate format. These included ICD for diseases, the British National Formulary (BNF) for drugs and the International Classification of Procedures in Medicine (ICPM). Later this was extended to include the UK national coding scheme for operations OPCS-4.

Dr James Read, a GP in Loughborough, undertook the editing task and developed new sections for examination findings, preventive care, administrative procedures and other subjects for which no suitable model could be found. Dr David Markwell developed the software. What was originally planned to take 3 months took almost 3 years and the scheme was finally launched as the eponymous Read Codes in 1986 [5].

As the work evolved, we found that we had improved on earlier classification and coding systems in several key respects. No paper version was ever published, facilitating regular updates and extensions. The simple position-dependent unidimensional hierarchy was easy to implement in software. The scheme was designed by GPs for clinical use by GPs in their surgery (rather than for secondary use such as epidemiology and international comparisons).

The first publication was in the British Journal of Healthcare Computing in May 1986 [6]. The next section is based on this original paper. The number of codes in the original version 1 (May 1986) is shown in Table 8.5.

Later developments greatly increased the number of terms, but there is no evidence that this increased usability to a significant extent – probably the reverse.

Hierarchical Codes

The structure of the classification hierarchy is mapped directly by codes. In the same way as a map grid reference specifies a position on a map, each code specifies its position within the classification. The original Read Clinical Classification has four-digit alphanumeric codes using the numerals 0–9 and the letters A–Z (letters I and O were not used to avoid confusion with 1 and 0). The first character relates to level 1, the second to level 2 and so on. Consider code B136; this is broken down as follows (Table 8.6):

Table 8.6 Example of Read
Code hierarchy

First level	B...	Neoplasm
Second level	B1..	Malignant neoplasm
Third level	B13.	Carcinoma stomach
Fourth level	B136	Ca. greater curvature-stomach

Table 8.7 Pick list for
rubella

0	H/O: rubella	1418
1	Rubella	A47.
2	Rubella + pregnancy	K2A3
3	Rubella-congenital	0251
4	Rubella health educ.	6794
5	Rubella antibody titre	439.
6	Rubella contact	65P5
7	Rubella damage-preg.	K364
8	Rubella screen	62J.
9	Rubella vaccination	65P.

The four-digit codes increase in detail from left to right. The alphanumeric coding system using four digit codes allows 1,679,616 possible entries (36^4). The scheme was later extended to allow lower-case letters (a-z), giving 60 options at each level, total 60^4 (about 12 million options). This is a code-dependent hierarchy.

Automatic Encoding

The classification incorporates automatic encoding. Entry of the first few letters of any term displays a list beginning with those letters from which the user chooses by line number. Consider the term 'rubella'. Entry of the letters 'rub' triggers a list (Table 8.7).

National and international medical classifications, such as ICD, have been developed to facilitate the production of statistics for epidemiology and research. None of these classifications covers the whole field of medicine, and none is suitable for clinical use because their coded content is not sufficiently specific.

James Read aimed to be comprehensive in both breadth of cover and also in the detail of the terms used in general practice. The Read Clinical Classification was based where possible on existing classifications, but large areas of medicine had not been classified before and Read extended the areas covered by existing schemes to include history, symptoms, examination findings, prevention and administration (and medication).

Table 8.8 Hierarch for
cerebral haemorrhage

Level 1	G...	Circulatory system diseases
Level 2	G7..	Cerebrovascular disease
Level 3	G71.	Cerebral haemorrhage
Level 4	G711	Subarachnoid haemorrhage
	G712	Intracerebral haemorrhage
	G713	Extradural haemorrhage
	G714	Subdural haemorrhage

Diseases

At the time, the International Classification of Diseases Ninth Revision (ICD-9) was
the standard statistical classification of diseases, used by hospitals throughout the
world. Sections of the Read Clinical Classification, which deal with diagnoses, injuries and death, are directly based on ICD-9. The Read first-digit codes A to Q correspond directly to ICD chapters, with the exception of chapter XVI (symptoms,
signs and ill-defined conditions) which is covered in greater detail elsewhere. Each
Read category is precisely cross-referenced to ICD.

This section of the Read Classification has 17 first level codes, 115 two-digit
codes, 728 three-digit codes, 2598 four-digit codes and 2575 synonyms. The level
of detail at each level is illustrated by an example (Table 8.8).

Procedures

The International Classification of Procedures in Medicine (ICPM) complemented
ICD-9 as a standard classification of surgical, diagnostic and therapeutic procedures. The Read Clinical Classification covers the whole of ICPM with the exception of the section on drugs, medicaments and biological agents.

In many cases the content and detail has been expanded to provide clinically
specific rubrics. For example the results of laboratory procedures are classified as in
Table 8.9.

The decision to include both the procedure (urine test for glucose) and the finding (Urine glucose test negative) in the same structure was probably a mistake
which has created problems ever since.

A change, made shortly after the publication of this paper, was to start sub-lists
at 0 rather than 1. The lists shown here are those in the original paper, not those
widely implemented.

Similarly in operative procedures, mastectomies for example, are classified as
(Table 8.10).

Two problems in the operations sections were the length of many operation
names, which required abbreviations that were sometimes hard to understand, and
the need for further levels of detail.

Table 8.9 Urine test for
glucose

Level 1	4...	Laboratory procedures
Level 2	46..	Urine examination
Level 3	466.	Urine test for glucose
Level 4	4661.	Urine glucose test not done
	4662.	Urine glucose test negative
	4663.	Urine glucose test=trace
	4664.	Urine glucose test=+
	4665.	Urine glucose test=++
	4666.	Urine glucose test=+++
	4667.	Urine glucose test=++++

Table 8.10 Mastectomies

Level 1	7...	Operative procedures
Level 2	7F..	Breast operations
Level 3	7F1.	Mastectomy
Level 4	7F11	Breast lump local excision
	7F12	Partial mastectomy
	7F13	Simple mastectomy
	7F14	Extended simple mastectomy
	7F15	Radical mastectomy
	7F16	Extended radical mastectomy
	7F17	Subcut mastect. + prosth implant
	7F18	Subcutaneous mastectomy

Sections of the Read Clinical Classification covering diagnostic procedures (including laboratory and X-ray) and therapeutic procedures (including surgery) comprise 6023 code categories and 2483 synonyms.

History/Symptoms

The history and symptoms section contains family, social and medical history as well as presenting symptoms. The relevant section in ICD-9 (Chapter XVI symptoms, signs and ill-defined conditions) is incomplete and reclassification was needed

Table 8.11 Smoking

Level 1	1. ..	History/symptom
Level 2	13..	Social/personal history
Level 3	137.	Tobacco consumption
Level 4	1371	Complete non-smoker
	1372	Trivial smoker
	1373	Light smoker
	1374	Moderate smoker
	1375	Heavy smoker
	1376	Very heavy smoker
	1377	Ex-smoker
	1378	Tobacco consumption unknown

to provide adequate clinical detail. Where any history/symptom factor has gradable variables each option is offered as a separate fourth level category. For example (Table 8.11).

Each term was defined. 'Heavy smoker' is 12–24 cigarettes a day or 80–160 per week, and 20 cigarettes is equivalent to 2 large cigars, 5 medium cigars, 10 small cigars or an ounce of tobacco.

The history/symptoms section had 1299 codes and 901 synonyms. History data is of cardinal importance in diagnosis and the prevention of disease and disability. "Listen to the patient, he is trying to tell you the diagnosis."

Occupations

The OPCS Classification of Occupations was the basis of this section of the Read Clinical Classification with 1749 coded occupational categories and 936 synonyms. Occupation is an important part of any patient database used for prevention or epidemiology.

Examination/Signs

The classification of examination findings and signs is organized by systems. This part was classified from scratch in the absence of any other recognized classification covering patient examination. This section comprises 19 second level, 282 third level and 1480 fourth level categories with 890 synonyms.

For example, retinal examination is classified as (Table 8.12):

Table 8.12 Retinal
inspection

Level 1	2. ..	Examination/Signs
Level 2	2B..	Central nervous system exam.
Level 3	2BB.	O/E – retinal inspection
Level 4	2BB1	O/E – retina normal
	2BB2	O/E – retinal vessel narrowing
	2BB3	O/E – retinal A-V nipping
	2BB4	O/E – retinal microaneurisms
	2BB5	O/E – retinal haemorrhages
	2BB6	O/E – retinal exudates
	2BB7	O/E – retinal vascular prolif.
	2BB8	O/E – vitreous haemorrhages

Prevention

Preventive procedures were classified from scratch. This is a key section of the classification particularly as computer-based prevention records and protocols could lead to major changes in the quality of patient care. This section includes:

- Contraception
- Pregnancy care and birth details
- Child healthcare
- Vaccination and immunization
- Chronic disease monitoring
- Health education and counseling
- Screening, etc.

The level of detail provided for cervical smear screening is shown in Table 8.13. The preventive procedures section had 1279 categories and 460 synonyms.

Administration

This section covers all aspects of practice administration. Examples include the stages of patient registration and de-registration, administrative details of patient encounters, processing of claim forms, staff administration, practice finance and audit reporting.

Table 8.13 Cervical
screening

Cervical neoplasia screening
Cx Screen - not offered
Cx Screen - offered
Cx Screen - not wanted
Cx Screen - wanted
Cx Screen - not needed
Cx Screen - up to date
Cx Screen - not attended
Cx Screen - not reached
Cx Screen - done
Cx Screen - no result yet
Cx Screen - normal
Cx Screen - abnormal
Cx Screen + fee claimed

Table 8.14 Contraceptive
FP1001 claims processing

Level 1	9...	Administration
Level 2	93..	Contraception administration
Level 3	931.	FP1001 claim status
	9311	FP1001 claim signed
	9312	FP1001 claim sent to FPC
	9313	FP1001 claim up to date
	9314	FP1001 claim due
	9315	FP1001 claim due next visit
	9316	FP1001 claim cancelled

There are 696 coded categories and 416 synonyms in the administration section. For example contraception FP1001 claim status is classified as Table 8.14.

Drugs

A significant extension, made later in 1986, shortly after the first paper was published, was to extend the scheme to cover medicines. Drugs were given codes starting with lower-case letters a-z, corresponding to the chapters in the first edition of the BNF.

Development

One of the reasons for changing the name from the Abies Medical Dictionary to the eponymous Read Codes was to encourage other suppliers to use them too. One of the first to take up this offer was EMIS.

In 1987, the Department of Health commissioned the Joint Computing Group of the BMA's General Medical Services Committee and the Royal College of General Practitioners to evaluate clinical coding systems for use by GPs. The working party considered the following morbidity coding schemes:

1. ICD-9
2. ICHPPC-2
3. ICPC
4. OXMIS
5. Read Clinical Classification
6. RCGP classification
7. Update morbidity dictionary
8. SNOMED

The final report (August 1988) listed the most important requirements to be:

1. Comprehensive in breadth and depth
2. Appropriate for GP usage
3. Provision for central maintenance
4. Amenable to statistical analysis
5. Compatibility with ICD-9
6. A hierarchical structure (second level requirement)
7. Accessibility of coding structure to the user (third level requirement)

The working group recommended the Read codes, with some qualifications:

1. Longer rubrics were needed for operations
2. Align to national coding schemes (ICD-9, OPCS-4, PPA Drug Index, SOC (standard occupational classification)
3. A fully resourced UK standing professional committee should be established to maintain and control the classification
4. Guidance should be provided on usage

The Department of Health set out to implement these recommendations and after almost 2 years of tortuous negotiations purchased the Read Codes for £1.25 million in April 1990, leading to the establishment of the NHS Centre for Coding and Classification [7].

Why Read Codes Were Successful

Features of the first generation Read codes that made them successful were.

1. A single responsible author/editor
2. Fit for purpose, written by a GP for GPs
3. Comprehensive (examination findings, history, administration etc.)
4. Modest evolutionary step (built on ICD-9 etc.)
5. Easy to implement in software and on screen
6. Understandable by users

The Read Codes improved on earlier classification and coding systems in several respects.

1. They were designed specifically for use by GPs in their surgery, not for epidemiology and international comparisons.
2. The simple position-dependent one-dimensional hierarchy was easy to understand by users
3. No paper version was ever published, facilitating multiple updates and extensions.
4. Easy to implement in software.

Problems with the Read Codes

One of the problems with Read-coded data has been the quality of information. It is easy to make a mistake when entering data, which seriously impacts data quality. For instance, entry of the term *physio* will give a list of options, the first being the occupation [03J1. physiotherapist]. It is easy to choose an occupation when what should be chosen is [8H77. refer to physiotherapist]. This sort of systemic misuse is not good for data quality.

The Read Codes combine the features of a classification and a coding scheme. However, no hierarchical coding scheme can ever be multipurpose, because they are built around a single hierarchical axis and each code is classified in one way only. The Read Codes have been highly successful in General Practice, for which they were designed. However, attempts to use the original versions in hospitals proved impracticable, primarily because the simple hierarchical scheme could reflect only one view, namely the general practice perspective. Hospital doctors did not understand why information retrieval in one dimension (which was of little interest) was easy, but in another dimension was difficult and slow.

Position-dependent coding schemes cannot be updated. Once a concept has been placed in the classification, it is not practicable to move it, even if it has been placed in a location that is later regarded as wrong. It is not possible to add in new codes in the middle of a sequence.

Another problem is the inherent multi-dimensionality of medicine. For example, tuberculosis meningitis is a type of tuberculosis, which is an infectious disease (and is given code A130.), but it is also an inflammatory disease of the central nervous system and has another code F004.. Having two separate codes creates code redundancy, which can cause inaccuracies in hierarchy-based analysis of clinical data stored using the codes.

Being restricted to only four levels (later extended to five levels) in the hierarchy causes another problem. Consider the mastectomy hierarchy (Table 8.15):

It is not possible to add a more detailed variant of this operation, such as subcutaneous mastectomy for gynecomastia (man boobs) in the appropriate position because there is no 6th level. A possible solution is to add it as a sibling alongside subcutaneous mastectomy in the 5th level with a code such as 71307. However this creates the danger that when retrieving cases of subcutaneous mastectomy (71304), those recorded using 71307 would be missed.

The NHS Clinical Terms project was started in 1992, as a major attempt to address all of the issues listed above. Expenditure on the Read Codes between 1990 and 1998 was £32 million [8]. The resulting scheme, which is known as Clinical Terms Version 3 (CTV3), was merged with the College of American Pathologist's SNOMED RT during 1999–2002 to create SNOMED CT (see Chap. 9). First we consider the early history of SNOMED.

SNOP and SNOMED

SNOMED has a long history. Back in 1955 the College of American Pathologists (CAP) established a committee to develop a nomenclature for anatomic pathology. In 1965, they published the *Systematized Nomenclature of Pathology* (SNOP), which describes pathology findings using four axes:

- Topography (anatomic site affected)
- Morphology (structural changes associated with disease)
- Etiology (the cause of disease) including organisms
- Function (physiologic alterations associated with disease).

SNOP was the first multi-axial coding system used in healthcare. By 1975 Roger Côté and colleagues had extended SNOP by adding additional dimensions covering diseases and procedures to give it a broader scope with the name *Systematized Nomenclature of Medicine* (SNOMED).

Table 8.15 Mastectomy hierarchy		
7....	Operations, procedures, sites
71. ..	Endocrine system and breast operations
713..	Breast operations
7130.	Total mastectomy operations
71304	Subcutaneous mastectomy

Table 8.16 SNOMED 3.5 axes

T	Topography – Anatomic terms (13,000 records)
M	Morphology – Changes found in cells, tissues and organs (6000 records)
L	Living organisms – Bacteria and viruses (25,000 records)
C	Chemical – Drugs (15,000 records)
F	Function – Signs and symptoms (19,000 records)
J	Jobs Terms that describe the occupation (1900 records)
D	Diagnosis – Diagnostic terms (42,000 records)
P	Procedure – Administrative, diagnostic and therapeutic procedures (31,000 records)
A	Agents – Devices, physical agents and activities associated with disease (1600 records)
S	Social context – Social conditions and important relationships in medicine (500 records)
G	General – Syntactic linkages and qualifiers (1800 records)

The original SNOMED was developed around a model of illness that started with normal structure (topography) and function. Sickness typically involves some abnormal physiology (function) or structure (morphology). This has some cause (etiology), which may be internal or external. Medicine seeks to reverse the process from the sick state to the healthy state by using administrative, diagnostic and therapeutic procedures, which act on function or body structure. Disease was added to give easy mapping to ICD. Occupations and organisms were added later. By 1998 SNOMED 3.5 had expanded to eleven axes and 157,000 records (Table 8.16).

The next generation, SNOMED RT (Reference Terminology) involved a major change. Spackman, Campbell and Côté (1997) describe the need for a reference terminology as follows:

> The need for a reference terminology can be illustrated by the situation facing many managed care organizations. They may have several different hospitals and clinics, each with an existing set of information systems. These organizations need to aggregate data from several systems in order to manage the quality and cost of care across the entire organization. Rather than totally replacing their existing information systems with one common system, they have a need to record data from each system using or referring to a common reference terminology. Aggregate data can then be grouped and analyzed using the various hierarchies of the reference terminology [9]

In developing SNOMED RT, the decision was made to adapt the KRSS (Knowledge Representation System Specification) as the description logic syntax for SNOMED RT.

SNOMED RT gave each concept code a semantic definition stated in description logic. SNOMED RT was completed in 2000 and provided one of the two key sources for SNOMED CT (see Chap. 9).

LOINC

Laboratory and clinical systems need to merge data for a variety of purposes, including clinical care, quality improvement and reporting, public health reporting, and research. While many systems use electronic messages to transmit results, most systems use their own internal, idiosyncratic codes to identify the results inside those messages. As a result, receiving systems cannot understand their contents without mapping every item to their own codes. This problem would be solved if each system used the same set of "universal" identifiers for clinical and laboratory observations. LOINC (logical observation identifiers names and codes) provides this set of universal identifiers [10]. LOINC provides codes for the observation names (eg eye color), not the observation finding (eg blue eyes). LOINC is like a bar-code for test data; LOINC provides codes for questions and where needed, other vocabularies provide codes for the answers.

LOINC is a community-built universal code system that facilitates the exchange, pooling, and processing of laboratory and other clinical observations. It is a controlled terminology that contains unique identifiers and "fully specified" names built using a formal structure, which distinguishes among tests and observations that are clinically different.

Researchers at the Regenstrief Institute set up the LOINC Committee and development of the database in 1994. Since then, the Regenstrief Institute and the LOINC Committee have published more than 50 versions of the standard that now contains around 80,000 terms (2016). Regenstrief Institute serves as the overall steward for the LOINC development effort, and works together with the LOINC Committee to define the overall naming conventions and policies for the development process. The Regenstrief Institute and the LOINC Committee have intentionally shaped LOINC development to be empirical, nimble, and open.

LOINC has been widely adopted. Today, there are more than 40,000 users from 170 countries. LOINC terms are available in 21 languages and dialects. LOINC has been adopted in both the public and private sector by government agencies, laboratories, care delivery organizations, health information exchange efforts, healthcare payers, research organizations, and within many exchange standards.

LOINC creates codes and a formal name for each concept that corresponds to a single kind of observation measurement or test result. The formal LOINC name is fully specified in the sense that it contains the features necessary to disambiguate among similar clinically distinct observations. The fully specified name is constructed using a six-part semantic model to produce a pre-coordinated expression. This does not capture all possible information about the procedure or result – just enough to unambiguously identify it.

The six parts of the model are:

1. Component or analyte: the substance or entity being measured or observed (eg potassium)
2. Property: the characteristic or attribute of the component or analyte (eg mass concentration g/L)
3. Time: the interval of time over which an observation was made
4. System: the specimen or thing upon which the observation was made (eg urine or blood)

5. Scale: how the observation value is quantified or expressed: quantitative, ordinal, nominal.
6. Method: a high-level classification of how the observation was made (optional). Only used when the technique affects the clinical interpretation of the results.

In addition to the fully specified name, LOINC also provides alternate names for use in other contexts.

The LOINC structure also contains codes for the atomic elements (parts) that make up the fully specified names. LOINC parts are used to construct hierarchies that organize LOINC terms, link synonyms and descriptions across many terms, and are the basis for an efficient mechanism for translating LOINC names. LOINC also contains a robust model for representing enumerated collections of observations (panels), which captures the hierarchical structure of the elements, attributes of individual data elements, value sets, and panel-specific attributes of data elements.

The laboratory section of LOINC covers the observations that can be made on a specimen, and includes most clinical laboratory testing, including chemistry, urinalysis, hematology, serology, microbiology (including parasitology and virology), toxicology, and molecular genetics. Presently, about 70 % of the terms in LOINC are for laboratory observations.

The clinical section of LOINC covers the observations that can be made on a whole person or population. Some of the domains covered by Clinical LOINC include vital signs, hemodynamic measurements, anthropomorphic measures, patient assessments, obstetrical ultrasound, radiology reports, and clinical documents and sections.

LOINC is distributed and made freely available from the LOINC website, with new releases published twice yearly. The main LOINC database is distributed in several file formats. In addition, Regenstrief develops and distributes a software program called the Regenstrief LOINC Mapping Assistant (RELMA) that provides tools for browsing the database and mapping local terminology to the LOINC terms. Additionally, Regenstrief develops the accompanying documentation and also distributes several accessory files to the main distribution, such as a file containing the full representation of enumerated collections and a hierarchy file.

LOINC terms and other resources have been translated by volunteers into many languages from its native English that are also made available from its website. Regenstrief has developed a refined process for enabling translation of the LOINC terms based on the parts from which they are constructed. By translating a list of parts for the terms of interest, translators only have to translate an element like "glucose" once, and it can then be applied to all of the terms that contain it.

Mapping local observation codes to LOINC provides a bridge across the islands of data that reside in isolated electronic systems. However, mapping is a resource-intensive and often rate-limiting step in interoperable data exchange. RELMA (Regenstrief LOINC Mapping Assistant) software helps with this process [11].

By design, LOINC covers a circumscribed content domain (observation identifiers) and is often used in conjunction with other terminology standards. LOINC is included as a source vocabulary in the National Library of Medicine's Unified Medical Language System (UMLS). LOINC has historically been used in conjunction with the Systematized Nomenclature of Medicine (SNOMED) with LOINC providing codes for the question and SNOMED providing codes for the answer or value. Recently, SNOMED has developed a model for representing observable entities based on LOINC's semantic structure and work is well underway to more closely coordinate terminology development between LOINC and SNOMED CT. The Unified Code for Units of Measure (UCUM) is a standard for electronic communication of units of measure with growing adoption that Regenstrief has included as way to represent units used for a LOINC term within the LOINC database.

As LOINC has been implemented in many settings, the adopting communities have uncovered different approaches that address unique features of those contexts. Lessons from nation-wide and regional health information exchanges, public health reporting, and implementation in resource-constrained settings can help inform our ongoing initiatives to advance the use of health information technology to improve the effectiveness and efficiency of care delivery.

UMLS

The National Library of Medicine has developed the Unified Medical Language System (UMLS) as an important terminology resource, intended for use mainly by developers of health information systems [12]. The UMLS contains a number of knowledge sources and tools, including the UMLS Metathesaurus and Semantic Network.

The UMLS Metathesaurus is a large multipurpose multilingual vocabulary database that contains over five million terms and one million concepts covering information about health and biomedical concepts, their names and relationships between them. The Metathesaurus is built from over 100 different source vocabularies and seeks to reflect and preserve the meanings concept names and relationships from these sources. For example, if two source vocabularies use the same name for different concepts or define the same concept in different ways, the Metathesaurus represents both sets of meanings and relationships, and indicates which meaning is used in which source vocabulary. The Metathesaurus preserves the many views of the world present in its source vocabularies because these views may be useful for different tasks.

The UMLS Semantic Network consists of a set of broad subject categories, referred to as Semantic Types, and a set of important relationships (Semantic Relations), which exist between Semantic Types. The Semantic Types are similar to, but not the same as the SNOMED CT hierarchies. There are 133 semantic types.

The Semantic Relations are similar to Relationships in SNOMED CT and includes the "isa" sub-type relationship for hierarchies and other non-hierarchical relationships grouped according to: physically related to; spatially related to; temporally related to; functionally related to; and, conceptually related to.

Another UMLS product is the CORE (Clinical Observations Recording and Encoding) Problem List Subset of SNOMED CT, which is a common list of SNOMED CT concepts for use in problem lists and patient summaries.

References

1. WHO. ICD-10. International statistical classification of diseases and related health problems. Geneva: WHO; 1992.
2. Fetter RB, et al. Case mix definition by diagnosis related groups. Med Care. 1980;18:1–53.
3. Mayes R. The origins, development, and passage of Medicare's revolutionary prospective payment system. J Hist Med Allied Sci. 2006;62(1):21–55.
4. Robinson D, Schulz E, Brown P, Price C. Updating the Read Codes: user-interactive maintenance of a dynamic clinical vocabulary. JAMIA. 1997;4(6):465–72.
5. Benson T. The history of the Read Codes: the inaugural James Read Memorial Lecture 2011. Inform Prim Care. 2011;19(3):173–82.
6. Read J, Benson T. Comprehensive coding. Br J Healthcare Comput. 1986;3(2):22–5.
7. Chisholm J. The read clinical classification. BMJ. 1990;300:1092.
8. The purchase of the Read Codes and the management of the NHS Centre for coding and classification select committee on Public Accounts Sixty-Second Report 1998. www.parliament.uk
9. Spackman K, Campbell K, Côté R. SNOMED RT: a reference terminology for health care. AMIA Proc. 1997: 640–4.
10. McDonald C et al (eds). Logical Observation Identifiers Names and Codes (LOINC) Users' Guide. Indianapolis: Regenstrief Institute. 2011.
11. Vreeman DJ, Hook J, Dixon BE. Learning from the crowd while mapping to LOINC. J Am Med Inform Assoc. 2015;22(6):1205–11.
12. National Library of Medicine (US). UMLS reference manual. Bethesda: National Library of Medicine; 2009.

Chapter 9
SNOMED CT

Abstract SNOMED CT is a comprehensive multilingual clinical terminology used in electronic health records and interoperability. Its components are concepts (codes), descriptions (terms) and relationships. Concepts are organized into hierarchies. SNOMED CT expressions are either a single concept (pre-coordinated) or more complex post-coordinated expressions using compositional grammar. Reference sets are used to provide subsets of the whole terminology for specific purposes.

Keywords SNOMED CT • Components • Concept • Description • Relationship • SctId • Defining relationship • Qualifying relationship • Hierarchy • Description logic • Expression • Compositional grammar • Pre-coordination • Post-coordination • Subtype qualification • Axis modification • Subsumption testing • Transitive closure table • Reference set • Language reference set • Navigation reference set • Map reference set • Release format 2 (RF2) • Delta release • SNOMED CT documentation

Introduction

SNOMED CT is the most comprehensive multilingual clinical healthcare terminology available. It is used in electronic health record systems to facilitate clinical documentation and reporting and to retrieve and analyse clinical data.

SNOMED CT is both a coding scheme, identifying concepts and terms, and a multi-dimensional classification, enabling concepts to be related to each other, grouped and analysed according to different criteria. IHTSDO (see Chap. 6) describes SNOMED CT as the global language of healthcare. While SNOMED CT has some of the attributes of a language, it is probably more useful to think of it as a coding system, where the codes used are unambiguous and are designed for computer processing.

Today most uses of SNOMED CT use single codes, although SNOMED CT provides a compositional grammar allowing complex expressions to be built up. Such expressions create challenges for reporting and analysis, but in future such post-coordinated expressions may become widely used.

© Springer-Verlag London 2016   155
T. Benson, G. Grieve, *Principles of Health Interoperability*,
Health Information Technology Standards, DOI 10.1007/978-3-319-30370-3_9

SNOMED CT has two key features that make it a major improvement over most other coding systems used in healthcare. First it is virtually future-proof, by being inherently evolvable. Concepts, terms and their codes can be added or deprecated (once added codes are not deleted) without limit and their relationships with others can be changed. Another aspect of its flexibility is support for multiple languages and dialects. Furthermore, SNOMED CT supports multiple relationships, including multiple parent-child hierarchical relationships. This simply reflects practical reality, but, as described in Chap. 8, many current coding schemes can only support a single axis of classification, which can be severely limiting.

At its heart, SNOMED CT is a model of meaning (see Chap. 11). The focus is on what is technically true and correct. For example in SNOMED CT, urine and glucose are *substances*, urine glucose concentration is an *observable entity*, a urine glucose test is a *procedure* and a urine glucose test result is a *clinical finding* and if a urine glucose test is not done it is a *situation with explicit context*. This can be quite confusing (we explain this in Chap. 10).

On the other hand, users are only interested in what is useful to them, and this is known as the model of use (see Chap. 11). However the model of use varies according to context. GPs, urologists and nurses are interested in different things about urine. In contrast, the model of meaning is always correct. Good implementations use a model of use for the user interface, supported by the universal model of meaning in the background, to provide the best of both.

On its own SNOMED CT does very little, but its value is realized when it is built into software, such as an EHR. The best implementations will be designed specifically for use with SNOMED CT.

Kent Spackman, the leader of the team that developed SNOMED CT postulated two golden rules: [1]

> *The first rule of coding is that yesterday's data should be usable tomorrow.*

Clinical data needs to be treated as being permanent. We have to be able to use yesterday's and today's data for the indefinite future. On the other hand, if nobody is going to re-use the data, there is no need to code it.

> *The first rule of data quality is that the quality of data collected is directly proportional to the care with which options are presented to the user.*

There is an enormous variety of medical activity and any attempt to impose a *one size fits all* approach is doomed to failure. This is why the model of use, implemented using context-specific reference sets is a key aspect of SNOMED CT implementations. Good implementation invariably takes full account of context.

SNOMED CT provides tools to record information about patients in a way that can be indexed and retrieved for reuse clinically at the point of care and subsequently for management, surveillance and research. It provides the clinical content and expressivity required for precise clinical documentation.

SNOMED CT has a broad coverage of routine clinical medicine, but still needs extension in areas such as social care and some specialist areas.

Origins of SNOMED CT

In 1999 the English National Health Service (NHS) and the College of American Pathologists (CAP) agreed to merge SNOMED RT (Reference Terminology) with the Read Codes Version 3, also known as the NHS Clinical Terms Version 3 (ctv3), to produce a single joint clinical terminology – SNOMED CT (Clinical Terminology). The merger was completed in 2002 with the first release of SNOMED CT.

SNOMED CT was a true merger. Every Read Code and previous SNOMED RT code ever released was incorporated into SNOMED CT so that migration to it would not result in loss of information. In 2007, the International Health Terminology Standards Development Organization (IHTSDO) acquired all rights to SNOMED CT (see Chap. 6).

Cimino's Desiderata (see Chap. 7) were key design criteria. Additional criteria were that all content should be understandable, reproducible and useful. Understandable means that definitions should be understood by average clinicians, given brief explanations. Reproducible means that retrieval and representation of the same item should not vary according to the nature of the interface, user preferences or the time of entry. Usable leads to the conclusion that we should ignore distinctions for which there is no use in healthcare.

SNOMED CT is large. The number of concepts, descriptions and relationships varies with every release. SNOMED CT contains more than 300,000 active concepts, about one million English descriptions and more than 1.4 million relationships. There is no paper version. It can only be accessed using specialised software, such as a SNOMED CT browser. Its sheer size is a significant issue in development, use and maintenance.

SNOMED CT cannot be used manually, partly because it is too big, but more importantly because it works in a way that is completely different from earlier coding schemes such as ICD or the Read Codes. The reference structure of SNOMED CT relies on explicitly defined relationships that need computer software to work. This is considerably more complex than code-dependent hierarchies, but is more powerful, flexible and future-proof. It allows any concept to be defined or qualified in as many ways as are needed.

SNOMED CT provides an extensible foundation for expressing clinical data in local systems, interoperability and data warehouses. The terminology is composed of *concepts*, *descriptions* and *relationships* that provide a way to represent clinical information across the broad scope of healthcare and can support analysis and clinical decision support.

SNOMED CT is organised into *hierarchies*. A node in a hierarchy represents each concept, with one or more subtype relationships to its parent(s). Understanding these hierarchies is important and their content is described in Chap. 10 SNOMED CT Concept Model. This chapter describes the structure and components of SNOMED CT.

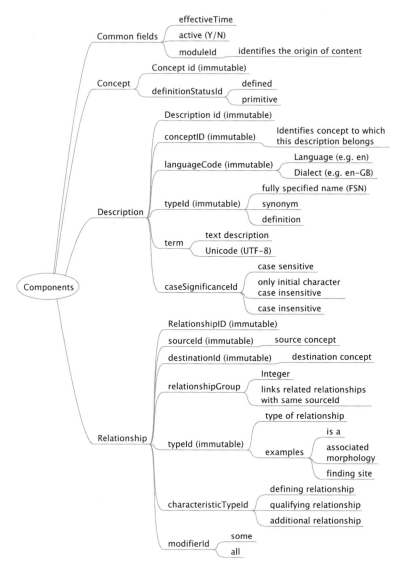

Fig. 9.1 SNOMED CT components

Components

SNOMED CT is composed of components, such as concepts, relationships, descriptions, reference sets and cross maps. A SNOMED CT Identifier (*sctId*) identifies every component. Figure 9.1 shows the main SNOMED CT components and their fields in RF2 release format.

All components have an *effectiveTime* field which states the time when this version of a component superseded the previous version or was created initially. Use of *effectiveTime* allows a single file/table to hold all released versions of SNOMED CT, computation of *snapshot* view for any specified date and *delta* release files of new rows supporting incremental updates.

All components carry an *active* field. An important principle of SNOMED CT is that of permanence. Once a component, such as a concept or description has been created it is never deleted, but the active flag may be set to inactive status. They also have a *moduleId*, which is used to identify the origin and organization responsible for maintaining this component.

Concept

SNOMED CT is concept-oriented. Concepts are clinical meanings that do not change. Each concept has a unique concept identifier (*conceptId*), which is an *sctId*. The conceptId is important because it is the code used to represent the meaning in clinical records, documents, messages and data.

SNOMED CT concept identifiers are a sequence of digits that do not reflect the meaning of the concept. There is seldom any value in displaying these to end users. Concept identifiers are simply unique identifiers used within computer systems and are not intended to be used by clinicians.

Description

Each concept is associated with a set of text descriptions, which provide the human-readable form of the concept. Every concept has at least two descriptions (terms) – the fully specified name (FSN) for that concept and a display term in the language being used.

The FSN is a phrase that names a concept in a way that is unique and unambiguous. Each FSN contains a suffix in parentheses that indicates its primary hierarchy, eg `myocardial infarction (disorder)`. The display term is often the FSN without its suffix (hierarchy tag), eg `myocardial infarction`.

All other descriptions are synonyms. Synonyms may be marked as being preferred or acceptable. The preferred term is a common phrase or word used by clinicians to name a concept and is used as default display term for that concept in a particular language or context.

Other synonyms may be marked as acceptable eg `heart attack` or `cardiac infarction`. A list of synonyms for a concept shows the various ways a concept may be described, rather like a thesaurus (see Fig. 9.2).

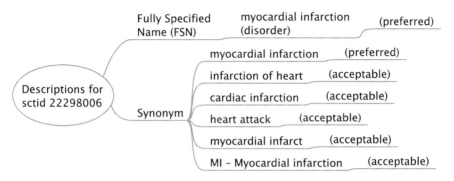

Fig. 9.2 Example descriptions for myocardial infarction

SNOMED CT terms can also be homonyms, where the same term is used for different concepts. The preferred terms and synonyms are not necessarily unique within a language or dialect. For example, the FSN `cold sensation quality (qualifier value)` has a preferred term of `cold`, but `cold` is also a synonym of `common cold (disorder )`. The two meanings of `cold` have different concept Ids.

Each description links a human-readable term with a concept. It has an associated unique numeric *descriptionId*. Terms are encoded using Unicode, which supports all languages. The language code of each description is recorded using ISO 2-character codes (eg `en` for English). The case of the first letter of the term can flagged as having case significance, as in `pH` or `Alzheimer's`.

Relationship

Relationships are the distinguishing feature of any reference terminology such as SNOMED CT. More than 1.4 million relationships have been defined in SNOMED CT and this number continues to grow.

Each relationship is defined as an Object-Attribute-Value triple, which can be processed by a computer. The Object is the source concept – the one that has the relationship, identified by a concept identifier (*sourceId*). The Attribute specifies the type of relationship (*typeId*), and is also a SNOMED CT concept. The Value is the target (*destinationId*).

All relationships are written using a notation known as Description Logic (DL), such as:

```
|concept|:|attribute|=|value|
```

The allowable attributes and values, which may be used to define or qualify concepts, are set out in the SNOMED CT Concept Model (see Chap. 10).

Four different types of relationship (*characteristicType*) are: defining relationship, qualifying relationship, historical relationship and additional relationship.

Defining Relationships are used to define each concept by its relationships with other concepts. Only relationships that are always true are used as defining relationships. Defining relationships specify the concept's supertypes (parents) or defining attributes. Defining relationships are specified as attribute-value pairs, where each attribute and value is a SNOMED CT concept. Supertypes, used in hierarchies are specified using the |is a| relationship, which has the conceptId 116680003.

Every active SNOMED CT concept except the SNOMED CT root has one or more supertypes. Supertype relationships allow users to identify whether a patient with a specific condition has a more general condition that subsumes the specific one. It lets you answer questions such as *"is angina pectoris a type of heart disease?"*

A concept is *defined* if its defining relationships are sufficient to distinguish it from all its supertype and sibling concepts. If a concept is sufficiently defined we can say that another concept, which is represented as a combination of the same defining characteristics, is equivalent to it or a subtype of it. This can be important in search.

Large parts of SNOMED CT are not yet sufficiently defined. *Primitive* concepts are not fully defined and do not have the unique relationships needed to distinguish them from their parent or sibling concepts. For example, pneumonia is a lung disease but unless defining characteristics are specified that effectively distinguish pneumonia from other lung diseases then it is a primitive concept.

SNOMED CT concepts are formally defined by their relationships with other concepts. These defining relationships may be either subtype relationships or attribute relationships. For example the concept |Appendicectomy| is a subtype of |procedure| and has defining attributes |method|=|excision| and |procedure site|=|appendix|.

Qualifying Relationships are used in post-coordinated expressions in health data. They are optional non-defining relationships that may be applied by a user or implementer. The range of possible values that can be used in qualifying relationships is constrained by the SNOMED CT Concept Model (see Chap. 10).

Additional Relationships allow non-definitional information to be distributed that may change over time or is specific to a particular national or organizational context (eg that a medicine is prescription only).

SNOMED CT Hierarchies

SNOMED CT is organised into hierarchies in which concepts are explicitly related by subtype relationships to parent concepts immediately above them in a hierarchy. A particular concept may have parents (immediate supertypes) and ancestors, as well as children (immediate subtypes) and descendants.

General concepts are at the top of the hierarchy. At each level down the hierarchy, concepts become increasingly specialized. Unlike a tree-structure, any SNOMED CT concepts can have more than one parent

SNOMED CT has about 19 top-level hierarchies (the number changes from time to time as the system evolves), which descend from a single *Root* concept | SNOMED CT concept| 138875005.

Some hierarchies have well-defined sub-hierarchies. For example, the Clinical Finding top-level hierarchy has a sub-hierarchy for Disease (or disorder); the Organism hierarchy has separate sub-hierarchies for Animals, Plants and Microorganisms.

The SNOMED CT hierarchies fall in three main groups: *object hierarchies*, which mainly comprise concepts that are likely to be qualified; *value hierarchies*, which are mainly concepts that act as values in object-attribute-value triples; and a *miscellaneous* group.

A *definitionStatusId* shows whether the concept is *sufficiently defined* (using relationships) or is a *primitive* concept.

SNOMED CT Identifier

The SNOMED CT identifier (*sctId*) is used to identify all types of component, including concepts, descriptions and relationships (see Fig. 9.3).

The is sctId is an integer between 6 and 18 digits long. One way of thinking of the sctId is as a 64-bit integer, although it has an internal structure. The internal structure of the sctId includes a meaningless *item identifier sctid* (between 3 and 8 digits), a 7-digit *namespace identifier* (which is only used in *extensions*), a 2-digit *partition identifier* and a single *check-digit*. The 8-digit item identifier allows almost 100 million items within any namespace.

Extensions are additions to SNOMED CT, usually specific to a single country or organization, and each is identified using a meaningless 7-digit *namespace identifier*, giving a theoretical potential of up to 10 million namespaces. An sctId including a namespace identifier is also known as a *long format sctId*. Namespaces are themselves defined as SNOMED CT concepts within a namespace hierarchy

If no namespace is identified in an sctId, this is a *short format sctId*, and it is assumed that the component is part of the International Release of SNOMED CT.

The *partition identifier* indicates the type of component referred to by that sctId. The *partition identifier* is a 2-digit number. If the first digit of the partition identifier is a zero (0), this component is part of the International Release; if it is a 1, then the component is part of an extension set. The second of the two digits in the partition identifier indicates which of the partitions of SNOMED CT the sctId is identifying, where:

- Concept (0)
- Description (1)
- Relationship (2)

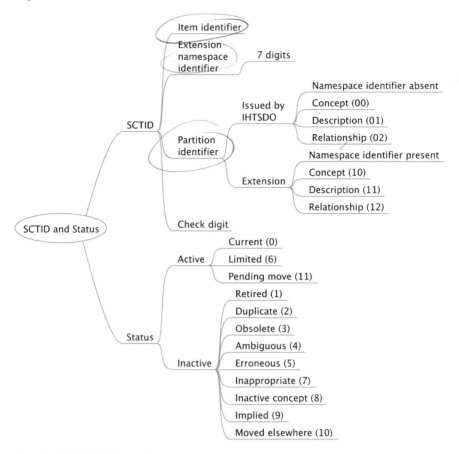

Fig. 9.3 SNOMED CT identifier

The *check digit* at the right hand end of the sctId uses *Verhoeff's Dihedral Group D5 Check* algorithm, which detects all single-digit errors and adjacent transpositions and about 95 % of twin errors and jump transpositions [2].

Expressions

A SNOMED CT expression is a collection of references to one or more concepts used to express an instance of a clinical idea (Fig. 9.4). It expresses an instance of a real world phenomenon (such as a headache) in a particular patient.

An expression is said to be *pre-coordinated* when a single concept identifier is used to represent a clinical idea. Including commonly used concepts in a pre-coordinated form makes the terminology easier to use. All of the 300,000 concepts in the SNOMED CT release are pre-coordinated, which allows a wide range of clinical information to be expressed in pre-coordinated form. It is generally easier

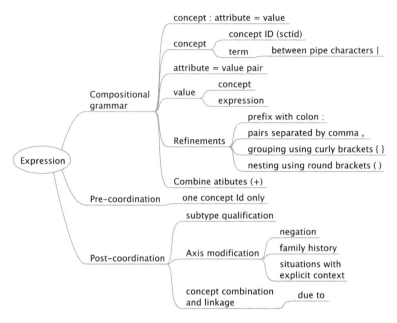

Fig. 9.4 SNOMED CT expressions

to handle pre-coordinated expressions than post-coordinated but this is at the cost of reduced flexibility.

SNOMED CT allows the use of *post-coordinated* expressions to represent a meaning using a combination of two or more concept identifiers. Post-coordinated expressions may be single level expressions or nested to any number of levels of detail. In a nested expression each attribute value is itself an expression, which can be nested. Nested post-coordinated expressions provide a powerful but complex means to allow SNOMED CT to describe things in great detail and cover unexpected requirements.

Compositional Grammar

SNOMED CT expressions are presented using compositional grammar [3]. This same compositional grammar is used to define SNOMED CT Concepts

At the simplest level, a single SNOMED CT concept identifier is a valid expression. Concept identifiers (conceptId) are shown as a sequence of digits. Other sctIds are not usually shown in compositional grammar. For example:

```
80146002
```

A concept identifier may be optionally followed by a term enclosed by a pair of pipe " | " characters. The term must be the term from a SNOMED CT description that is associated with the concept identified by the preceding concept identifier. For example, the term could be the preferred description, or the preferred description associated with a particular translation. The term may include any valid UTF-8 characters except for the pipe "|" character. Whitespace before or after the concep-tId is ignored as is any whitespace between the initial "|" character and the first non-whitespace character in the term or between the last non-whitespace character and before the second "|" character. For example:

```
80146002 | appendectomy |
```
set id   Term (handwritten annotation)

A concept identifier (with or without a following term) can be followed by a *refinement*. A colon (:) is used as a refinement prefix, between the concept to be qualified and the qualifying expression. A refinement consists of a sequence of one or more attribute-value pairs. The attribute and the value are both represented by concept identifiers (with or without a following term). The attribute is separated from the value by an equals sign *Refinement prefix* (handwritten annotation)

```
80146002|appendectomy| : 260870009|priority| = 25876001|emer
gency|
```
CONCEPT *QUALIFYING EXPRESSION* (handwritten annotations)

If there is more than one attribute-value pair, the pairs are separated by commas, representing a logical AND *ATTRIBUTE - VALUE PAIR* (handwritten annotation)

```
80146002|appendectomy|:260870009|priority|=25876001|emer
gency| , 425391005|using access device|=86174004|laparoscope|
```

Curly braces represent grouping of attributes within a refinement, for example to indicate that the method applies to a specific site *ATTRIBUTE   VALUE* (handwritten annotations)

```
80146002|appendectomy|: { 260686004|method|=129304002|ex
cision-action|, 405813007|procedure site-direct|=181255000|entire
appendix| }
```

The ungrouped attributes, if any, are all listed first, followed by all the grouped attributes.

Round brackets represent nesting to allow the value of an attribute to be refined

```
161615003|history of surgery|:363589002|associated proce-
dure|= ( 80146002|appendectomy|: 260870009|priority|=25876001|emer
gency )
```

Subtype Qualification

Subtype qualification is where the concept is linked with an attribute concept in such a way that the post-coordinated expression is equivalent to a subtype of the unelaborated concept. For example, the concept |asthma| could be qualified with the attribute concept | severe | to produce an expression for | severe asthma |, which is a subtype of the concept | asthma |. Where expressions have been post-coordinated and saved in this way the application can compute equivalence and hence subsumption when retrieving the stored expression.

There are four types of subtype qualification:

- Qualification

```
|fracture of femur|:|laterality|=|right|
```

- Refinement of a defining attribute
- Addition of unsanctioned qualifiers
- Addition of nested qualifiers.

Concepts can also be linked together to indicate causality or temporal relationships. For example, |Anaemia| can be linked by the attribute |Due to| to |Ascorbic acid deficiency|. The resulting post-coordination is equivalent to |Anaemia due to ascorbic acid deficiency|.

Axis Modification

Axis modification occurs where elaboration fundamentally changes the meaning of the concept, rather than simply refines it. Such an elaboration of the concept means that it is no longer subsumed as a subtype of a parent concept (subsumption is discussed in the next section). For example, if we elaborate the concept | asthma | to associate it with the mother of a patient, the meaning of the resulting expression has a different meaning from the concept of | asthma | by itself, and therefore different clinical implications. If a clinician runs a query of all instances of asthma in their practice, they would not expect get back instances of asthma linked to a family member.

Concepts that are negated by being post-coordinated with a negation concept (such as | known absent |) have their meanings fundamentally shifted. For instance, |asthma not present| is not a subtype of |asthma|. To say a person does not have |status asthmaticus | (an acute exacerbation of asthma that does not respond to standard treatments) is not to say that they don't have | asthma |.

Negated concepts subsume in the opposite direction to their positive counterparts. Whereas positive expressions are subsumed by more general instances, negated expressions are subsumed by more specific negative expressions. Concepts with axis modification, such as negation and family history must be treated differently from concepts that have been refined through subtype qualification. A special hierarchy of | Situations with Explicit Context| is used to express this type of concept that overrides the standard contextual defaults of SNOMED CT.

Subsumption Testing

Subsumption testing is used in information retrieval because most research, audit and decision support applications usually assume that a supertype includes all its subtypes (children and descendants). For example, a project may need to identify all patients with diabetes, which implicitly assumes that all types of diabetes should be included.

In any subsumption test there are two SNOMED CT expressions (or codes), one of which is tested for subsumption by the other. The *candidate expression* is tested to see if it is subsumed by (is a descendent of) another expression. The *predicate expression* is tested to see if it subsumes (is an ancestor of) another expression.

A **transitive closure table** is a list of all the ancestors of each concept. Transitive closure tables provide a fast direct way of checking whether one concept is a subtype (child) of any other and provides a means of high-performance subsumption testing. High-speed subsumption testing is essential for clinical decision support and is very useful in all types of analysis and reporting.

Reference Sets

Reference sets (refsets) are important in the practical application of SNOMED CT. Reference sets provide a single mechanism for referencing and adding information to SNOMED CT components (Fig. 9.5). All reference sets have common metadata, but the fields (columns) used vary according to use and purpose. When using a system, a user is only interested in a tiny proportion of the whole of SNOMED CT and reference sets provide a way of enabling this.

The number of members in a reference set may vary enormously. A language reference set may have hundreds of thousands of members. A realm reference set containing the concepts commonly used in a clinical specialty may contain several thousand members, but the set of concepts or descriptions in context reference sets for a specific clinical protocol, template or data entry field may only contain a few members.

One way to think of a reference set is as an index entry pointing to a set of pages relevant to a topic. Each reference set member is uniquely identified by a UUID, can be inactivated using the *active flag* and is versioned using an *effectiveTime*. Reference sets are extensible with additional columns and are not limited by fixed specifications.

Language Reference Sets allow descriptions to be set for a language, dialect or context of use. SNOMED CT can be translated into any language or dialect. Each translation uses existing concepts with new language-specific descriptions. A language Reference set is a set of references to the descriptions that make up that language or dialect edition. For example, British English (en-GB) and US English (en-US) are different dialects of English in which many medical terms have different spellings; English, French and Spanish are different languages.

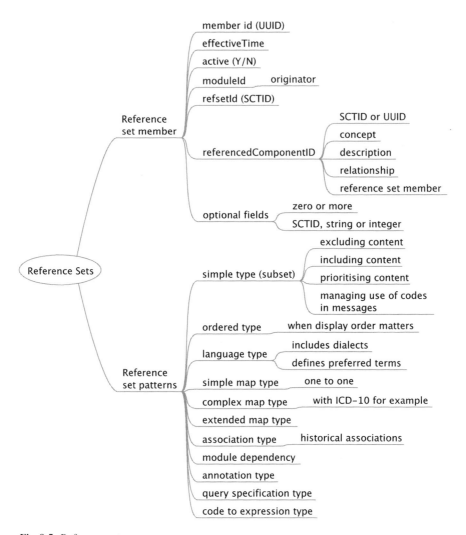

Fig. 9.5 Reference sets

Realm Reference Sets cover the terminology used in a specific area of expertise or locality. Examples of realms include: a specialty, a professional discipline, an organization, a country, or a specialty within a country (eg US dentists).

Context Reference Sets identify components that are included in or excluded from the set of values that can be used in a particular context. Simple reference sets are not necessarily mutually exclusive and the content of different reference sets may overlap. For example this type of reference set may be used to limit the content of a field to those permitted by an interoperability message standard.

Navigation Reference Sets provide an alternative tree-view of a set of terms in a specified order. Navigation hierarchies can reflect the way that people think when entering data, sometimes referred to as the model of use. Navigation hierarchies are useful for display, navigation and data entry. These are usually handcrafted, to limit the number of levels, the number of choices at each level, to list terms in a sensible order and ensure consistency over time. However, large numbers of handcrafted hierarchies are difficult to maintain. Each reference set member includes a reference to the parent concept, a child concept and the sequence order of that child.

Map Reference Sets are used to reference other terminologies and classifications, such as the International Classification of Diseases (ICD). Mapping is needed to allow data collected for one purpose such as clinical care to be used for another purpose such as reimbursement, avoiding the costs and errors of having to re-enter data. It is also needed when data needs to be migrated to newer systems. Ideally, computer programs will use the mapping tables to translate codes automatically, but unfortunately the rules of many coding systems, such as ICD-9 CM and ICD-10 are such that fully automated coding is not yet feasible.

A single SNOMED CT concept may need to be mapped to one or more target codes. The map from SNOMED CT to ICD-10 CM involves two main reference sets: a Descriptor reference set and a Complex Map reference set. Other reference sets may also be involved but are less important.

The Descriptor reference set contains metadata that describes the attributes of all publication reference sets and their information content. The ICD-10 map is one item in this file. The metadata items are themselves SNOMED CT concepts in the foundation metadata concept hierarchy.

The complex map reference set contains one or more map records for each source concept mapped, including the ICD-10 target codes. Each SNOMED CT concept may have none, one or more mappings to ICD-10 CM.

History Reference Sets SNOMED CT includes component history files, which maintain a record of changes to existing components, in line with the principle of permanence. The reasons for why it is inactive may be declared in a reference set. These reasons may include being: retired, duplicate, obsolete, ambiguous, erroneous, inappropriate, inactive concept, implied, or moved elsewhere.

Reference Set Development

The process of reference set development and maintenance is challenging and time-consuming. It is no easier to develop and maintain a reference set that has the support of a large clinical community than it is to develop any other consensus standard. Clinicians want reference sets that meet their particular needs, complete and yet focused. Reference set development is likely to remain a growth area for many years to come.

A number of tools have been developed to help with the tasks of building, maintaining and using reference sets.

Creating a new reference set requires access to a namespace in order to generate the sctId needed. Within that namespace, at least one module ID concept (with an FSN and preferred term) is required under the *module* sub-hierarchy (within the Core Metadata) for each authoring organization. The steps required to create a new reference set include:

- Create the reference set concept in the foundation metadata hierarchy.
- Create the descriptor for the reference set (by adding members to the reference set descriptor reference set).
- Add members to the reference set.

For each reference set, a formal document should record the rules, principles and approach used to determine the members of that reference set.

A typical reference set development project is likely to involve the following steps:

- Establish scope and team and identify which existing pattern (if any) can be used.
- Identify relevant terms from existing records and evidence base (literature)
- Compare to SNOMED CT content.
- Derive reference sets, including hiding some of the complexity of SNOMED CT, allocating priorities and the sequence order of terms.
- Validate using panels and in practice for comprehensiveness, relevance, reliability and usability.
- Implement and deploy software that enables users to achieve their goals.
- Maintenance.

Reference sets need to be maintained and the content re-examined when new releases of SNOMED CT are made available. Processes need to be established to address any concepts that have become inactive and new concepts added in each new release.

Releases

The SNOMED CT International Edition is released twice a year by IHTSDO. This may be supplemented by national Extensions. All releases use the same file formats, known as RF2 (Release Format 2), which was introduced in 2011. Three release types are supported.

A *snapshot* release contains the most recent version of every component. This is useful for installing SNOMED CT.

A *delta* release contains only those components that have been changed in any way since the previous release. This is useful for updating a system.

The *full* release contains every version of every component that has ever been released. It is voluminous and provides a complete historical record and can be used to obtain a view of the state of any component at a particular time.

The core terminology data files are a Concepts file, Descriptions file and Relationships file. Reference set files have common metadata but a number of different structures.

The pattern for release file names consist of five elements, each separated by an underscore "_" and followed by a full stop "." and a file extension:

```
<FileType>_<ContentType>_<ContentSubType>_<Country|Namespace>_<Ve
rsionDate>.<Extension>
```

The FileType specifies the type of file and the release format. For example the main terminology files have File type *sct2*, where the *2* tag refers to Release Format 2.

Documentation

The SNOMED CT Starter Guide, the Technical Implementation Guide and the Editorial Guide are three key reference documents, available from IHTSDO, which describe SNOMED CT in detail. These are aimed at different audiences and contain a good deal of overlap.

SNOMED CT Starter Guide [4] (56 pages) provides a good overview of a range of topics associated with SNOMED CT, organised into 15 Chapters.

SNOMED CT Technical Implementation Guide (TIG) [5] (757 pages) is intended for SNOMED CT implementers, such as software designers who need an authoritative point of technical reference and advice to support their involvement in designing, developing, acquiring or deploying software applications. It includes sections on: implementation, structure and content, release file specifications, the Concept Model, terminology services, record services, change management and extension services.

SNOMED CT Editorial Guide [6] (171 pages) describes editorial policies regarding the purpose, scope, boundaries, requirements, concept model, hierarchies, terming, and other policies related to the content in SNOMED CT. It is primarily intended to guide those who are responsible for editing the content of the International Release, but secondarily is important for those creating extensions.

References

1. Spackman K. SNOMED User Group Meeting. Chicago. 2005.
2. Wagner NR, Putter PS. Error detecting decimal digits. Commun ACM. 1989;32(1):106–10.
3. SNOMED CT Compositional Grammar Specification and Guide. The International Health Terminology Standards Development Organisation, Copenhagen. Version 2. 03 July 2015.
4. SNOMED CT Starter Guide. IHTSDO. December 2014.
5. SNOMED Clinical Terms Technical Implementation Guide. IHTSDO. January 2015.
6. SNOMED CT Editorial Guide. IHTSDO. January 2016.

Chapter 10
SNOMED CT Concept Model

Abstract The SNOMED CT Concept Model is the set of rules that governs how concepts are permitted to be modeled using relationships to other concepts. SNOMED CT concepts are organised in hierarchies and this chapter describes these hierarchies and what attributes and values can apply to each type of concept.

Keywords SNOMED CT hierarchies • Attributes • Clinical finding • Procedure • Situation with explicit context • Observable entity • Event • Staging and scales • Specimen • Body structure • Organism • Substance • Pharmaceutical/biologic product • Physical object • Physical force • Social context • Environments and geographic locations • Qualifier value • Special concept • Record artifact • SNOMED CT model components

The SNOMED CT Concept Model is the set of rules that governs how concepts are permitted to be modeled using relationships to other concepts. The Concept Model is a key part of SNOMED CT and provides a way of cross checking what you are doing.

These rules assert which attributes and values can be applied to each type of concept. For example, the concept model asserts that any subtype of the |Clinical finding| hierarchy can be related using the attribute |FINDING SITE| to a concept that is a subtype of either |Anatomical structure| or |Acquired body structure|. Mention of |Acquired body structure| provides a reminder that findings may be associated with prostheses.

For clarity the names of concept preferred terms are shown in lower case enclosed by pipe characters, e.g., |term| and the names of attributes are shown in uppercase, e.g., |FINDING SITE|.

SNOMED CT Hierarchies

SNOMED CT concepts and their hierarchies fall into three main groups, which are used in expressions.

Object hierarchies include all concepts that apply directly to patients and may be further qualified:

© Springer-Verlag London 2016

T. Benson, G. Grieve, *Principles of Health Interoperability*,
Health Information Technology Standards, DOI 10.1007/978-3-319-30370-3_10

|Clinical finding|
|Procedure|
|Situation with explicit context|
|Observable entity|
|Event|
|Staging and scales|
|Specimen|

Value hierarchies include concepts which are used to represent the values in relationships, where a relationship can be thought of as an object-attribute-value triple

|Body structure|
|Organism|
|Substance|
|Pharmaceutical/biologic product|
|Physical object|
|Physical force|
|Social context|
|Environment or geographical location|

Miscellaneous hierarchies

|Qualifier value|
|Record artifact|
|Special concept|
|SNOMED CT Model Component|

All hierarchies are based on subtype relationships specified explicitly by ISA attributes. Concepts can have more than one parent, giving a poly-hierarchical structure.

The concept model is a key part of SNOMED CT and compliance with its rules requires users to understand these hierarchies and how to use about 50 attributes, which are used to define concepts and in post-coordinated expressions. The examples used in this chapter are all defining relationships (Fig. 10.1).

Attributes

Each attribute has a domain and a range. The domain is the hierarchy to which this attribute may be applied, eg |Clinical finding|. Each attribute can take values only from a particular value hierarchy. For example, values for the |FINDING SITE| attribute may only come from the |Acquired body structure| or |Anatomical structure| hierarchies. The set of allowable values are referred to as the attribute's range, eg

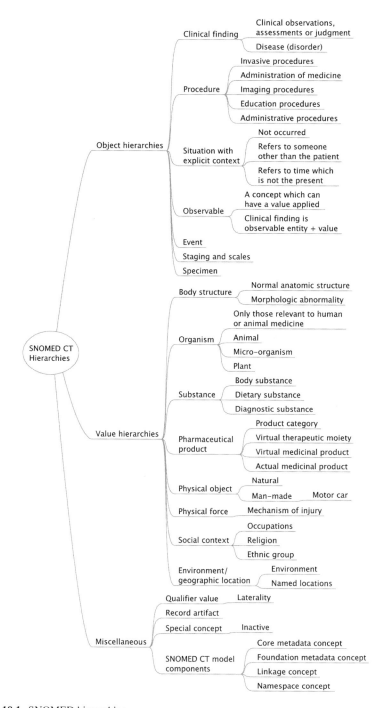

Fig. 10.1 SNOMED hierarchies

```
|Pneumonia|:|FINDING SITE|=|Lung structure|
```

In this example, the attribute |FINDING SITE| is allowable for a concept in the Clinical Finding hierarchy such as |Pneumonia|. The value |Lung structure| is a valid value for this attribute since it is in the |Anatomical structure| hierarchy.

The |Qualifier value| hierarchy may be used when other more explicit domains are not appropriate.

The range is the set of values that can be applied to each attribute and is defined in the Concept Model using the notation shown below:

(<<) this code and descendants,
(<) descendants only,
(<=) descendants only (stated) except for supercategory groupers,
(==) this code only,
(< Q) descendants only when in a qualifying Relationship,
(< Q only) descendants only, and only allowed in a qualifying Relationship.

A supercategory grouper is sufficiently defined by reference to a value that is at the top of the value hierarchy, resulting in a very general meaning, such that the code is not useful for record entry, but is useful as an organizer of the hierarchy.

Attributes are used to indicate a relationship between two concepts and are used to logically define a concept, and as qualifying attributes in post-coordinated expressions. Within SNOMED CT, attributes are classified into:

Concept model attributes: About 50 defining attributes, which are used to model concept definitions and are in the Concept Model, plus the |IS A| attribute.

Unapproved attributes: more than 1000 other attributes that may be used to model concept definitions or in post-coordinated expressions, but which have not yet been used in modeling pre-coordinated concepts as part of the SNOMED Concept Model. The term *unapproved attribute* is perhaps a bit misleading, because these are full SNOMED concepts.

Concept history attributes: about 7 attributes used for tracking history such as

|REPLACED BY|
|MAY BE A|
|SAME AS|
|MOVED FROM|
|MOVED TO|

Object Hierarchies

Clinical Finding

Clinical findings in SNOMED represent the result of clinical observations, assessments or judgments, and include both normal and abnormal clinical states; this covers a very broad range of concepts, with a range similar to that of HL7 Observation (Fig. 10.2).

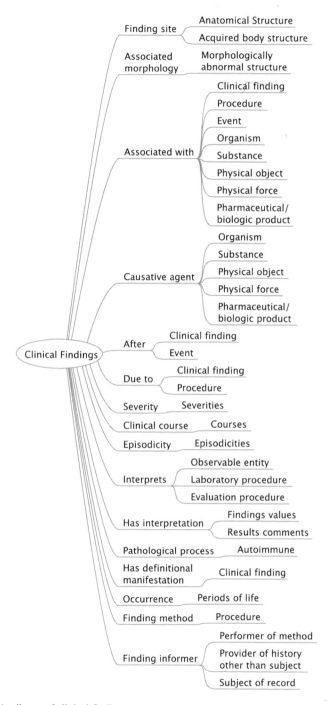

Fig. 10.2 Attributes of clinical findings

The default context for a clinical finding is that the finding has actually occurred, it relates to the subject of record (the patient), currently or at a stated past time.

Concepts within the |Disorder| sub-hierarchy of |Clinical Findings| are always abnormal clinical states.

Clinical findings allow the following attributes:

|FINDING SITE| specifies the body site affected by a condition and has values from the body structure hierarchy, eg

        |appendicitis|:|FINDING SITE|=|appendix|

|ASSOCIATED MORPHOLOGY| specifies morphologic changes seen at the tissue or cellular level that are characteristic features of a disease; it has values from the morphologic abnormality hierarchy, eg

        |appendicitis|:|ASSOCIATED MORPHOLOGY|=|inflammation|

|ASSOCIATED WITH| asserts that a clinical finding is associated with another clinical finding, procedure, pharmaceutical product, substance, organism, physical object, physical force or event without asserting or excluding a causal or sequential relationship between them. It has three subtypes: |AFTER|, |DUE TO| and |CAUSATIVE AGENT|.

|AFTER| is used when a clinical finding occurs after another clinical finding or procedure, showing the sequence of events.

|DUE TO| relates a clinical finding to its cause, which may be another clinical finding or an event.

|CAUSATIVE AGENT| identifies the cause of a disease such as an organism, substance, pharmaceutical product, physical object or force. It does not include vectors such as mosquitoes, which transmit malaria.

|SEVERITY| is used to represent the severity level of a clinical finding, such as mild, moderate or severe. This should be used with caution because these concepts are relative to other values in the value set presented to the user.

|CLINICAL COURSE| represents the course and/or onset of a disease, such as acute or chronic. Note that the term acute may mean any combination of rapid onset, short duration or high severity.

|EPISODICITY| represents episodes of care provided by a physician or other care provider, such as a general practitioner. This attribute is not used to represent episodes of disease experienced by the patient.

|INTERPRETS| may refer to observable entity, laboratory procedure or evaluation procedure, eg

        |abnormal glucose level|:|INTERPRETS|=|glucose measurement|

|HAS INTERPRETATION| designates the judgment aspect being made, eg

```
        |abnormal glucose level|:|HAS INTERPRETATION|=|outside ref-
erence range|
```

|PATHOLOGICAL PROCESS| provides information about the underlying pathological process, such as infectious, parasitic or autoimmune, eg

```
        |autoimmune    parathyroiditis|:|PATHOLOGICAL    PROCESS|=|
autoimmune|
```

|HAS DEFINITIONAL MANIFESTATION| links disorders to the observations (manifestations) that define them, eg

```
        |hypertension|:|HAS DEFINITIONAL MANIFESTATION|=|finding of
increased blood pressure|
```

|OCCURRENCE| refers to the specific period of life during which a condition first presents, such as childhood.
|FINDING METHOD| specifies the procedure by which a clinical finding was determined, eg

```
        |finding by palpation|:|FINDING METHOD|=|palpation|
```

|FINDING INFORMER| specifies the person or other entity from which the finding was obtained, eg

```
        |patient-reported outcome|:|FINDING INFORMER|=|subject of
record|
```

Procedure

In SNOMED a procedure is broadly defined as any type of action done intentionally as part of the process of delivering healthcare, including history taking, physical examination, testing, imaging, surgical procedures, disease-specific training and education, counseling and administrative procedures (Fig. 10.3).

The definition of a procedure in SNOMED is much broader than the definition in the HL7 RIM, where the term is limited mainly to surgical procedures.

Allowed attributes of procedure include:

|METHOD| represents the action being performed to accomplish the procedure, including surgical, drug administration, education, manipulation and therapy. Method can be regarded as the anchor of each relationship group that defines a procedure. Attributes are grouped within the method to which they apply. Typical values of |METHOD| are incision, excision, removal, injection.

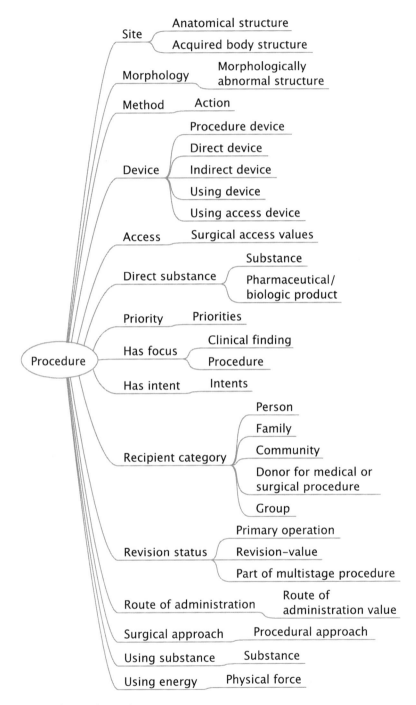

Fig. 10.3 Attributes of procedures

|PROCEDURE SITE| describes the body site acted on or affected by a procedure. A procedure may have a direct site or an indirect site.

|PROCEDURE MORPHOLOGY| specifies the morphology or abnormal structure involved in a procedure. It also uses the idea of direct and indirect morphology.

```
|excision  of  benign  neoplasm|:|DIRECT  MORPHOLOGY|=|benign
neoplasm|
    |removal of stitches from wound|:|INDIRECT MORPHOLOGY|=|wound|
```

|PROCEDURE DEVICE| describes the devices associated with a procedure.

|ACCESS| describes the route used to access the site of a procedure, such as open, closed and percutaneous.

|DIRECT SUBSTANCE| substance or pharmaceutical product on which the procedure's method acts directly.

|PRIORITY| refers to the urgency assigned to the procedure (eg emergency).

|HAS FOCUS| clinical finding or procedure.

|HAS INTENT| specifies the intent of the procedure (eg diagnosis).

|RECIPIENT CATEGORY| is used to specify the recipient, where the recipient is not the subject of the procedure, such as when the subject is a donor.

|REVISION STATUS| primary, revision or part of a multi-stage procedure.

|ROUTE OF ADMINISTRATION| specifies the route by which a procedure introduces a given substance into the body.

|SURGICAL APPROACH| specifies the directional, procedural or spatial access to the site of a surgical procedure.

|USING SUBSTANCE| specifies a substance used to execute the action of a procedure, such as contrast media in radiography.

|USING ENERGY| describes the energy used to execute an action (eg gamma radiation).

Situations with Explicit Context

In SNOMED, findings and conditions can appear either as subtypes of |Clinical finding| or can be subtypes of |Situation with explicit context| (Fig. 10.4). Concepts under |Situation with explicit context| are used use when it is important to make clear:

- who is the subject of the record (the patient or someone else such as a family member)
- when the event took place (past, present or future)
- whether a finding was present, absent, unknown or is a potential risk
- whether a procedure was done, not done or planned.

|Situations with explicit context| express information about situations that override the standard context defaults of SNOMED and in many cases cause axis modification. This is one of the most confusing issues in SNOMED, but it needs to handled properly or else patients may be classified as having conditions that they do not have.

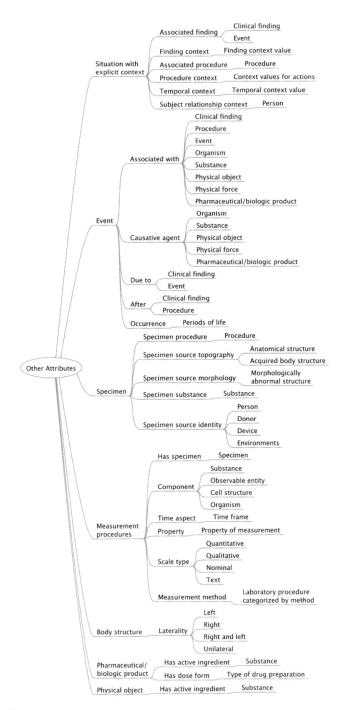

Fig. 10.4 Other attributes

The standard defaults in SNOMED are that the procedure has actually occurred or the finding is present, the data refers to the patient (not, for example, a family member) and the procedure or finding is occurring now or at a specified time. For example, a statement noting maternal history of breast cancer refers to the mother of the patient, not the patient and is in the past not the present.

The top-level situations include |Finding with explicit context|, |Procedure with explicit context| and |Family history with explicit context|. The attributes associated with |Situations with explicit context| include:

|ASSOCIATED FINDING|
|FINDING CONTEXT| is used to represent a situation in which a clinical finding
    may not be known, may be absent or may not yet be present, eg
|ASSOCIATED PROCEDURE|
|PROCEDURE CONTEXT|
|TEMPORAL CONTEXT|
|SUBJECT RELATIONSHIP CONTEXT|

Observable Entity

Observable entities (usually referred to as **observables**) are variables or properties which can have values applied to them. Most observables can be represented by a question, such as "What is the patient's height?"

When given a value, observables provide a specific finding or assertion about health related information. Examples include the names of lab tests, physical exam tests and dates of significant events. A clinical finding can be modeled as an observable entity plus a value. One use of observable entity is as headings on a template.

For example the concept |hair color| is an observable entity, while the concept |gray hair| is a finding.

Event

Events are occurrences that happen, which are not healthcare procedures or interventions, such as travel, earthquake, and death.

Staging and Scales

This hierarchy contains the names and components of assessment scales and tumor staging systems.

Specimen

Specimens are entities that are obtained, usually from a patient, for examination or analysis. They can be defined by attributes such as:

- Body structure, normal or abnormal, from which they are obtained
- Procedure used to collect the specimen
- Source from which it was collected
- Substance of which it is comprised.

Value Hierarchies

Body Structure

Body structure includes normal as well as abnormal anatomical structures. Normal anatomical structures can be used to specify the body site involved in a disease or procedure. Abnormalities of body structure are represented in a sub-hierarchy |Body structure, altered from its original anatomical structure (morphologic abnormality)|. Body structure has one attribute:

|LATERALITY| (eg |left|, |right|, |left and right|).

Organism

The organism hierarchy includes organisms of significance in human and animal medicine, including causes of diseases and required for public health reporting of the causes of infectious diseases. Sub-hierarchies of organism include: |animal|, |microorganism| and |plant|.

Substance

The substance hierarchy contains concepts for recording the active chemical constituent of drugs, foods, allergens, poisons and many other substances. Sub-hierarchies include |body substance|, |dietary substance| and |diagnostic substance|.

Pharmaceutical/Biologic Product

The |pharmaceutical/biologic product| hierarchy is separate from the |substance| hierarchy in order to clearly distinguish between manufactured drug products and their active chemical constituents.

Medicinal products have four distinct types:

- **Product category** describes common drug categories used in prescribing, such as |beta-blocking agent|
- **Virtual Therapeutic Moiety** (VTM) is product name only (eg aspirin). The VMP is linked to the active ingredient substance(s). Some subtypes of VTM include form as well as name.
- **Virtual Medicinal Product** (VMP) includes the product name, strength and form as used on a drug prescription (eg aspirin 75 mg tablet). This level is used to support CPOE and e-prescribing.
- **Actual Medicinal Product** (AMP) is a single unit dose of a marketed medicinal product and includes product name (trademarked brand name), strength, dosage form, flavor (when applicable), and manufacturer. AMP does not include packaging information. AMPs are often country specific and are therefore found in national extensions, such as the NHS dictionary of medicines and devices (dm + d).

Physical Object

Physical objects may be natural and man-made (eg a motor car) and include medical devices.

Physical Force

Physical force is used to represent forces, which play a role in causing injuries.

Social Context

The Social context includes social conditions and circumstances relevant to healthcare such as ethnic group, occupation, religion, education, housing, care provision, family relationships and life style.

Environments and Geographic Locations

The Environments and geographic locations hierarchy includes types of environments as well as named places such as countries, states and regions.

Miscellaneous Hierarchies

Qualifier Value

Includes concepts used as qualifying values in attributes, which are not defined elsewhere; for example, |left| and |right| used with the |laterality| attribute.

Special Concept

One sub-hierarchy of Special concept is Inactive concept, which is the supertype for all concepts that have been retired. Another is for special navigational concepts.

Record Artifact

Used to refer to parts of electronic patient records, and different types of document.

SNOMED CT Model Components

The SNOMED CT Model Component hierarchy contains metadata concepts in four sub-hierarchies.

Core Metadata Concept

These codes are used to describe the structural information for the core release data such as concepts, descriptions and relationships.

Foundation Metadata Concept

These concepts provide supporting metadata and structural information for derivative release structures including refsets.

Linkage Concept

All concepts that can be used as a relationship type are subtypes of |linkage concept|, which has two sub-hierarchies:

|Attributes|
|Link assertion|

Attributes have been discussed in some detail above.

Link assertions allow SNOMED concepts to be used to assert relationships between statements, rather than SNOMED concepts; for example in HL7 messages. A link assertion is primarily used to link information items together in patient records, messages and other documents. For example, link assertions have been used in HL7 V3 messages to link clinical statements, such as linking a clinical finding or procedure with one or more problems in a problem list. Subtypes of link assertions include:

|HAS EXPLANATION|
|HAS REASON|
|HAS SUPPORT|
|IS ETIOLOGY FOR|
|IS MANIFESTATION OF|
|HAS PROBLEM MEMBER|
|HAS PROBLEM NAME|

Namespace Concept

This is where globally unique namespaces are defined.

Chapter 11
Implementing Terminologies

Abstract Implementation of terminologies involves making choices about when to use codes and when to structure information in separate fields. Terminology binding is the process of specifying in archetypes and templates what codes belong in which fields. The model of use is how information appears to the end user. This often differs from the model of meaning, which is always true, but sometimes harder to use. It is important to distinguish between coded text, which refers to a code, and codeable text where a code reference is optional. Value sets represent the permissible codes, while a coding scheme may represent all codes in a scheme. A number of different scenarios are explored for when codes may be known or not known. The task of the receiver is critical, to ensure information is not lost.

Keywords Reference terminology • Terminology binding • Information model • Model of use • Model of meaning • Structured data entry • Views into EHR • Common user interface • Structural models • Reference model • ISO 13606 • Archetype • Template • Coded data types • Coded text • Codeable text • Null flavor • Code system • Version • Original text • Display name • Translation • Value set • Complex coded expressions • Common scenarios • Coded data types in v2 • V3 and FHIR • Receivers • Expression storage

In this chapter we discuss some of the implementation issues of using SNOMED CT in the context of interoperability. This is an area where development is taking place quite rapidly and it is difficult to make definitive statements about how each should be used. Instead we focus on the core principles. Most of this chapter is written with SNOMED CT in mind, but the principles also apply to other terminologies.

SNOMED CT is a reference terminology, and works in a way that differs substantially from traditional position-dependent coding schemes and enumerations. The value of using it depends on how well it is implemented and used. There is no one best or right way to implement SNOMED CT, although there are wrong ways [1].

Architects and designers will introduce alternative approaches suited to their own circumstances. The golden rule is to select the simplest approach that meets the requirement. Einstein is attributed as saying: *Everything should be made as simple as possible, but no simpler*. In any specification we need to be as stringent as possible, to reduce disorder and degrees of freedom, while still meeting the legitimate user requirement.

© Springer-Verlag London 2016

T. Benson, G. Grieve, *Principles of Health Interoperability*,
Health Information Technology Standards, DOI 10.1007/978-3-319-30370-3_11

SNOMED does not require that all applications in a network use it for the internal representation of their data. This is clearly impractical because few legacy applications were designed around it. The choice as to whether or not to use SNOMED CT as part of the internal record structure depends on the type of data being recorded and the external political environment.

The description logic used in SNOMED expressions has the virtue of being universally consistent, flexible and relatively easy to learn. In many cases it may be easier to use SNOMED description logic to qualify concepts than to create additional data elements or HL7 V3 ActRelationships to achieve the same end.

Terminology Binding

Terminology binding is the process of establishing links between elements of a terminology such as SNOMED and an information model [2]. This is particularly important in interoperability, primarily because interoperability involves two translations, from the source system to the wire format and from the wire format to the destination system, and the information models used in the source and destination systems are likely to be substantially different. The wire format is the common link, which is why it needs to be standardised.

Issues of information structure are closely related to those of terminology. It is simply not possible to slot any terminology into any data structure and expect it to work. After all, in every spoken language, the grammar (syntax) and words (terminology) have evolved together. The idea that syntax and semantics are independent dimensions is a gross over-simplification.

However, for reasons that seemed good at the time, standards development organizations responsible for healthcare terminologies (such as SNOMED CT), message syntaxes (such as the HL7 V2 and V3) and information models developed standards independently and in parallel. Their justification was the perceived need to interoperate with a wide range of legacy schemes and to meet the needs of different national requirements and languages. For these reasons, terminology coding schemes such as SNOMED CT and ICD-10 were designed to be syntax-neutral, so as to work with any syntax. Similarly, information reference models such as the HL7 RIM and EN 13606 were also designed to be terminology neutral. An additional complication is that the flexibility built into both HL7 and SNOMED CT means that there are usually several possible ways to perform the binding.

Bindings can be expressed in a variety of ways [3]. The simplest way is to use explicit specified value sets. A more complex approach is to specify rules that determine how a definition can be formed. For example a rule could be defined that specifies that any injury involving a long bone shall be qualified by its laterality (left or right).

In reality, system architects are well advised to select both the message syntax and the terminologies to be used at the outset of their work.

Furthermore, electronic information systems operate at two different levels, which Rector [4] describes as the *Model of Use* and the *Model of Meaning*.

Model of Use

The Model of Use describes how a system such as an EHR is used in practice and how data is captured and displayed. The Model of Use represents the human interface. The same data could be captured in any number of different ways and the skill of the designer is to make this as easy and efficient as possible for each work process. Different use cases have different Models of Use (Fig. 11.1).

Data capture (the Model of Use) has to be designed to make each activity as quick and easy as possible. Healthcare workflow is made up of a relatively small number of common high volume activities, such as requesting tests, prescribing medication and making referrals, together with many less common activities that are specific to the specialty of the clinician, the patient's problems and their progress along the care pathway. The specifics of tasks such as patient assessment, diagnosis, monitoring and planning treatment, tests and follow-up, differ considerably according to what is the matter with the patient.

The two main methods of recording coded data are ad hoc data entry and structured data entry.

Ad hoc Data Entry is one method of data coding, in which the user simply types in the first few letters of the term in mind and the system responds with a list of matching terms and synonyms. These terms are often organised hierarchically in a tree-structure, as exemplified by the Read Codes (see Chap. 8). The user can scan up and down the hierarchy, moving between levels of greater or lesser detail to find just the right term. The term selected is shown, providing the user with an opportunity to validate the choice. This method has been used successfully by almost all GPs in the UK for over 20 years, although there remain problems with the completeness and consistency of data captured in this way.

The practicality of ad hoc data entry depends on keeping the size of the list of matching terms to a size that can be scanned quickly and easily. Research has shown that the human eye can read up to six lines of text of about 15 characters without moving the eye muscles. Anything more takes longer to read and increases the errors.

The number of options displayed can be limited limit by providing specific value sets for each task to restrict the number of options available, or by displaying the most commonly used items first (known as velocity coding).

Structured Data Entry removes the requirement to type in the first few letters of the term; the user simply points and picks with a mouse or on a touch-screen. This works well when the task has a narrow scope and clearly defined path. However, the designer needs to do a lot of work to structure each step to fit in with each user's way of working. This is even more difficult when working with patients with several conditions. The computer protocol must follow the natural clinical order of the task

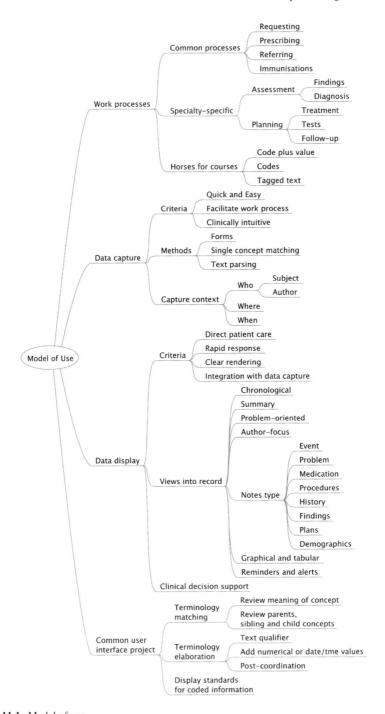

Fig. 11.1 Model of use

and items on picking lists need to be grouped in a natural order and use appropriate terms. A further constraint is that each item needs to be recorded in a form suitable for information retrieval.

Data collection needs to be integrated with the display and review of previous records and clinical decision support warnings and alerts. At any one moment the user's screen can only show a limited amount of information, although a patient's EHR may contain thousands of separate items of information.

One approach is to offer multiple views into the record with instant switching between views. Such **views** into an EHR include:

- Chronological. The most natural way to display information is in date/time accession order. Reverse chronological order shows the most recent first. Data may be grouped by the clinically relevant date, so tests may be linked both to the date of sample and to the report date.
- Author views, so that users can see quickly the last entries that they or another person made for this patient.
- Clinical documents such as discharge and referral letters.
- Care events such as admissions, discharges and clinic visits.
- Problem list including diagnoses and allergies.
- Medication, distinguishing between present, previous and proposed medication.
- Operations and major procedures.
- Clinical progress notes.
- History including presenting history, advance directives, mental capacity and social history.
- Findings, with sections for examination findings, laboratory results, vital signs, imaging findings. Flow charts and graphs can be provided for specific types of data, which need to be monitored and such as blood pressure, blood chemistry and assessments.
- Plans indicating who is to do what and when.
- Demographics.

Common User Interface

The NHS Common User Interface project, undertaken in collaboration with Microsoft set out to deliver of a consistent user experience across applications and devices to improve patient safety, reduce the cost of end user training and support, and increase the productivity and effectiveness of clinicians.[1] The work is presented in a number of documents covering.

- Terminology Matching [5]
- Terminology Elaboration [6]
- Display Standards for Coded Information [7]

---

[1] The results of the Common User Interface project are available at www.mscui.net.

Three main ways of entering SNOMED coded data are to use *options, single concept matching* and *text parser matching*.

Options are used where the user selects from a set of options rather than entering text.

In single concept matching, the user enters a note for a single clinical concept and selects an appropriate match returned by the SNOMED server and elaborates this as required.

In text parser matching, the user enters notes as unconstrained free text and the system matches words and phrases against the SNOMED database. This requires the system to identify and match SNOMED concepts as well as build post-coordinated expressions based on sanctioned attribute relationships from within the text. The first step in the process is to identify the context, so as to identify the appropriate form, navigation or other subset to constrain the options and simplify the task. Having entered text, the system may present a set of choices, which can be refined or elaborated in various ways:

- Adding unstructured text to the expression to give the expression further meaning.
- Browsing alternative matches and refining selected matches; this may include reviewing the parents, siblings and child concepts in a navigation hierarchy or qualifying attributes.
- Matching a SNOMED expression from within a passage of text and leaving some of the text itself un-encoded but associated with the encoded expression.
- Adding a qualifier to a SNOMED expression, using the qualifiers offered by the system, such as the severity of a condition. For example, the concept Asthma can be qualified with the attribute concept Severe to produce an expression that is the subtype of the concept Asthma.
- Adding or selecting numerical or date and time values for a SNOMED expression. For example, the user should be able to add the value 38.9 and the unit Degrees centigrade to the concept of Body temperature.

Model of Meaning

Each Model of Use needs to be convertible into a Model of Meaning to make it computable. The Model of Meaning is a representation for reporting and statistical analysis purposes, which represents our understanding of the world, so that we can reason about it in general, and individual patients in particular. Computers are limited in their ability to process data and require information in a common, standardized format. The Model of Meaning provides such a format for data processing and reasoning. There is a great advantage for reporting and clinical decision support in working with a common Model of Meaning. This is a key rationale for the development of reference models such as the HL7 RIM and ISO 13606 reference models (see Fig. 11.2).

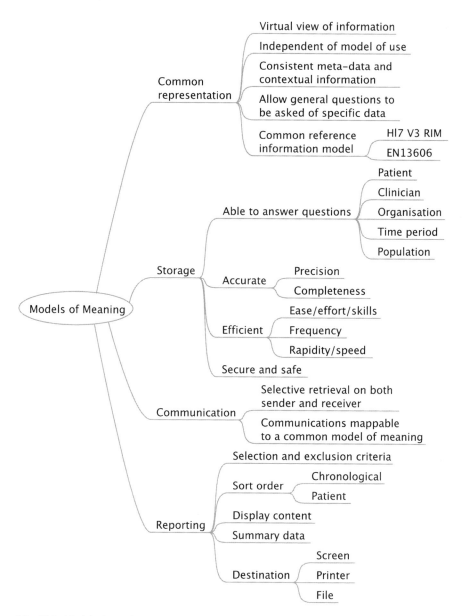

Fig. 11.2 Model of meaning

The HL7 TermInfo project provides a set of recommendations on how to bind SNOMED with HL7 Version 3, and expresses this requirement as follows:

Every application has its own data entry screens, workflow, internal database design, and other nuances, and yet despite this, we talk of semantic interoperability. In order to achieve interoperability, and enable a receiver to aggregate data coming from any of a number of applications, it must be possible to compare data generated on any of these applications. In order to compare data, it helps to imagine a canonical or normal form. If all data, regardless of how it was captured, can be converted into a common form, it becomes possible to compare.

This issue applies not only when we wish to exchange data between applications, (semantic interoperability), but also within EHR systems where a single system is used to support different models of use (semantic operability).

Some suppliers have chosen to use a standard reference model as the basis of their applications architecture. The use of the HL7 RIM for this purpose has been termed **RIMBAA** (RIM-based Application Architecture). The ISO 13606 and OpenEHR reference models are also used in this way. It is also practicable to use a proprietary architecture, which has a direct mapping to standardized reference models.

For storage and analysis, we need information to be in a form that allows it to be re-used in a wide range of different ways. The best way to do this is to store the data in a form that reflects the Model of Meaning (rather than that of the Model of Use). The storage system needs to keep information safe and secure, but the value of a system depends on what you can get out of it. The primary purpose is to enable questions to be answered accurately and efficiently. Accurate reports are precise and complete; efficient reports are quick and timely.

Users need to be able to count and extract groups of patient record data for innumerable reasons. The process of specifying a report involves:

- Selection and exclusion criteria to identify the records required and what codes to search for.
- The sort order in which to display the results.
- The content and format of each record displayed.
- The summary data at the top and bottom of the report (headings, totals, percentages).
- Where to send the report – screen, printer or file.

Data retrieval is hard if the underlying structure of the data is not the way that the user thinks it is, yet this is often the case. If the database is organised using a common Model of Meaning, then this model needs to be understood by everyone who wants to interrogate the data. This requires education and training.

A second requirement is that users become familiar with the layout of their data collection screens and picking lists and may well think of this as the way the system works. They reasonably expect to use the same lists and groups for their reports. The ability to report data in the way it is structured on data collection screens is a basic requirement. However, this can be difficult to deliver if the data is stored using a Model of Meaning, which differs considerably from the Model of Use; system designers may need to go to some trouble to resolve this issue.

Another difficulty is the need for multi-dimensional analysis. A trauma surgeon may want to know:

- How many fractured shaft of femur did I see last year?
- How many open reductions and fixations did I perform (on any bone)?
- How many times did I use a locked intramedullary nail?

A code with a meaning such as `open reduction and fixation of fracture of the shaft of femur using a locked intramedullary nail` may appear at first sight to allow these questions to be answered, but in practice it is difficult to answer this sort of question in this way. This is why multi-axial coding systems such as SNOMED are needed, but they have to be implemented properly in electronic patient record systems to meet this requirement.

A practical difference between the Model of Use and the Model of Meaning is that the former is usually context specific, while the latter aims to be universally true. At the point of care, the context is implicitly understood, while for big data analytics the information needs to be unambiguously explicit.

Structural Models

There are fundamental differences between structural models such as HL7 RIM and clinical terminologies such as SNOMED, in spite of the overlap between them.

A structural model provides a framework that represents clinical information in a consistent standardised way, and relates each entry to common metadata such as its subject, author, date/time and location. Structural models can handle the basic structure of each entry quite easily, but become increasingly complex as the granularity of the data becomes finer and more detailed.

Structural models may be built around standardised reference models, such as the HL7 V3 RIM or the EN13606 Reference Models (including OpenEHR).

The basic structure of the HL7 V3 RIM, with its backbone of Act, Role and Entity, linked with ActRelationship and Participation association classes, is quite simple, although this top-level simplicity covers up a lot of complexity. The HL7 Clinical Statement pattern provides a more refined and complex model for representing clinical information, as used in CDA Level 3.

The **ISO 13606 reference model** covers the same domain as the HL7 Clinical Statement pattern, but is based on the traditional structure of medical records, with record components such as folder, composition, section, item, cluster and element, as well as participants such as subject of care, healthcare professional, organisation and software or device. Many people find it easier to work with these traditional concepts, rather than having to translate these every time into the more abstract HL7 concepts of Act, Entity and Role. ISO 13606 terms can be used at the analysis level and mapped to HL7 RIM-based artifacts as a subsidiary step.

Two key ideas in ISO 13606 are archetypes and templates. **Archetypes** provide a standardised approach for representing and sharing clinical data specifications. Each archetype defines how the EHR reference model hierarchy is organised to represent the data for one clinical entry or care scenario. These archetype definitions are represented in a standardized form, using a formal language known as Archetype Definition Language (ADL), and so they can be shared and used across record-sharing communities to define how locally organised clinical data should be mapped consistently (even if the data originate from multiple systems).

Templates are constrained archetypes, for a specific purpose. A template contains just the functionality that is required, while an archetype usually contains a lot of detail that is not required in every use case.

The HL7 Clinical Statement pattern and the ISO 13606 reference models both contain many pre-defined attributes, which provide a structure to the model, but both allow the same information to be structured in more than one way. They are deliberately terminology-neutral. For example, any number of different coding schemes can be used and there is a good deal of flexibility in the way that data is handled. For example, both models allow the term "family history of asthma" to be handled either as a single concept, `family history of asthma`, or as a composite of `family history` plus `asthma`.

These methods do not recognise that some things can be done better using terminology than in the information model. In particular SNOMED description logic provides a powerful way to qualify any concept, to almost any level of detail. The advantage of this is that such post-coordinated SNOMED expressions all use a simple standardised compositional grammar, which can be computer processed. The alternative in HL7 v3 is to create multiple ActRelationships, which are more complex and potentially ambiguous.

Coded Data Types

In many specifications, it is useful to distinguish between "Coded Text" and "Codeable Text". Both specify that the content may be represented by a code, but the rules around each differ. When they are used, the coded data types may have a *value set* assigned to them – this defines the list of codes that are supposed to be used for this data element.

Coded Text

Coded Text is a reference to a concept – the intent is that this is a code picked directly from a list of possible codes. If the field is mandatory, then a code from the specified value set must be provided – it is not valid to just provide text. If there is no known code from the list of possible codes, then there is no proper value. Text may be provided in addition to the code. Because of this, Coded Text is only used when a value set is assigned, and generally for status or workflow options. For people familiar with HL7 v2, Coded Text=CNE; for v3 and CDA, Coded Text=CV CNE; for FHIR, this is code, Coding or CodeableConcept with a binding with strength=*required*.

Codeable Text

Codeable Text is a reference to a concept, where the concept might be represented as either a code or text. Text is a proper alternative, but a code should be provided. It is usually appropriate to provide text with the code as well. Codeable Text data types may have a fixed value set assigned to them, or they may be open (any code or text), though the specification should specify value sets for all codeable text elements. For people familiar with HL7 v2, Codeable Text=CWE; for v3 and CDA, Coded Text=CV CWE; for FHIR, CodeableConcept with a binding with strength=*extensible*.

Both the Coded Text and Codeable Text data types are represented using complex data types in HL7 standards. For example, the CD and CE data types used in HL7 v3 and CDA are closely related and share the aspects in Table 11.1.

None of the attributes have any length limits, nor does the originalText text content. Translations have the same type as the containing class (CD or CE). CD is differs from CE because it allows post-coordination. The displayName is provided so that an end-system that doesn't know the coding system can still display something useful to its users. It can be difficult at times to determine what the original text is. Sometimes it has the same value as the displayName, but this does not make it redundant. In XML, this looks like:

```
<x nullFlavor="[NF]" code="[code]" codeSystem="[oid]"
    displayName="[display]"/>
    <originalText>[text]</originalText>
    <translation nullFlavor="[NF]" code="[code]" codeSystem="[oid]"
    displayName="[display]"/>
</x>
```

Table 11.1 Data type components used in both Coded Text and Codeable Text

Group	Attributes	Meaning
Code	code : string codeSystem : string codeSystemVersion : string	Identifies the code system and code defined by it
Display	displayName : string	One defined display representation for the code
Text	originalText : ST (element)	Provides the text that the user said/typed/chose when picking the code or in place of the code
Translations	Translation (element)	Recursive reference to more of the same type.

Note that here and elsewhere in this section, <x>is used as the element name for the coded element, as it may vary depending on the specification. <x>is most usually <code>or <value>.

There is a tricky relationship between the nullFlavor, code, codeSystem and text. The interplay between various aspects of the coded data types in specifications can be complex, particularly once translations come into play. The rest of this section works through some common scenarios explaining how to represent them in a coded data type.

nullFlavors

The CD data type includes an attribute called "nullFlavor". This attribute is used to indicate why the coded value is unknown. It can have one of the values in Table 11.2.

Note the indenting of codes – it denotes that there are relationships between the codes; eg ASKU is a special type of UNK – if something is ASKU, it is also UNK.

In addition to this table, the value of the nullFlavor can be "OTH" – this is a special value that means that the "concept" – the meaning – is known, but it is not a

Table 11.2 HL7 v3 codes for nullFlavor

Code	Name	Definition
NI	No information	The value is missing for some unknown reason
		Note that is exactly the same as not including at all
UNK	Unknown	The value is not known
ASKU	Asked but unknown	Information was sought but not found (eg, patient was asked but didn't know)
NAV	Temporarily unavailable	Information is not available at this time but it is expected that it will be available later
NASK	Not asked	This information has not been sought (eg, patient was not asked)

Table 11.3 Codes for indigenous status with overlap between them

Code	displayName
1	Aboriginal but not Torres Strait Islander origin
2	Torres Strait Islander but not Aboriginal origin
3	Both Aboriginal and Torres Strait Islander origin
4	Neither Aboriginal nor Torres Strait Islander origin
9	Not stated/inadequately described

valid code in the context of the specified value set (see below). OTH should only be used as described below.

Sometimes coding systems include codes that overlap with the nullFlavors. Table 11.3 shows codes for indigenous status used in Australia.

The "9" code overlaps with the nullFlavor values, though not clearly. As a rule of thumb, using a code is preferable to a nullFlavor if a suitable code exists.

Code Systems and Versions

The codeSystem is an attribute that identifies the coding system by providing an OID (2.16.840.1.113883.6.96) or a UUID (441D40AF-0A07-426C-96AA-00E9D4C4A713). UUIDs are also known as GUIDs. These are opaque identifiers that uniquely identify a coding system. They are opaque because when you look at them, you cannot determine what they mean.

The primary use of the codeSystem attribute is to distinguish between codes, so that the code "X" used by one system isn't accidentally confused with the code "X" used by another system for another use. When vendors are assigning identifiers to internal code systems, this is the first thing to keep in mind – codes must never clash within a code system.

The second use of the codeSystem attribute is to recognise the coding system, so that the code can be interpreted correctly. For instance, 2.16.840.1.113883.6.96 identifies the SNOMED CT terminology. HL7 maintains an OID registry at http://www.hl7.org/oid/index.cfm?ref=common where OIDs must be registered. So it is possible to resolve any OID in a CDA document using this registry. The second thing to keep in mind when assigning identifiers to internal coding systems is that if they are going to be used in interoperability, they need to be registered with a coherent useful description of the code system.

In addition to the codeSystem attribute, there is also a codeSystemVersion attribute, which exists to handle changes in meaning of a code over time. In principle, the meaning of a code should never change over time – the definition of "X" should always mean the same thing.

However in practice, the meaning of codes may change over time – occasionally the definitions are revised to clarify meaning. Because of this, it is beneficial to supply a code system version if it is known.

Original Text

OriginalText has two uses:

- When there is no right code to pick, the original text is the meaning of the concept.
- Even when there is a code, the code does not always capture all the details and/ or nuances that the user had in mind. In this case, the originalText may be (should be) closer to the user's meaning.

Some applications do not have the ability to code data immediately and rely on "post coding" often by another person. In this case, the original text captures the concept as expressed by the data enterer prior to coding. It can be difficult to determine what the correct original text is. Table 11.4 provides a guide to common scenarios.

In this example (Fig. 11.3), it would be possible to assign several codes to the narrative; the different codes are different CodeableText values in the structured data. In the context of CDA, the text would be rendered narrative for the section containing the (in this case) radiology report. The data entries include the CD data types that correspond to the Codeable Text data elements. In these cases, the CD data types can refer to the content in the narrative directly instead of duplicating the text.

Table 11.4 Scenarios for picking original text

Scenario	Original text
User picks a code from a list of codes, displayed as the codes themselves (usually this only works with small lists of well known terms, particularly where the codes are meaningful)	None
User picks a code from a list of codes, displayed as text	Display text
User typed some text which was processed in the background	Text user typed
User typed some text which started a code look up	The text description of the code they picked
User chose a code from a list and typed more text to clarify further (see image below)	The display name for the code, with the clarifying text appended
User typed some text which was processed into a suggested list of codes, and then the user typed more text to further narrow the suggested list	The choice of "original text" becomes a little arbitrary; in the case where the original text stands as part of a report (see image below), the first original text applies

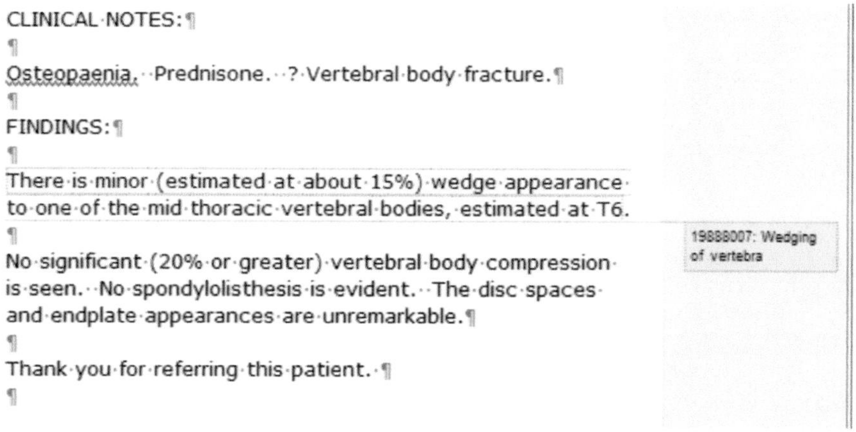

CLINICAL·NOTES:¶
¶
Osteopaenia.··Prednisone.··?·Vertebral·body·fracture.¶
¶
FINDINGS:¶
¶
There·is·minor·(estimated·at·about·15%)·wedge·appearance·
to·one·of·the·mid·thoracic·vertebral·bodies,·estimated·at·T6.
¶   19888007: Wedging
No·significant·(20%·or·greater)·vertebral·body·compression·   of vertebra
is·seen.··No·spondylolisthesis·is·evident.··The·disc·spaces·
and·endplate·appearances·are·unremarkable.¶
¶
Thank·you·for·referring·this·patient.·¶
¶

Fig. 11.3 Coding free text

In practice, the CDA narrative content would look like this:

```
<text>
  <paragraph>CLINICAL NOTES:</paragraph>
  <paragraph>
    Osteopaenia. Prednisone. ?Vertebral body fracture.
  </paragraph>
  <paragraph>FINDINGS: </paragraph>
  <paragraph>
    <content id="e23">
      There is minor (estimated at about 15%) wedge appearance
      to one of the mid thoracic vertebral bodies, estimated at T6.
  </paragraph>
  <paragraph>
    No significant (20% or greater) vertebral body compression is
    seen. No spondylolisthesis is evident.The disc spaces and
    endplate appearances are unremarkable.
  </paragraph>
  <paragraph>Thank you for referring this patient.</paragraph>
</text>
```

A CD value referring to the text as shown above would be constructed like this:

```
<value code="19888007" codeSystem="2.16.840.1.113883.6.96"
  displayName="Wedging of vertebra">
  <originalText><reference value="#e23"/><originalText>
</value>
```

Table 11.5 Source of displayName for common code systems

Code system	Source of displayName
SNOMED CT	Preferred name
ICD-10	Preferred name
HL7 code systems and v2 tables	The Print name for the code

Conformant CDA implementations must always be able to resolve the original text by following a reference instead of expecting the text to be provided directly in the originalText.

displayName

The CD data type includes a displayName attribute, which is the text that is designated for use to represent the code/concept by the coding system. Table 11.5 summarizes the source of the displayName for common code systems.

Translations

Translations are used to allow in place mappings between different code systems, for instance, if a sender is using ICPC, and a receiver expects SNOMED CT. They also allow for a community of systems (including senders) to gracefully transition from one coding system to another. Because we often find multiple systems using different coding systems, perhaps with gradual migration to SNOMED CT, translations are important. The translation structure is recursive (CDs contain translations which are CDs, which can contain translations...), so the translations should not have originalText (there's only one "text" for the entire concept) and nested translations should be avoided.

Generally, if the root code is a LOINC or SNOMED CT code, there is no need for translations. Translations should not be used with data elements that have a type "Coded Text".

Value Set vs Code System

Data elements may have a value set assigned, which specifies the set of allowed values for the codes. Simple value sets specify a list of possible codes.

When value sets are usually based on a single coding system (which is usually the case) a reference to the code system and value set are the same and usually reference the code system.

However more complicated value sets are possible that control how complex coded expressions are used.

Complex Coded Expressions

SNOMED CT allows for post-coordination using a defined expression language:

```
<value code="128045006:{363698007=56459004}"
    codeSystem="2.16.840.1.113883.6.42">
  <originalText>Cellulitis of the foot</originalText>
</value>
```

The SNOMED CT expression syntax allows additional "display text" to be included in the expression (following each code) surrounded by pipes (|) as in: 15574005 | foot fracture |: 272741003 | laterality |=7771000 | left |. This form is useful for explanation, but should not be used in interoperability. The display name should go in the displayName attribute.

ICD-10 also allows for dual coding, where one code clarifies the other. Here's an example:

```
<value code="J21.8 B95.6" codeSystem="2.16.840.1.113883.6.260">
  <originalText>Staph aureus bronchiolitis</originalText>
</value>
```

Expression-based coding systems such as post-coordination create problems; while the need for such capability arises innately and obviously to clinical users, all aspects of their implementation are difficult, and support for them is not generally available within most clinical systems.

Many interoperability specifications recommend against use of post-coordination and reference sets are usually enumerations of pre-existing codes.

The HL7 v3 CD and the CE data types differ in that the CD data type allows for qualifiers – additional qualifiers that modify the meaning of the primary code. These qualifiers are intended to support representation of these complex coded expressions in HL7 CDA documents, but are complex to use. Instead of using the CD qualifiers, implementers should use expressions inside codes as shown within the examples above. Implementers should also be aware that for some SNOMED expressions the code and displayName attributes can be quite long, and must not be truncated.

Code systems are identified by an OID or a UUID, which uniquely identifies the coding system. Any coding systems that are identified by an OID should be registered in the HL7 International OID registry at http://www.hl7.org/oid/index.cfm?ref=common. UUIDs should not be registered. Table 11.6 summarises the OIDs for common coding systems.

Note that local Implementation Guides contain many small terminologies, which are documented in place where they are used. Table 11.6 focuses on the main commonly used clinical coding systems

Coding system	OID
SNOMED CT	2.16.840.1.113883.6.96
LOINC	2.16.840.1.113883.6.1
ICD-10	2.16.840.1.113883.6.3
ICPC 2+	2.16.840.1.113883.6.140.1

Table 11.6 OIDs used for common coding systems

Common Scenarios for Coding

There are two possible approaches to coding, depending on whether the type of the data element is "Coded Text" or "Codeable Text".

For Coded Text, these are the possible scenarios:

1. The correct code is known
2. The correct code is not known

For Codeable Text, these scenarios apply:

3. The value (Coded or not) is not known at all
4. User picks code directly from the value set
5. User enters text
6. User picks a code provided by some other code system (eg ICPC2+, ICD-10, etc.).
7. User picks a code from another code system and then provides additional clarifying text
8. User chooses a code they have defined themselves
9. The message or document is being prepared on an interface engine from a v2 CWE type, and it is not known which of processes #4–#8 applied.

Note that in cases 5 through to 8, a code in the expected code system could be determined by either consulting a mapping table, or using some form of linguistic/statistical analysis. At present, the generally available linguistic/statistical mapping processes are far from ready for production. This means that the primary reliance will be on mapping tables. The section below describes how to code the scenarios above both with and without such mapping tables on the grounds that they will gradually become available.

The following checklist assists in determining the applicable scenario:

• Is the type of the data element Coded Text or Codeable Text?
• What value set is assigned to the data element?
• What value set and/or code system does the application use?

What Happens When the User Cannot Find an Appropriate Code?

#1: Coded Text – The Correct Code Is Known

Coded text is simple – either the correct code is known, or it is not. If the correct code is known, then it is used directly

```
<x code="01" codeSystem="1.2.36.1.2001.1001.101.104.16299"
    displayName="None known">
</x>
```

If desired, an originalText can be provided.

```
<x code="01" codeSystem="1.2.36.1.2001.1001.101.104.16299"
    displayName="None known">
  <originalText>There are no known medications</originalText>
</x>
```

It is not usually appropriate to provide an originalText for a Coded Text data element; the choice lists are usually small and infrastructural. In the specific case above, the original text would correspond to the caption/label on the radio button that the user checked to choose none known, but this should not imply anything different to the meaning of the code.

#2: Coded Text – The Correct Code Is Not Known

If the correct code is not known, then a nullFlavor is used:

```
<x nullFlavor="UNK" codeSystem="2.16.840.1.113883.3.879">
</x>
```

This says that the value of the indigenous status is unknown.

It may be appropriate to provide additional text if some additional information is known that cannot be coded correctly:

```
<x nullFlavor="UNK" codeSystem="2.16.840.1.113883.3.879">
  <originalText>Chinese Malay / Aboriginal</originalText>
</x>
```

Note that the value is still unknown. Many Coded Text value sets contain codes for unclear concepts such as these (1 or 9 in this case), and use of originalText in this context should always be reviewed.

#3: Codeable Text – The Value (Coded or Not) Is Not Known at All

For data elements of type Codeable Text, if the correct value is not known at all, then a nullFlavor is used:

```
<x nullFlavor="NASK">
</x>
```

This indicates that the value of the data element is unknown because the patient was not asked. In some cases, it might not be known why the data element is missing. In these cases:

```
<x nullFlavor="NI">
</x>
```

This is equivalent to simply omitting the whole element "x" from the CDA document altogether (which is also valid).

Note that for CodeableText, you should not provide a nullFlavor and an original-Text – if any text is known, then the concept is not null.

#4: Codeable Text – User Picks Code Directly from the Expected Value Set

If the user picks code directly from the expected value set, the correct code system is being used. For example, if this is SNOMED CT, and the user chose the code 263063009 (Fracture dislocation of joint), and there is no applicable value set, or the code is in the value set, the code would be represented as:

```
<x code="263063009" codeSystem="2.16.840.1.113883.6.96"
   displayName=" Fracture dislocation of joint">
   <originalText>Fracture dislocation of joint</originalText>
</x>
```

In the unlikely case that the user picked the code "263063009" directly without seeing any display text, the code would be represented as:

```
<x code="263063009" codeSystem="2.16.840.1.113883.6.96"
   displayName=" Fracture dislocation of joint">
</x>
```

This form of representation is more likely for coding systems other than SNOMED CT (particularly smaller code systems where the codes are meaningful to humans). Here is a simple example:

```
<x code="M" codeSystem="oid for gender"
   displayName=" Male">
</x>
```

#5: Codeable Text – User Enters Text

The user may enter text – because the user application only has a text field for this value, or the user couldn't find the code that said what they wanted.

Continuing with the dislocation example, and assuming that the user has written "fracture/dislocation", text would be represented like this:

```
<x>
   <originalText>Fracture/dislocation</originalText>
</x>
```

If a code in the target coding system is later generated based on some linguistic/statistical process, it can be added as a translation:

```
<x>
   <originalText>Fracture/dislocation</originalText>
   <translation code="263063009" codeSystem="2.16.840.1.113883.6.96"
      displayName=" Fracture dislocation of joint"/>
</x>
```

In advanced use cases, it may be useful to indicate that the user did look for a code before entering text. Note that whether this is known depends on the application workflow. The following example demonstrates the correct way to represent that the coding was not possible:

```
<x nullFlavor="OTH" codeSystem="2.16.840.1.113883.6.140.1">
   <originalText>Fracture/dislocation</originalText>
</x>
```

The next example uses ICPC2+, which has the OID 2.16.840.1.113883.6.140.1, as the original coding system. If the text is later mapped to SNOMED CT:

```
<x nullFlavor="OTH" codeSystem="2.16.840.1.113883.6.140.1">
   <originalText>Fracture/dislocation</originalText>
   <translation code="263063009" codeSystem="2.16.840.1.113883.6.96"
      displayName=" Fracture dislocation of joint"/>
</x>
```

#6 Codeable Text – User Picks Code Directly from the Expected Value Set

The user may pick a code directly from the expected value set. As an example, assume that the user picks the ICPC2+ code "L76013" (Fracture):

```
<x code=" L76013" codeSystem="2.16.840.1.113883.6.140.1"
    displayName=" Fracture">
  <originalText>Dislocation or fracture</originalText>
</x>
```

This assumes the user picked from a list that includes the text, not just the ICPC 2+ codes – in which case there would be no originalText. This representation also holds for the situation where the user typed the text first, and some additional process followed that led to picking the code.

If a code in SNOMED CT is available from either mapping or a linguistic/statistical process, it is added as a translation:

```
<x code="L76013" codeSystem="2.16.840.1.113883.6.140.1"
    displayName=" Fracture: other">
  <originalText>Fracture dislocation of joint</originalText>
  <translation code="263063009" codeSystem="2.16.840.1.113883.6.96"
    displayName=" Fracture dislocation of joint"/>
</x>
```

Note that even if the expected code is not available when the document is written, because the code/codeSystem that the user picked is correctly coded, when the mappings become available later (or the linguistic/statistical processes improve to become useable later), systems can convert to the desired code system.

There is an unusual variation to this case – where the user picks a SNOMED CT code, but it is not in the correct value set (the SNOMED CT Problem/Diagnosis Reference Set in this case). However by the rules of Codeable Text, this is still a valid concept:

```
<x code="209393006" codeSystem="2.16.840.1.113883.6.96"
    displayName="Other open fracture dislocation"/>
  <originalText>Fracture dislocation of joint</originalText>
</x>
```

If this gets mapped into the right reference set later:

```
<x code="209393006" codeSystem="2.16.840.1.113883.6.96"
    displayName="Other open fracture dislocation"/>
  <originalText>Fracture dislocation of joint</originalText>
  <translation code="263063009" codeSystem="2.16.840.1.113883.6.96"
    displayName=" Fracture dislocation of joint"/>
</x>
```

Fig. 11.4 Addition of clarifying text

This could be done automatically based on the definitions within SNOMED CT itself.

#7 Codeable Text – User Picks a Code from Another Code System and Then Provides Additional Clarifying Text

The user picks some code from another coding system, and then provides some further clarifying/qualifying text (see Fig. 11.4).

In this case, the "text" is the displayName of the code + the extra text. Usually a separator is used in the original text, so it looks like this:

```
Aneurysm;artery;cerebral - minimum deficit
```

This modified original text swallows up all the other possibilities as the "text that the user intended", and the code would be represented like this:

```
<x code="K90001" codeSystem="2.16.840.1.113883.6.140.1"
    displayName="Aneurysm;artery;cerebral">
  <originalText> Aneurysm;artery;cerebral - minimum deficit
</originalText>
</x>
```

If the code is mapped to the expected code set, then it would be represented like this:

```
<x code="K90001" codeSystem="2.16.840.1.113883.6.140.1"
    displayName="Aneurysm;artery;cerebral">
  <originalText> Aneurysm;artery;cerebral - minimum deficit
</originalText>
    <translation code="128608001" codeSystem="2.16.840.1.113883.6.96"
    displayName="Cerebral arterial aneurysm"/>
</x>
```

In this case, SNOMED CT does not appear to have a more specific code for:

```
Cerebral arterial aneurysm with minimum deficit
```

but if such a code existed, and the tooling was capable of performing the mapping, it could also be used.

#8 Codeable Text – User Chooses a Self-Defined Code

In some clinical systems, when a user cannot find a code that represents their intent, they can simply define their own code that only they see and use. Note that this process has obvious dangers, and the various clinical systems exert different levels of control over the appropriateness of this action. These considerations are out of scope here.

As an example, assume that the user encountered the situation above (Aneurysm;artery;cerebral – minimum deficit), and instead of offering the ability to provide extra qualifying text, the system allows the user to create their own code. If, the user creates a code AA1001, which means Cerebral arterial aneurysm with minimum deficit, and the user picks this new code, this would be represented as:

```
<x code="AA1001" codeSystem="441D40AF-0A07-426C-96AA-00E9D4C4A713"
   displayName=" Cerebral arterial aneurysm with minimum deficit">
   <originalText>Minimal deficit Cerebral arterial aneurysm</ </
   originalText>
</x>
```

The code system here is a UUID that scopes the code AA1001 so that it could never be confused with any else's AA1001 code, should they use that particular code. In practice, the codeSystem could be an OID, but this would require some kind of external system to distribute unique identifiers to the installed base of the application; UUIDs are much easier in this case (and may be generated by some system API such as coCreateGUID on Windows). In these cases, systems must track and store the UUID so that it is consistently used for this purpose.

In the long term, it is possible that systems to gather and map these custom codes to national code systems will be put in place (this is not possible now, but people are already interested in the idea). For this reason, vendors should keep appropriate local records over the local codes so that this might be possible in the future.

#9 Codeable Text – CDA Generated on an Interface Engine from HL7 v2

In the short term, many CDA documents will be generated by an interface engine on the perimeter of an organisation from existing exchanges. In practice, this means converting v2 messages to CDA documents, and in most cases, the user process around the coding will not be known.

In these cases, the CD data type is generated from a CWE or CNE data type. Ideally there would be consistency between the v2 and other specifications, so that CNE maps to Coded Text and CWE maps to Codeable Text, but this is not always the case. In general, though, there is not much difference between the CWE and CNE data types, and what difference there is often misunderstood.

Table 11.7 shows a mapping between the v2 CWE data type, the v3/CDA CD data type. This table is based on HL7 v2.7 and is indicative; actual usage of the CWE data type varies widely and implementers should consult their message specifications and sources carefully.

Advice for Receivers

When receiving codes, you can reverse engineer to be sure about the exact circumstance that applied. However, it is generally not required to do so. The following advice suffices for most uses:

- Displaying the concept to the user:

  - If you get an originalText, display this to the user
  - Otherwise, if you get one, show the displayName
  - Otherwise, if you can, look up the code and show its meaning
  - Otherwise, show the code, if you get one
  - Otherwise the nullFlavor description in brackets
  - If you don't get anything then show "blank" or "—") or equivalent
  - It is sometimes useful to display the code in brackets if assigned (this alerts the user that the concept is coded, if the work flow depends on the code)

- Storing the concept:

  - Codes, displayNames, and originalText may be arbitrarily long. (>255 chars is possible)
  - They should never be truncated
  - Some unlimited type storage is appropriate. In practice this is challenging; in the end most implementations choose some variation of storing the entire document as a blob, indexing the parts of the document that are used for searching/matching, and marking in those indexes where content has been truncated.

Table 11.7 Coded data types alignment between HL7 v2, v3/CDA and FHIR

V2	CDA	FHIR
CWE-1 Identifier	CD.code	CodeableConcept.coding.code[2]
CWE-2 Text[1]	CD.displayName	CodeableConcept.coding.display[2]
CWE-3 Name of Coding System[3]	CD.codeSystem(Name?)	CodeableConcept.coding.system[2]
CWE-4 + 10 Identifier	CD.translation.code	CodeableConcept.coding.code
CWE-5 + 11 Text	CD.translation.displayName	CodeableConcept.coding.display
CWE-6 + 12 Name	CD.translation.codeSystem(Name?)	CodeableConcept.coding.system
CWE-7 Version ID	CD.codeSystemVersion	CodeableConcept.coding.version[2]
CWE-8 + 13 Version ID	CD.translation.codeSystemVersion	CodeableConcept.coding.version
CWE-9 Original Text[4]	CD.originalText[5]	CodeableConcept.text
CWE-14 System OID	CD.codeSystem	CodeableConcept.coding.system[2]
CWE-15 Value Set OID	CD.valueSet[6]	Extension on Coding[7]
CWE-16 Value Set Version	CD.valueSetVersion[6]	Extension on Coding[8]
CWE-17 + 20	CD.translation.codeSystem	CodeableConcept.coding.system
CWE-18 + 21	CD.translation.valueSet[6]	Extension on Coding[7]
CWE-19 + 21	CD.translation.valueSetVersion[6]	Extension on Coding[8]
	CD.codingRationale[6, 9]	CodeableConcept.coding. userSelected

Notes:

1. CWE-2/5/11: This is sometimes used as original text, and sometime as display name

2. If the data type is a CodeableConcept, this is customarily the first coding, but order doesn't matter

3. The rules for 'name of coding system' are much looser than for v3 CD.codeSystem and FHIR Coding.system. Name is usually coded, but not always

4. The definition of original text forces implementers to tighten up the usage of CWE2 etc (see note 1). This is one of the biggest changes in 2.3, but was not followed well

5. CD.translation.originalText should never be used

6. Defined in ISO 21090 (post CDA release 2 of data types), can be used as an extension in CDA if necessary

7. The extension URL is pre-defined by the FHIR specification: http://hl7.org/fhir/StructureDefinition/valueset-reference

8. No extension is pre-defined for this one

9. Coding Rationale covers more than just 'user selected'

- Making decisions based on the code:

    – Check the root and the translations for the preferred code
    – It may not matter whether the code is an expression or not (need to consult documentation on terminology service/library)

Other Implementation Issues

One of the key decisions is whether the information items really needs to be coded, or whether free text would be more appropriate. Data coding requires additional work by programmers and users, so there is little merit in coding just for the sake of it, without good reason.

When thinking about the level of coding required, it may be useful to consider data in three categories.

1. **Data likely to be coded and complete** so that we can be confident about making comparisons between both the numerator and the denominator of any variable. Examples include: care processes such as clinic attendances and admissions, invasive procedures such as surgical operations, medicines prescribed through a computer and laboratory test results. These data can be used for comparisons, because you can usually trust both the numerator and the denominator.
2. **Data likely to be coded but often not complete**. Examples include history, examination findings and problem lists. If a record does not mention a finding, you cannot assume it was absent in the patient. This type of data cannot be used reliably for comparisons.
3. **Data unlikely to be either coded or complete** Most free text, dictated notes and images fall into this category. It is difficult to reliably extract data or make valid comparisons.

In some cases an external authority may specify that SNOMED, or some other coding scheme, must be used. For example, a national government agency may require data to be coded in a particular way, for example, to provide comparable information for management or research. In this situation, the architect needs to decide whether or not there is more value in using the specified coding scheme internally, or in supporting a translation table between a private internal representation and the specified coding scheme.

Many designers continue to use proprietary structures and codes for the internal representation of their data, which are then mapped unambiguously to SNOMED or to other coding schemes such as ICD-10, using translation aids such as the National Library of Medicine's UMLS (Unified Medical Language System).

The **scope** of uses of SNOMED CT is wide and includes:

- Entries in problem lists, admission and discharge diagnoses, provisional, working and differential diagnoses.
- History of current condition, presenting symptoms and other symptoms
- Allergies, adverse events and propensities to adverse reactions
- Operative, diagnostic and therapeutic procedure requests, delivery and outcomes
- Medications including prescriptions, dispensing records, drug administration charts, current, discontinued and proposed medication
- Past medical, surgical, family and social history
- Clinical examination findings and vital signs

- Laboratory, diagnostic and other investigation requests, procedures and results
- Other clinical information such as plans, goals, risks, progress notes and assessment scales
- Administrative information such as admission, discharge, transfer and referral events
- Other values including drugs, organisms, substances and body structure.

Data Entry The ease of use of the user interface is of particular importance in any clinical system. This depends on many factors, including the speed, usability, and relevance of searches. The approach used may depend on the number of options applicable at any point. For example, different approaches may be best when entering yes/no answers, selecting from a short list with less than 10 options, or from a long enumerated list with possibly thousands of options, such as a list of possible medications. Different approaches are also needed to describe complex situations such as the victim of an accident with multiple injuries, which can only be described accurately using post-coordinated expressions or using free text.

Data Retrieval A system's capability to perform information retrieval can be judged in terms of:

- Query expressivity – ability to support pre-coordinated and/or post-coordinated expressions
- Subsumption testing – ability to test whether a patient with a specific condition can be assumed to have a more general condition that subsumes the specific one. Subsumption testing allows the computer to answer questions such as: "does this patient who has angina pectoris have a heart disease?"
- Concept equivalence – the ability to retrieve equivalent information, when it has been recorded using different but equivalent terms and expressions
- Context awareness – the ability to take account of context data, recorded either in the record or in SNOMED expressions, when interpreting and evaluating results
- Performance.

Communication The work required to use SNOMED CT in interoperability messages and other communication differs for outbound and inbound communications, according to whether the information is being sent or received. The level of sophistication depends on:

- Level 0 Mapping-based support for SNOMED expressions
- Level 1 Native SNOMED support for communication of pre-coordinated expressions
- Level 2 Native SNOMED support for communication of post-coordinated expressions

As a general rule it is considerably easier to send outbound communications than to process incoming ones.

Expression Storage refers to the extent that EHR systems support the storage and use of pre-coordinated and post-coordinated SNOMED expressions. A simple EHR system may only support the use of pre-coordinated concept identifiers (sctId). A more sophisticated system would support the storage of post-coordinated SNOMED expressions. This could be an expression stored using SNOMED description logic resulting in strings of indeterminate length. An alternative is to use an expression reference table, which enables a fixed length reference, such as a UUID, within the records (Fig. 11.5).

Fig. 11.5 Terminology binding

When to Use HL7 and SNOMED

One of the best ways to simplify a situation, which is inherently complex, is to fix whatever can be fixed. This is why many countries have resolved to use SNOMED CT as the standard terminology used in interoperability and HL7 V3 CDA as the standard interchange format.

However there is a substantial overlap between HL7 and SNOMED. The same information can be expressed in different ways using SNOMED and HL7 CDA structures. This forms an obstacle to semantic interoperability.

It is impossible to draw a clean dividing line between the what should be done using HL7 CDA structures and what should be done using SNOMED description logic. This problem was investigated in the TermInfo Guide which recommends: [8]
SNOMED should be used for specifying:

- Specific concepts and value sets, for example, diseases, symptoms, signs, procedures, drugs, etc.
- Representation of constraints on use of terminology, such as reference sets including navigation hierarchies and value sets.
- Simple semantic relationships, such as laterality or the relationship between 'viral pneumonia', 'lung', 'virus', 'infectious disease'.
- Constraints on combination of concepts, for example, restrictions on 'finding site' refinement of 'appendicitis', or conventions on representing laparoscopic variants of a procedure.
- Post-coordinated expressions at various levels of nesting.

The HL7 information model should be used for specifying:

- Instance information and meta-data for any clinical statement such as dates and times, numbers and quantities.
- Identifiable instances of real-world entities such as people, organizations, places.
- Representation of relationships between distinct instances of record entries and other classes. For example, assertions of causal relationships between entries, grouping of entries related by timing, problem or other organizing principles.
- Overall record and communication architecture, such as EHR compositions, CDA documents and HL7 messages, showing the way that items should be grouped together and anchors for terminology components, such as codes.
- Differences due to the work process for a specific use case.

There remain some grey areas, where the choices are not clear-cut, such as how best to handle issues such as context, negation and uncertainty. An important criterion is that the proposed bindings should not involve a proliferation of pre-coordinated codes. For example it would not be a good idea to require codes to represent every possible cross product of adverse effects of combinations of medicines or substances.

Where more than one approach appears to be viable and broadly equal in impact, then we need to avoid unnecessary divergence by selecting a single approach, perhaps based on precedence. If one method has already been used successfully and an alternative has not, then we should prefer the approach that has already been demonstrated to work. As always in interoperability the goal is clarity, to minimise ambiguity and disorder.

References

1. Vendor Introduction to SNOMED CT. IHTSDO. May 2015.
2. SNOMED CT IHTSDO Glossary. IHTSDO. January 2015.
3. Markwell D, Sato L, Cheetham E. Representing clinical information using SNOMED Clinical terms with different structural information models. In: Cornet R, Spackman K, editors. Representing and sharing knowledge using SNOMED. Proceedings of the 3rd international conference on Knowledge Representation in Medicine (KR-MED 2008). 2008: 72–9.
4. Rector AL, Qamar R, Marley T. Binding ontologies and coding systems to electronic health records and messages. Appl Ontol. 2009;4(1):51–69.
5. NHS CUI Design Guide Workstream – Design Guide Entry – Terminology – Matching. 2007.
6. NHS CUI Design Guide Workstream – Design Guide Entry – Terminology – Elaboration. 2007.
7. NHS CUI Design Guide Workstream – Design Guide Entry – Terminology – Display Standards for Coded Information. 2007.
8. HL7 Version 3 Implementation Guide: TermInfo – Using SNOMED CT in CDA R2 Models, Release 1 Draft Standard for Trial Use. December 2015.

Part III
HL7 and Interchange Formats

Chapter 12
HL7 Version 2

Abstract Hl7 v2 is the most widely used healthcare interchange format. Messages are sent in response to trigger events. Messages comprise a set of segments, defined using an abstract message syntax table. Segments contain fields and fields contain components; components may contain sub-components, which are separated by delimiters. Segments are specified in segment definition tables. Z-segments are locally defined. Data types are the basic building blocks in each element. Some are simple, others are complex. Codes are defined using tables.

Keywords HL7 v2 • Trigger event • Segment • Abstract message syntax table • Field • Component • Sub-component • Data type • Delimiter • Message header (MSH) • Event type (EVN) • Patient identification details (PID) • Patient visit (PV1) • Request and specimen details (OBR) • Result details (OBX) • Z-segment • Coded no exceptions (CNE) • Coded with exceptions (CWE) • HL7 tables

HL7 Version 2 (v2) is the most widely used healthcare interoperability standard in the World. It is used in over 90 % of all hospitals in the USA and is widely supported by healthcare IT suppliers worldwide.

At first sight, the HL7 v2 documentation may appear to be large and formidable, but it is based on a few basic principles, which are quite easy to grasp that account for its success.

To understand some of the features of HL7, we need to go back to its origins in 1987. The initial focus of HL7 was on exchanging information about admissions, discharges and transfers (ADT) within hospitals. The first version, HL7 V1.0 was issued a few months later. The next year, 1988, HL7 v2.0 was published, and this included a major extension to add in messages for exchanging orders and reports for tests and treatment, based closely on the ASTM (American Society of Testing and Materials) E.1238.88 standard. Version 2.1, which was the first widely used version, was published in 1991.

The HL7 v2 standard has been in continuous development for more than 25 years. At the time of writing, the latest version is Version 2.8.2, which was approved as an ANSI standard in April 2015 [1].

Due to the way that HL7 v2 is designed, it is not possible to understand a v2 message without access to the standards documentation and a detailed implementation guide for that specific implementation (Table 12.1).

T. Benson, G. Grieve, *Principles of Health Interoperability*,
Health Information Technology Standards, DOI 10.1007/978-3-319-30370-3_12

Table 12.1 HL7 v2.8
chapters

Chapter	Title
1	Introduction
2	Control
2A	Control – Data Types
2B	Control – Conformance
2C	Control – Code Tables
3	Patient Administration
4	Order Entry
4A	Order Entry: Pharmacy/ Treatment, Vaccination
5	Query
6	Financial Management
7	Observation Reporting
8	Master Files
9	Medical Records/Information Management
10	Scheduling
11	Patient Referral
12	Patient Care
13	Clinical Laboratory Automation
14	Application Management
15	Personnel Management
16	eClaims
17	Materials Management
Appendix A	Data Definition Tables
Appendix B	Lower Layer Protocols
Appendix C	BNF Message Descriptions
Appendix D	Glossary
Appendix E	Index
	HL7 Version 2 XML Encoding Syntax, Release 2

The structure of the HL7 v2 standard documentation is shown in Table 12.1, which shows the contents of HL7 v2.8. The most important chapters are Chap. 2, Control and Chapter 2A Data Type Definitions. The full documentation has almost 2500 pages and almost one million words. It contains an enormous amount of knowledge and experience about health informatics.

During its long development period the scope and size of v2 has increased greatly. However the basic ideas have hardly changed. One of the core principles has been the preservation of backward compatibility, while the standard has evolved by addition. The idea being that a system, which can understand a new message in a new version, should also be able to understand a previous version. Ideas, which have been superseded, are flagged as being deprecated, but not replaced.

Older versions are still widely used because there is minimal return on investment achieved by replacing a working interface with a later version and a significant risk of hitting unexpected problems. However, interface engineers may need to work with several different versions and recognise the differences between them. There are important differences between versions, so it is always important to know what version is being used. In this chapter, we focus on principles, and most of the examples used here are applicable to version 2.4. It is beyond the scope of a single chapter in a book of this sort to document the differences between releases.

To understand the HL7 v2 documentation, you need to know about the message syntax and data types. Message syntax describes the overall structure of messages and how the different parts are recognised. Each message is composed of segments in specified sequence, each of which contains fields also in a specified sequence; these fields have specified data types. Data types are the building blocks of the fields and may be simple, with a single value, or complex, with multiple components. These components themselves have data types, which can be simple or complex, leading to sub-components (Fig. 12.1).

Message Syntax

HL7 v2 messages are sent in response to trigger events. The message name is derived from the message type and a trigger event. The message type is the general category into which a message fits. For example, patient administration messages are ADT. Examples of message types are shown in the Table 12.2, which also shows the HL7 v2 Chapter where they are described in detail.

The trigger event indicates what happened to cause a message to be generated. Trigger events are specific to a message type. For example some of the ADT trigger events are shown in Table 12.3.

The full message name for an admit notification is ADT^A01 (the "^" is the HL7 field component separator). The message name is always entered in the ninth field of the message header segment (MSH-9).

Each HL7 v2 messages comprises a set of segments. For example, a simple message, noting that a patient has been admitted to the hospital contains the following segments in the order shown in Table 12.4.

The overall structure and allowable content of each message is defined in an abstract message syntax table, which lists segments in the order in which they occur (Table 12.5). The abstract message syntax also shows which segments are optional and which can be repeated. A segment listed on its own is mandatory and may not repeat. Optional segments are surrounded by square brackets [XXX]. Segments that are allowed to repeat are indicated using curly braces {XXX}. If a segment is both optional and repeatable, it has both brackets and braces [{XXX}]. Note that the order is not important: [{XXX}] and {[XXX]} are equivalent.

Segments can be grouped into logical groupings containing more than one segment and may be nested such as: OBR, [NTE], {OBX, [{NTE}]}. This specifies one OBR segment with an optional NTE segment, plus any number of OBX, each with any number of NTE segments. Abstract message syntax allows choice of seg-

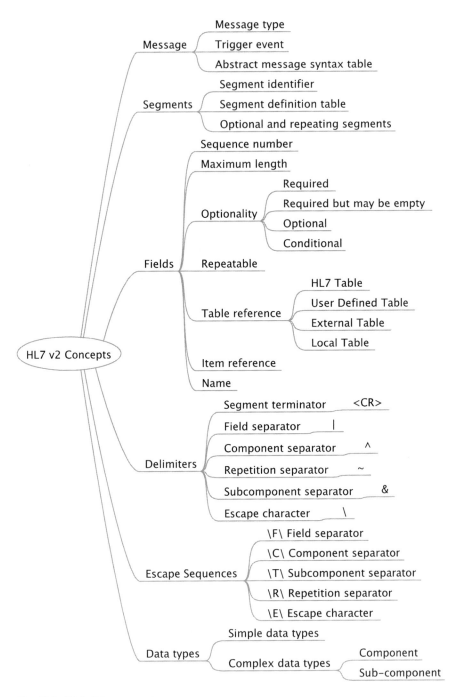

Fig. 12.1 HL7 v2 key concepts

Table 12.2 HL7 v2 message types

Value	Description	v2 Chapter
ACK	General acknowledgment message	2
ADT	ADT message	3
ORM	Order message	4
ORU	Observation result unsolicited	7

Table 12.3 Hl7 v2 trigger events

Value	Description
A01	Admit/visit notification
A02	Transfer a patient
A03	Discharge/end visit
A04	Register a patient

Table 12.4 List of segments

MSH	Message Header
EVN	Event Type
PID	Patient Identification
PV1	Patient Visit

Table 12.5 Example of HL7 v2 abstract message syntax table

ADT^A01	ADT message	Chapter
MSH	Message Header	2
EVN	Event Type	3
PID	Patient Identification	3
[PD1]	Additional Demographics	3
[{NK1}]	Next of Kin / Associated Parties	3
PV1	Patient Visit	3

ments. For example: <OBR | RQD> means that either OBR or RQD may be used. The abstract message syntax table (Table 12.5) also shows which chapter of the HL7 v2.x standard contains the segment definition.

Table 12.5 also shows that segments MSH, EVN, PID and PV1 are mandatory. PD1 is optional. It is also indented, which indicates that it is nested under the PID segment, creating a group. NK1 is both optional and repeatable. Each segment has a three-character identifier, the segment ID (eg MSH). In a message the segment ID is always the first three characters of the line.

Segments contain fields and fields contain components; components may contain sub-components, which are separated by delimiters.

Delimiters

Delimiters (such as field separators, component separators and sub-component separators) are used to indicate the boundaries between these elements. The term element is used to refer to a field, a component or a subcomponent.

Segment Separator Within the HL7 syntax, the size of messages transmitted is reduced by truncation. If fields at the end of a segment or component are not needed, the appropriate terminator or separator character truncates them. The segment terminator (carriage return) truncates segments. In the same way, field separators truncate components and component separators truncate subcomponents. Each segment is ended with an ASCII carriage return < CR > character.

Most HL7 v2 implementations use default encoding with the delimiters to terminate segments and to separate components and subcomponents (Table 12.6). The delimiters are defined in the first two fields of the MSH segment (MSH-1 and MSH-2). There is also an XML representation (not described here).

The **field separator** (| usually referred to as pipe) is always the fourth character of each segment. In HL7 v2 fields are named according to their sequential position within a segment. For example MSH-9 is the ninth field in the MSH segment and is preceded by nine field delimiters. Two adjacent field separators (| |) indicate an empty field. If an application wishes to state that a field contains null and expects the receiving system to act on this, then an explicit null is represented as | *""* | .

The **component separator** (^ usually referred to as hat) separates the components of a field. Components are referred to by the segment, field and position in the field (eg MSH-9.1) For example the MSH-9 field (Message Type) contains two components: MSH-9.1 (message type) and MSH-9.2 (trigger event) and might be represented as ADT^A01. The field separator truncates any components, not needed at the end of a field. For example, the following two data fields are equivalent: |ABC^DEF^^| and |ABC^DEF|.

The **repetition separator** (~ usually referred to as tilde) is used to separate the first occurrence or repetition of a field from the second occurrence and so on.

Table 12.6 Default HL7 v2 delimiters

Symbol	Usage
|	Field separator
^	Component separator
~	Repetition separator
\	Escape character
&	Subcomponent separator
<CR>	Segment terminator

Table 12.7 HL7 v2 Escape characters and sequences

Symbol	Escape sequence
|	\F\
^	\S\
~	\R\
\	\E\
&	\T\ eg |Marks \T\ Spencer|

The **escape character** (\ back-slash) is used mainly in text elements to bracket text for special processing. The escape character can be used to send delimiters within a message (Table 12.7).

For example, the term *Barnes & Noble* must be written `Barnes \T\ Noble` or could be treated as two separate sub components *Barnes* and *Noble*.

The escape character may also be used to indicate certain formatting commands, such as `\.br\` to indicate line break, or `\.sp 3\` to skip 3 spaces in the formatted text (`FX`) data type.

The **subcomponent separator** (& ampersand) is used to separate sub-components within components, providing an additional level of granularity.

Segment Definition

Each segment is defined in a table such as that shown below for the MSH Message Header segment. All HL7 v2 messages begin with a single MSH segment. Table 12.8 provides an example of how segments are defined.

The columns of this table show:

SEQ:	Field sequence number
LEN:	Maximum field length
DT:	Data type
Usage:	Optionality (see below)
Cardinality:	Repeatable field. If Y can repeat any number of times; a number, such as Y/3 indicates a maximum number of 3 repeats.
TBL#:	The reference number of the HL7 table, which contains a controlled vocabulary from which values can be taken.
Item#:	HL7's internal database item number
Element Name:	Human readable name of the field.

The usage column shows the optionality of each field in the segment attribute table. The usage codes mean:

R   Required
RE   Required but may be empty; dependent on values in the patient's record. This field is required to be completed unless no data has been collected for it.
O   Optional

Table 12.8 Segment definition table for MSH segment

SEQ	LEN	DT	Usage	Cardinality	TBL#	Item #	Element name
1	1	ST	R			00001	Field Separator
2	4	ST	R			00002	Encoding Characters
3	180	HD	O			00003	Sending Application
4	180	HD	O			00004	Sending Facility
5	180	HD	O			00005	Receiving Application
6	180	HD	O			00006	Receiving Facility
7	26	TS	O			00007	Date/Time Of Message
8	40	ST	O			00008	Security
9	7	CM	R			00009	Message Type
10	20	ST	R			00010	Message Control ID
11	3	PT	R			00011	Processing ID
12	8	ID	R		0104	00012	Version ID
13	15	NM	O			00013	Sequence Number
14	180	ST	O			00014	Continuation Pointer
15	2	ID	O		0155	00015	Accept Acknowledgment Type
16	2	ID	O		0155	00016	Application Ack Type
17	2	ID	O			00017	Country Code
18	6	ID	O	[0..3]	0211	00692	Character Set
19	60	CE	O			00693	Principal Language Of Message

C   Conditional on the trigger event or on some other field(s). The field definitions following the segment attribute table should specify the rules of conditionality for this field.

CE   Conditional but may be empty

X   Not supported

B   Left in for backward compatibility with previous versions of HL7, and not to be used with this specification.

Segments

This section describes several of the most important segments found in HL7 v2 messages, including the message header (MSH), event type (EVN), patient identification details (PID), patent visit (PV1) request and specimen details (OBR), result details (OBX) and Z-segments (Fig. 12.2).

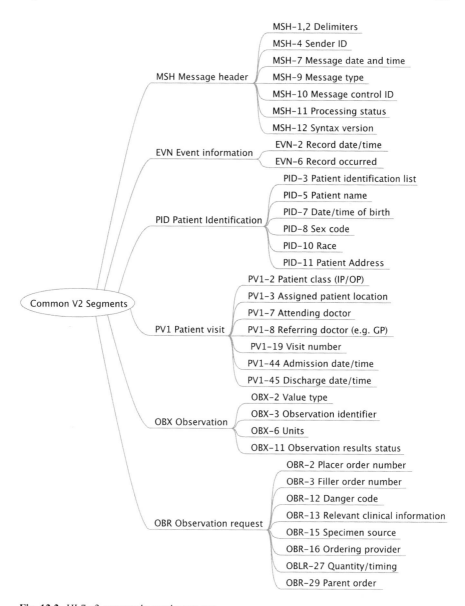

Fig. 12.2 HL7 v2 commonly-used segments

Message Header MSH

The report header (MSH) contains common metadata found in most messages, irrespective of subject. The first two fields of the MSH segment specify the delimiters used (see above).

Other required fields in MSH in addition to the Field Separator and Encoding Characters are:

SenderID (MSH-4) is a unique identifier for the sender, expressed as the combination of an identification code for the sender and a code for the naming authority that controls the assignment of these identification codes. The only constraint is that the combination of MSH-4.2 and MSH-4.3 is unique.

For example: `|^12345^Labs|`

DateTime (MSH-7) of message is the exact date/time, that the sending system created the message. For example `|20080805183015+0000|` indicates Aug 5, 2008 6.30 pm and 15 s GMT.

MessageType (MSH-9) is used to name each message. This field contains components for message type code, trigger event and message structure ID. A typical example is:

```
|ADT^A04^ADT_A01|
```

The first component `ADT` is the message type code. This represents the core function of the message.

The second component `A04` contains the trigger event type.

The third component is the abstract message structure code defined by HL7 (`ADT_A01`). Each message structure (eg `ADT_A01`) may be used with a number of trigger events (eg `A01`, `A04`, `A08` and `A13`).

MessageID (MSH-10) is used to uniquely identify the message. The sending system must assign an identifier, which is unique to the extent in combination with the SenderID it is globally unique. One way of ensuring uniqueness is to use a globally unique identifier such as a GUID, which is produced on the fly by software. However, GUIDs are longer than the 20 characters prescribed by HL7.

ProcessingStatus (MSH-11) shows whether the message is production (`P`) or for some other use such as debugging (`D`) or training (`T`).

SyntaxVersion (MSH-12) indicates the HL7 version with which this message claims compliance. Compliance with HL7 v2.4 is shown by `|2.4|`.

Event Type (EVN)

The EVN segment is used in all ADT messages to capture detailed time information about the trigger event. When used it is the second segment of each message. The most used fields are:

Recorded Date/Time (EVN-2) is the date and time that the trigger event data was recorded. The data type is TS (time stamp).

Event Occurred (EVN-6) is the actual time of the event, rather than the time that a message was triggered. It can help to ensure that data is reported in a logical order. For example, information transcribed from paper records may be recorded in the computer out of sequence.

Patient Identification Details (PID)

PatientID (PID-3) refers to the patient identifiers (one or more), which are used by the healthcare facility to uniquely identify a patient (eg hospital number, NHS number). In HL7 v2 these identifiers are sent in field PID-3, with the identifier in the first component (PID-3.1), an optional identifier for the issuing authority in the fourth component (PID-3.4) and an identifier type code (required) in the fifth component (PID-3.5).

For example a patient with hospital number 123456 at St Mary's Hospital (SMH) may be entered as |123456^^^SMH^PI|, where PI indicates that this is a Patient internal identifier.

If the sender only uses the NHS number, eg 9999999904, this could be exchanged as |9999999904^^^NHS^NH|.

The repetition separator, ~, separates the combination of both hospital number and NHS number, together: |123456^^^SMH^PI~9999999904^^^NHS ^NH|

PatientName (PID-5) includes the first (given) and last (family) name of the patient. These are provided in fields PID-5.1 and PID-5.2 respectively. Mary Smith would become |Smith^Mary|.

DateOfBirth (PID-7) is recorded as a date in format YYYYMMDD, eg |19620114| for 14 January 1962.

SexCode (PID-8) using an agreed coding system, such as M=Male and F=Female, eg |M|.

Patient address (PID-11) uses the components in Table 12.9.

For example, |14 Pinewood Crescent^Hermitage^^^RG18 9WL| shows two lines of address and a postcode.

Patient Visit (PV1)

The PV1 (patient visit) segment is used in this example for both the patient's GP and the patient location at which the sample was taken.

Patient Location (PV1-3) is the patient's location.

General Practitioner (PV1-8) is the person responsible for the Patients Health in the Community. This data is desirable but not mandatory.

Table 12.9 Components of patient address (PID-11)

Street address	PID-11.1
Second line of address	PID-11.2
City	PID-11.3
State, province or county	PID-11.4
Zip or postal code	PID-11.5
Country	PID-11.6

Request and Specimen Details (OBR)

The laboratory allocates each specimen an accession number, which is used to identify that specimen and any derivatives. In HL7 this is referred to as the Filler Order Number and is provided in field OBR-3.1.

Lab Test Code records what was requested to be done. An agreed code system, such as LOINC should be used. It is provided in field OBR-4, component OBR-4.1 with the text name in component OBR-4.2 and the name of coding system in OBR-4.3.

The date and time that the specimen was collected from the patient is provided in field OBR-7, using format YYYYMMDDHHMM. The time is optional.

The specimen source is provided using an agreed code or controlled vocabulary in field OBR-15.1. (eg WOUND SWAB)

Body Site (desirable) states the part of the body from which the specimen is taken. This is provided as a string in field OBR-15.4 (eg FOOT)

Site Modifier (optional) is sometimes reported, to provide additional information about the body site. If used it is provided in field OBR-15.5 (eg Right)

The doctor who ordered the test is recorded in field OBR-16, using an agreed identifier in OBR-16.1.

Result Details (OBX)

Each separate result is entered as a separate OBX segment, which relates to a single observation or observation fragment. It represents the smallest indivisible unit of a report.

Each result is represented as an attribute-value pair.

The data type of the value is specified in OBX-2. In HL7 terminology the attribute being measured is specified in OBX-3 (Observation Identifier) and the value is in OBX-5 (Observation Value). Internal references are specified in OBX-4.

In Microbiology, organisms, or the presence of an organism, are identified by either isolating the organism on a medium, or testing for the presence of an organism using a variety of tests. Isolates generally have associated antibiotic susceptibilities.

Observation Identifier (OBX-3) is the test that is being done (the attribute being measured) and typically uses LOINC or locally defined codes. Field OBX-3.1 contains the code; OBX-3.2 contains the human-readable display text; OBX-3.3 contains the coding scheme identifier if used. For example: |9999-9^Test name^LN|

(OBX-5) is the value of the result and is typically a numeric value or a code such as a SNOMED code.

The value type – the data type of the observation value – is specified in the Value Type (OBX-2).

The code value is OBX-5.1, display text is OBX-5.2 and code system identifier is OBX-5.3. Text strings can be transmitted in OBX-5.2 (eg `|^This is a result|`). Numeric values are represented as strings to allow non-numeric characters to be used (such as ">").

```
OBX|1|CE|5182-1^Hepatitis  A   Virus   IgM   Serum   Antibody
EIA^LN||G-A200^Positive^SNM|
```

Some microbiology results have an extra complication. The first stage is to identify the various isolates (such as bacteria), which are present in the specimen. The second stage is to test each of these isolates for susceptibility to treatment by various antibiotics. The solution is to use internal references to link all of the results for the same isolate together using the Observation Sub-ID (OBX-4). Each OBX segment for the same isolate contains the same integer value in OBX-4.

The Abnormal Flag (OBX-8). If the observation is an antimicrobial susceptibility, the interpretation codes are: S = susceptible; R = resistant; I = intermediate; MS = moderately susceptible; VS = very susceptible.

The observation result status is a required to indicate whether the result is Final, Preliminary or otherwise and should be present in OBX-11.

Z-Segments

HL7 v2 provides a facility for any users to develop their own segments, message types and trigger events using names beginning with Z. Z-segments are widely used and are one of the main reasons why there are so many different variants of HL7 v2 messages.

Z-segments can be placed anywhere in a message. Some message designers place all Z-segments at the end of a message, whilst others place them adjacent to related information.

A Simple Example

The following example is from a simple feed of laboratory test reports from a microbiology laboratory to an infection control monitoring system. Each report includes:

- A header stating the type, origin and date time of the message
- A single patient with ID number, name, sex, date of birth, address and General Practitioner identifier
- Specimen details of the laboratory accession number (ID), source, body site, time of collection and requester

- A set of test results, including the test name and result and abnormality flag.

The abstract syntax of the HL7 v2 message is:

```
MSH          Message header
PID          Patient Identification Details
PV1          Patient Visit
OBR          Results header
{OBX}        Results detail (repeats)
```

All segments are required.

The structure of an HL7 v2 message, which meets these requirements, is:

```
MSH|delimiters||sender|||dateTime||messageType|messageID
    |processingStatus|syntaxVersion
PID|||patientID^^^source^IDtype||familyName^givenName||d
    ateOfBirth|sex|||streetAddress^addressLine2^^^postC
    ode
PV1|||patientLocation|||||patientsGP
OBR|||accessionNumber|testCode^testName^codeType|||speci
    menDate|||||||||specimenSource^^^bodySite^siteModifier|
    requester
OBX||valueType|observableCode^observableName|observation
    SubID|valueCode^valueText^valueCodeType|||abnormalF
    lag
OBX …
```

A populated example is:

```
MSH|^~\&||^123457^Labs|||200808141530||ORU^R01|12345678
    9|P|2.4
PID|||123456^^^SMH^PI||MOUSE^MICKEY||19620114|M||||14
    Disney Rd^Disneyland^^^MM1 9DL
PV1||||5 N|||||G123456^DR SMITH
OBR|||54321|666777^CULTURE^LN|||20080802|||||||||SW^^^FOO
    T^RT|C987654
OBX||CE|0^ORG|01|STAU||||||F
OBX||CE|500152^AMP|01||||R|||F
OBX||CE|500155^SXT|01||||S|||F
OBX||CE|500162^CIP|01||||S|||F
```

Note that the OBX segment repeats. Information about the susceptibilities of organism detected (STAU – staphylococcus aureus) is linked to that organism finding by using the OBX-4 Observation Sub-ID field.

This could be rendered as in Table 12.10

Table 12.10 Rendering of
example microbiology report

Report from Lab123457, 15:30 14-Aug-2008, Ref 123456789
Patient: MICKEY MOUSE, DoB: 14-Jan-1962, M
Address: 14 Disney Rd, Disneyland, MM1 9DL
Specimen: Swab, FOOT, Right, Requested By: C987654
Location: 5 N
Patients GP: Dr Smith (G123456)
Organism: STAU

Susceptibility: AMP R
                          SXT S
                          CIP S

Data Types

Data types are the basic building blocks used to construct or constrain the contents of each element. Every field, component and subcomponent in HL7 v2 has a defined data type, which governs the information format in the element, what sub-elements it can contain and any vocabulary constraints. Some data types are Simple others are Complex (Fig. 12.3).

HL7 v2 has 89 data types in all, but most applications use only a small number of common datatypes.

Simple data types contain just a single value, while complex data types may contain more than one sub-element, each of which has its own data type. The data type of a component can also be a complex data type. In this case, that component's components are subcomponents of the original data type. No further recursion is allowed.

Complex data types reflect associations of data that belong together, such as the parts of a person's name, address or telephone number, or linking identifiers with their issuing authority.

Simple Data Types

Simple data types include:

DT (date) represents a date in format: YYYYMMDD. For example, 2 August 2008 is represented as 20080802. The months and days

DTM (date/time) is used to represent an event date and time including time zone if required. YYYYMMDDHHMMSS.SSSS+/-ZZZZ where +/-ZZZZ indicates the time zone.

TS (time stamp) has a similar format.

FT (formatted text) allows embedded formatting commands, bracketed by the escape character.

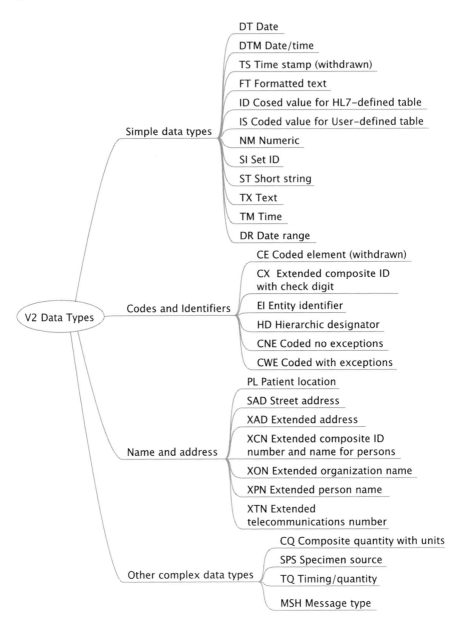

Fig. 12.3 HL7 v2 data types

ID represents a value from a HL7-defined table. Users are not allowed to add their own values.

IS represents a value from a user-defined table.

NM (numeric) is used for numeric values. It may be preceded by a sign and may contain a decimal point.

SI (set ID) gives the order of a segment instance within a message that may contain more than one segment with the same segment ID.

ST (string) is used for short strings up to 200 characters.

TX (text) is used for longer texts up to 64 K characters. In the TX data type the repetition separator (~) is used to indicate a hard carriage return (line break).

Complex Data Types

HL7 v2 supports a variety of complex data types to handle items such as coded values, identifiers, names, addresses and so on. The most commonly used complex data types fall into three broad categories: Codes and Identifiers; Names and Addresses, and Other Complex Data Types.

Codes and identifiers are particularly important in interoperability and HL7 v2 supports both internally defined (by HL7) and externally defined coding schemes. The commonly found complex data types for codes and identifiers include:

CE (coded element) can be used to represent and external code set or a non-coded text value.

CX (extended composite ID with check digit) is used for identifiers, including associated administrative detail such as context and optional check digit information. The CX data type includes the following components, which are optional except where stated (Table 12.11).

Table 12.11 Components of the CX data type

Component	Description
CX.1	ID number (required) – the value of the identifier. This may be alphanumeric, but if so the check digit and check digit scheme are null.
CX.2	Identifier check digit or digits
CX.3	Check digit scheme from HL7 v2 Table 0061 (eg Mod11)
CX.4	Assigning authority that creates the identifier, typically an OID (conditional)
CX.5	Identifier type code (required) from HL7 v2 Table 0203 (eg PN person number))
CX.6	Assigning facility
CX.7	Effective date
CX.8	Expiration date
CX.9	Assigning jurisdiction
CX.10	Assigning agency or department
CX.11	Security check
CX.12	Security check scheme PN

EI (entity identifier) is used to specify identifiers

HD (hierarchic designator) is used to represent a code value or an identifier. It is useful for elements that some systems may treat as a code and other systems may treat as an identifier.

Coding schemes are either internal, which means defined by HL7, or external, which means defined by some other party. There are two main data types, CNE and CWE, which are similar.

CNE (Coded with No Exceptions) is used when a required or mandatory coded field using a specified internal or external coding system must be used and may not be extended with local values. Components are optional unless stated. Table 12.12 shows the components available in the CNE data type.

CWE (Coded With Exceptions) is used when the set of allowable code values may vary on a site-specific basis or no code value is available for transmission, just a

Table 12.12 CNE components, description and version added

Component	Description	Version added
CNE.1	Identifier (required)	2.1
CNE.2	Text name of identifier	2.1
CNE.3	Name of coding scheme, from HL7 v2 Table 0396, or can use CNE.14	2.1
CNE.4	Alternate identifier (NB this should have exactly the same meaning as CNE.1)	2.1
CNE.5	Alternate text	2.1
CNE.6	Name of alternate coding system (table 0396)	2.1
CNE.7	Coding system version ID (this relates to CNE.1-3)	2.3
CNE.8	Alternate coding system version ID	2.3
CNE.9	Original text, this is the text seen by the person who selects the code	2.3
CNE.10	Second alternate identifier	2.7
CNE.11	Second alternate text	2.7
CNE.12	Name of second alternate coding system	2.7
CNE.13	Second alternate coding system version ID	2.7
CNE.14	Coding system OID	2.7
CNE.15	Value set OID	2.7
CNE.16	Value set version ID (this is a date and is required if CNE.15 is used)	2.7
CNE.17	Alternative coding system OID	2.7
CNE.18	Alternative value set OID	2.7
CNE.19	Alternative value set version ID (this is a date required if CNE.18 is used)	2.7
CNE.20	Second alternative coding system OID	2.7
CNE.21	Second alternative value set OID	2.7
CNE.22	Second alternative value set version ID (date required if CNE.21 is used)	2.7

2ext string (CWE.2). CWE is very similar to CNE, but the identifier component (CWE.1) is optional. The CNE and CWE data types have a similar structure:

In most applications only the first three components of coded element are used. However, both CNE and CWE allow the expression of a single concept in two different coding schemes. This may be useful when a sending system holds data using a different coding system than that required by the destination. The original text may also be transmitted.

Names and Addresses

FN (family name) surname.

PL (patient location) within an institution, may include bed, room, ward, floor, building, facility, status and type.

SAD (street address) house number and street.

XAD (extended address) the full location address. Street address (SAD), city, state, postal code, country; also allows start and end dates.

XCN (extended composite ID number and name for persons) is used for clinical staff. This is the largest data type with 23 components, combining the features of both CX and XPN into a single field.

XON (extended organization name) is used for healthcare organizations

XPN (extended person name) is used for patients and their relatives. Includes family name (FN), given name(s), title, suffix, type and date range.

XTN (extended telecommunication number) is for electronic addresses including telephone and email. It includes optional codes for use (eg home or work) and type (eg direct line or mobile)

Other Complex Data Types

CQ (composite quantity) has sub-components quantity and units.

SPS (specimen source) covers information about specimen type, body site, collection method, additives etc.

TQ (timing/quantity) allows the specification of the number, frequency, priority etc. of a service, treatment or test.

MSH (message type) is the data type used for the message type in field MSH-9.

HL7 v2 Tables

Coded values need to be uniquely identified, but there is always the problem that two different coding schemes use the same code value. The solution is to explicitly identify both the coding scheme and the code value. HL7 v2 defines four types of Table.

HL7 Tables are normative content and values provided should be interpreted as being required by the HL7 standard. The ID data type is most often used to encode values for HL7 Tables.

User Defined Tables are not normative content; values where provided are simply suggestions and are not intended to be interpreted as required by the standard. The name of the table is fixed by HL7, but the contents will vary from institution to institution. IS data type is often used to encode values for these tables.

External Tables are sets of coded values defined an published by other organizations. Examples include ICD, SNOMED CT and LOINC.

Local Tables have a non-HL7 assigned table identifier and contain locally or site-defined values.

References

1. Health Level Seven Standard Version 2.8.2 – an application protocol for electronic data exchange in healthcare environments. ANSI/HL7 V2.8.2-2015.

Further Reading

The following books only cover HL7 v2.
Heitmann K, Bloebel B, Dudeck J. HL7 communication standard in medicine, short introduction and information. Cologne: Verlag Alexander Mönch; 1999.
Henderson M. HL7 messaging, version 2, vol. 2. Aubrey: O'Tech Inc; 2007.
Bhagat R, Hui C. HL7 for busy professionals: your no sweat guide to understanding HL7. Anchiove. 2015.

Chapter 13
The HL7 v3 RIM

Abstract HL7 v3 was designed to address most of the problems with v2. It is based on a reference information model (RIM), with six main classes, Act, Entity, Role, ActRelationship, Participation and RoleLink. Structural attributes are used to specify each class in more detail. These include classCode, moodCode and typeCode. Data types in v3 are based on those in v2.

Keywords RIM • Strucural attributes • Act • Entity • Role • Act relationship • Participation • Role link • id • statusCode • negationIndicator • classCode • moodCode • activityTime • effectiveTime • Observation • Procedure • SubstanceAdministration • Supply • PatientEncounter • Person • LivingSubject • Player • Scoper • instanceIdentifier • Universally Unique Identifier (UUID) • ObjectIdentifier (OID) • Code data types • nullFlavor • realCode • TypeId • templateId

The Health Level Seven (HL7) Reference Information Model (RIM) is a static model of health and healthcare information as viewed within the scope of HL7 standards development activities. The RIM is the ultimate source from which all HL7 Version 3 (v3) protocol specification standards draw their information-related content.

Origins of v3

Even its supporters accept that HL7 v2 was developed in an ad hoc and unplanned way. For example, when an additional element is needed, it is added in the next available spot. Perhaps more importantly, v2 provides multiple ways of doing the same thing, leading to the well-founded jibe: *"when you have seen one implementation of v2, you have seen one implementation; every one is different"*.

Work on HL7 v3 began in 1992 with the establishment of the HL7 Version 3 Task Force. As with many things, many of its characteristics are best understood by considering its origins as the planned successor to Version 2. The HL7 web site explains the rationale for v3 as follows:

© Springer-Verlag London 2016

T. Benson, G. Grieve, *Principles of Health Interoperability*,
Health Information Technology Standards, DOI 10.1007/978-3-319-30370-3_13

243

*Offering lots of optionality and thus flexibility, the v2.x series of messages are widely imple-
mented and very successful. These messages evolved over several years using a "bottom-
up" approach that has addressed individual needs through an evolving ad-hoc methodology.
There is neither a consistent view of that data that HL7 moves nor that data's relationship
to other data.*

The success of v2 is largely attributable to its flexibility. It contains many optional
data elements and data segments, making it adaptable to almost any site. While
providing great flexibility, its optionality also makes it impossible to have reliable
conformance tests of any implementation and forces implementers to spend more
time analyzing and planning their interfaces to ensure that both parties are using the
same optional features.

Version 3 addresses these and other issues by using a well-defined methodology
based on a reference information model. Using rigorous analytic and message
building techniques and incorporating more trigger events and message formats
with very little optionality, HL7's primary goal for v3 was to offer a standard that
would be definitive and testable, and provide the ability to certify vendors'
conformance.

HL7 v3 was designed to be comprehensive in scope, complete in detail, extensible as requirements change, up-to-date and model-based, conformance testable
and technology independent. It uses an object-oriented development methodology
and a Reference Information Model (RIM) to create messages. The RIM is an
essential part of the HL7 Version 3 development methodology and provides an
explicit representation of the semantic and lexical connections that exist between
the information carried in the fields of HL7 messages [1].

The RIM was conceived as a universal reference model for healthcare interoperability, covering the entire healthcare domain. Each message specification would be
a view into this common model. The RIM is at the core of HL7 Version 3. You cannot understand v3 without understanding the RIM.

The effort to develop the RIM took place in two distinct phases. During the first
phase, from about 1992 to 1999 a complex class model was developed with more
than a hundred classes and several hundred attributes and associations, supported
by extensive documentation. In many ways, this was a rationalized super-set of the
content of v2. However, many people considered this model to be just too large to
learn and use.

During 1998–1999, a radical approach, known as the Unified Service Action
Model (USAM), was proposed to simplify the problem [2]. After a heated debate,
HL7 resolved to adopt USAM with effect from January 2000. USAM is based on
two key ideas, which lead directly to the structure of the RIM, as we know it today.

The first idea is that most healthcare documentation is concerned with "happenings", in which human and other things participate in various ways. Happenings
have a natural lifecycle such as an intent for it to happen, the event of the happening
and consequences of its happening. These are like the moods of a verb.

The second idea is that the same people and things can perform different roles
when participating different types of happening. For example, in different contexts
the same person can be either a care provider or the subject of care (patient).

Overview

HL7 v3 is a lingua franca used by healthcare computers to talk to other computers, to help provide information when and where needed. Healthcare communication is complex and any language needs to accommodate this complexity and also handle future needs. HL7 v3 is designed to handle most, if not all, healthcare communications in an unambiguous way, using a relatively small set of constructs, which can be learnt relatively easily.

The RIM specifies the grammar of v3 messages, the basic building blocks of the language (nouns, verbs etc.), their permitted relationships and data types. The RIM is not a model of healthcare although it is healthcare specific nor is it a model of any message although it is used in messages.

At first sight the RIM is quite simple. The RIM backbone has a small number of core classes and permitted relationships between them. However it presents quite a steep learning curve. The good news is that, once you reached the plateau, the ground becomes much less steep.

The RIM defines a set of pre-defined attributes for each class and these are the only ones allowed in HL7 messages. Each attribute has a specified data type. These attributes and data types become elements and attributes in HL7 XML messages. (Note that RIM attributes should not be confused with XML attributes). Message specifications, to do a particular task, the message designer uses a sub-set of the available RIM attributes; listing each element used and how many repeats are allowed. This is known as refinement. Each data type is constrained to the simplest structure that meets the requirements of the task.

HL7 v3 uses a graphical representation, called Refined Message Information Model (RMIM) to display the structure of a message as a color-coded diagram (see Chap. 14). Most RMIMs can be shown on a single sheet of paper or PowerPoint slide and these RMIM diagrams are used to design messages and to explain what each HL7 message consists of. The actual interchange (the wire format) is usually XML, validated by schemas.

All of the XML tags and attributes used in v3 messages are derived from the HL7 RIM and the HL7 v3 data types. The structure of each HL7 message is set out in an XML schema, which specifies which tags and attributes are needed or allowed in the message, their order and the number of times each may occur, together with annotations describing how each tag shall be used. HL7 message schemas are lengthy, detailed and verbose. The RIM itself is shown in Fig. 13.1.

The RIM Backbone

The RIM is based on a backbone structure, with three main classes, Act, Role and Entity, linked together using three association classes: ActRelationship, Participation and RoleLink. In the RIM, every happening is an Act, which is analogous to a verb

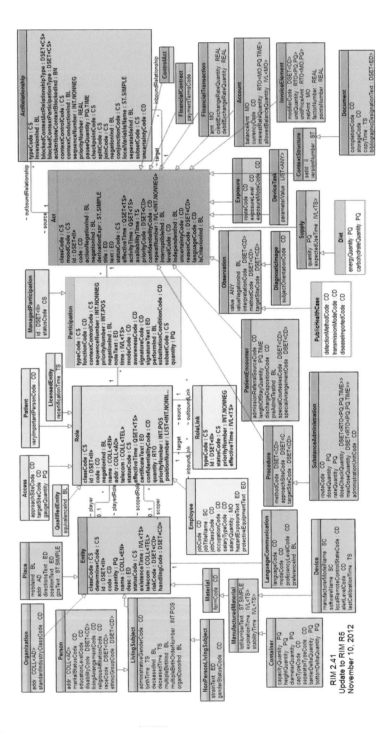

Fig. 13.1 HL7 v3 RIM normative content

in English. Each Act may have any number of Participations, which are Roles, played by Entities. These are analogous to nouns. Each Act may also be related to other Acts, via ActRelationships. Act, Role and Entity classes have a number of specializations. For example, Entity has a specialization called LivingSubject, which itself has a specialization called Person, which inherits the attributes of both Entity and LivingSubject.

Structural attributes provide a way to reduce the size of the original RIM from over 100 classes to a simple backbone of just six main classes. The primary use of structural attributes is in the design of messages or other services, which are then implemented in computer applications. Message designers select the values of structural attributes when designing messages, or groups of related messages. These values are then frozen and may not be changed by application programmers or any-one else down stream. In a very real sense, each class is named by its structural attributes. The semantic meaning of every class in an HL7 v3 message specification is specified by its structural attributes. The actual name of the class does not really matter; what matters is the meaning of its structural attributes.

The structural attributes for each of the backbone RIM classes are shown in Table 13.1:

Each of the main backbone classes (Act, Role and Entity) has a number other classes linked to it, using a line with an open triangle arrowhead at the backbone-class end. This is the UML symbol for specialization. The class that is pointing is a specialization of the class that is being pointed towards, which is a generalization.

The specialization inherits all of the properties of the generalization, while adding any specific attributes of its own. For example, the class Patient, at the top center of the RIM, is a specialization of Role with the addition of the optional attribute veryImportantPersonCode. The convention is that only a class, which has one or more additional attributes specific to it, is shown on the RIM.

Common Attributes

A number of frequently used attributes are found in more than one class. These include: id, code, and statusCode.

Table 13.1 v3 RIM structural attributes

Class	Structural attributes
Act	classCode, moodCode, negationInd, levelCode
Entity	classCode, determinerCode
Role	classCode, negationInd
ActRelationship	typeCode, inversionInd, contextControlCode, ContextConductionInd, negationInd
Participation	typeCode, contextControlCode
RoleLink	typeCode

The conventional way to denote each attribute is to use the XPath notation for classes and attributes, so Act/id is the attribute id in Act, and Role/id is an id in Role. Attributes may be prefixed with to show that id is an attribute.

id is used to identify classes and has the II (Instance Identifier) data type, which may be a universally unique identifier (UUID) or and object identifier (OID). id is used to give unique identity to people, persons, organisations, things and information objects.

There are also two main types of code used in HL7 v3. The first type covers the specialised codes used for structural attributes and are defined by HL7 itself. The second covers externally defined terms and codes such as SNOMED CT (Clinical Terms). While **classCode** is a structural attribute used to indicate the name of an Act, Role or Entity, the **code** attribute is used to specify precisely what the class means at a leaf level of granularity. Unlike classCode, code is not mandatory nor is it a structural attribute.

The classCode and code attributes are related in so far as code should be subsumed by classCode. This means that if Act/classCode is a procedure, then the Act/code should be a type of procedure and may not be anything else. The code attribute is usually populated from an external coding scheme. External coding schemes are identified using an OID. The combination of the OID and code value is unique. Each class may only contain a single code. If it is necessary to apply several attributes to a class, which are best done with code, then each code has to be in a separate class, which must be linked to the parent class using an ActRelationship.

The **statusCode** attribute is used to indicate the current sate of a class according to the appropriate state model. Acts may have statusCodes of new, active, completed, cancelled, aborted and so on. Entities may be active or inactive; Roles may be active, terminated, suspended, pending or cancelled.

The **negationIndicator** attribute is used to reverse the meaning of a class. The main features of the HL7 v3 RIM are summarized in Fig. 13.2.

Act

The Act class represents a record of something that has happened or may happen. Full representation of an Act identifies the kind of act, what happens, the actor who performs the deed, and the objects or subjects (e.g. patients) that the act affects. Additional information may be provided to indicate location (where), time (when), manner (how), together with reasons (why), or motives (what for). Acts can be related to many other Acts using the ActRelationship class. For example one Act may contain, cause, lead to, update, revise or view information about other Acts. Information in a document is treated as an Act – the act being the creation of the document content. Each transaction is a kind of act. An account is a record of a set of transactions.

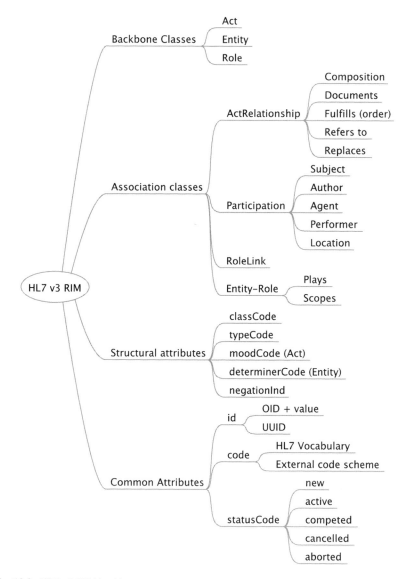

Fig. 13.2 HL7 v3 RIM backbone

Act/classCode is a structural attribute, which specifies whether an Act is an observation, an encounter or a procedure. Acts can include an enormous range of happenings such as:

- events, such as encounter, visits and appointments;
- observations such as tests, diagnoses and examination findings;
- notifications such as alerts, confirmation and consent;
- the supply and administration of medicines and other consumables; clinical, administrative and financial procedures.

Table 13.2 Mood code values and their meanings

EVN	Event (occurrence)	A service that actually happens, may be an ongoing service or a documentation of a past service
RQO	Request	A request or order for a service is an intent directed from a placer (request author) to a fulfiller (service performer)
PRMS	Promise	Intent to perform a service that has the strength of a commitment. Other parties may rely on the originator of such promise that said originator will see to it that the promised act will be fulfilled
PRP	Proposal	A non-mandated intent to perform an act. Used to record intents that are explicitly not Orders
DEF	Definition	Definition of a service

Act has another important structural attribute called **moodCode**, which is similar to the tense of a verb. The term mood is a grammatical term represents a category of verb use typically expressing: fact (indicative mood), command (imperative mood), question (interrogative mood), wish (optative mood), or conditionality (subjunctive mood). The moodCode indicates whether an Act has happened (an event), is a request for something to happen, a goal or a criterion. For example, *weight = 100 kg* is an observation event; *measure weight daily* is a request or order; *reduce weight to 80Kg* is a goal and *if weight is greater than 80Kg* is a criterion. Examples of mood-Code values are shown in Table 13.2.

In orders and observations, moodCode is used to distinguish between an order (RQO), which is something you want to happen, and a report (EVN), which is something that has happened. In clinical guidelines, moodCode distinguishes between the definition of the guideline as originally authored (DEF), the intent that it should be followed for a particular patient (PRP) and compliance (EVN).

statusCode specifies the state of an Act, such as (Table 13.3):
The full state-machine diagram for the Act class is shown in Fig. 13.3.

The Act class has two important time attributes: **activityTime** and **effective-Time**, which have rather different meanings. activityTime states when the Act itself occurs, but effectiveTime states the clinically relevant time of the Act. The difference is best explained by examples.

- The activityTime for an appointment booking is the time of making the appointment, while the effectiveTime is the appointment date/time.
- For a laboratory request, the activityTime is the time the request is made, while the effectiveTime is the time that the sample is requested to be taken (in hospitals, doctors often order blood tests with instructions for the sample is to be taken at some future time).
- For a laboratory test result, the activityTime is the time the test was performed, but the effectiveTime is the time the sample is taken from the patient.
- For a contract, the activityTime is the date of the contract itself, while the effectiveTime is the time that the contract holds good.
- The activityTime for a prescription is the date of the prescription, while the effectiveTime is how long the medication is to be taken.

Table 13.3 Some Act status codes and their meanings

new	Act is in preparatory stages and may not yet be acted upon
active	The Act can be performed or is being performed
completed	An Act that has terminated normally after all its constituents have been performed
cancelled	The act has been abandoned before activation
aborted	The act has been terminated prior to the originally intended completion

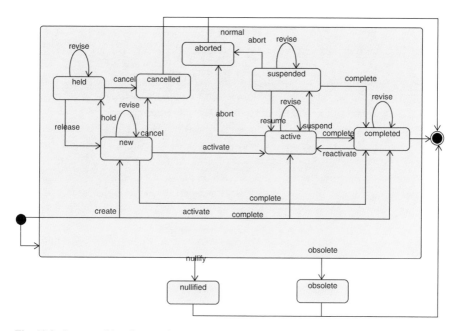

Fig. 13.3 State-machine diagram for Act class

Important Act specializations are Observation, Procedure, SubstanceAdministration, Supply and PatientEncounter.

Observation is defined as an Act of recognizing and noting information about the subject, and whose immediate and primary outcome (post-condition) is new data about a subject. Observations often involve measurement or other elaborate methods of investigation, but may also be simply assertive statements, such as a diagnosis. Many observations are structured as name-value pairs, where the Observation/code (inherited from Act) is the name and the Observation/value is the value of the property. value contains the information determined by the observation action. The value attribute is unique in the RIM in that it has the data type ANY, which is to say it can be any data type. In messages, the data type needs to be constrained to a specific data type, such as physical quantity (PQ) or a code. This works in a way, which is similar to the OBX segment in v2, which can also contain any data type.

In XML representations, the `xsi:type` attribute is used to specify the specific data type used, for example:

```
<value xsi:type="PQ" value="100" unit="mg"/>
```

Procedure is defined as an Act whose immediate and primary outcome (post-condition) is the alteration of the physical condition of the subject. Note that this definition of procedure is rather more limited than the definition of procedure used in SNOMED CT, although it includes most surgical procedures and physical treatment such as physiotherapy. It does not cover imaging or laboratory investigations, administrative procedures, counseling or medication.

SubstanceAdministration is defined as the act of introducing or otherwise applying a substance to the subject. This class is used when prescribing a medicine, with a moodCode of INT (intent), because the intent of a prescription is to administer medication. SubstanceAdministration is also used in EVN (event) mood to record that a medication has been administered to a patient. SubstanceAdministration requires a Participation of Consumable (CSM) to the Role of ManufacturedProduct (MANU) and hence to the Entity Material to identify the material or medicine involved. ManufacturedProduct may also be scoped by the manufacturerOrganization Entity.

Supply is defined as an Act that involves provision of a material by one Entity to another. For example, dispensing a medicine is a Supply, while prescribing and administration are both SubstanceAdministration. Prescriptions are Supply with a moodCode of RQO (request), while medicine administration has moodCode of EVN (event). Supply requires a participation of Product (PRD) to the Role of ManufacturedProduct.

PatientEncounter is defined as an interaction between a patient and a care provider for the purpose of providing healthcare-related services. Examples of PatientEncounter include inpatient and outpatient visits, home visits and even telephone calls. An appointments, which has been booked, is PatientEncounter in the mood PRO (promise) until it has taken place, when the mood is changed to EVN (event) (Fig. 13.4).

Entity

Entity is the second main backbone class in the RIM. An Entity is any living or non-living thing, which has or will have existence. It can also represent a group of things or a category or kind of thing.

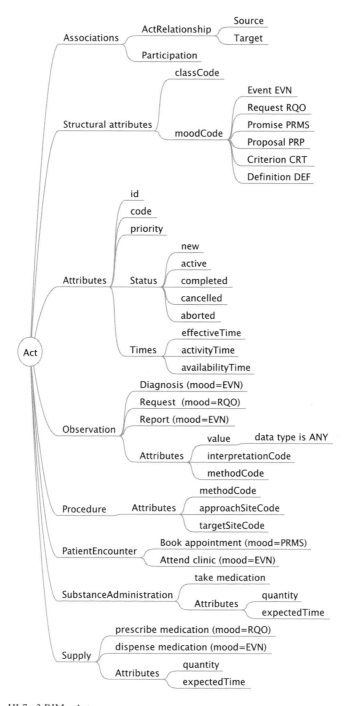

Fig. 13.4 HL7 v3 RIM – Act

Entity covers the whole universe of:

- Living things, such as people, animals plants and microorganisms;
- Non-living things such as places, manufactured items and chemical substances;
- Abstract things such as organizations.

Entity has two structural attributes: classCode, which states what type of thing it represents, and determinerCode, which is used to distinguish between an individual instance such as a person, a collection of instances such as a herd, or the generic class of that entity such as a particular type of micro-organism.

Entity has four main specializations, LivingSubject, Material, Place and Organisation.

Person is a specialization of LivingSubject, which is a specialization of Entity. In XPath notation a Person is:

        Entity/LivingSubject/Person

Person has the attributes inherited from Entity and LivingSubject as well as its own. For example, name is an attribute of Entity, while administrativeGenderCode, birthTime (date of birth) and deceasedTime (date of death) is each an attribute of LivingSubject.

LivingSubject has a second specialization, NonPersonLivingSubject, which is mainly used for veterinary subjects (animals, birds, fishes etc.), but also includes bacteria, plants, fungi etc.

Some attributes of an Entity, are also found in the Role class. These include id, code, name, addr (address), telecom, statusCode and quantity. The primary rule for determining whether to use an Entity attribute or a Role attribute is whether or not an attribute value is permanent. If it is permanent, then use Entity, if it is not permanent, and in particular, if it is related to how a thing is used or what a person does, then it is a Role attribute (Fig. 13.5).

Role

Role is the third main backbone class. Role is defined as a competency of an Entity playing the Role. Roles for people are usually positions or jobs, which they are qualified to do. Roles for inanimate objects such as places and machines are what they are normally used for.

There is also a wide variety of Roles, which can be played by different Entities. Each Entity can play multiple Roles. Examples of Roles include:

- People, such as patient, practitioner or employee
- Places such as hospital, home, clinic or place of birth
- Organisations such as care provider, employer or supplier

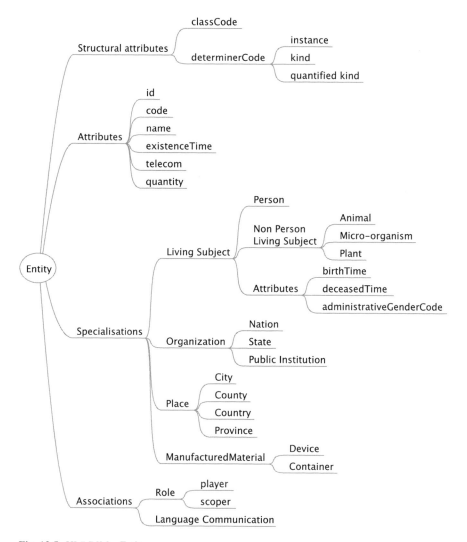

Fig. 13.5 HL7 RIM – Entity

- Things such as drug, instrument or computer system
- Responsible entities, such as parent, employer or manufacturer.

The most important Role specialization is Patient. However, it is important to remember that specializations are only shown explicitly in the RIM when they add additional attributes to the general class.

Patient is defined as a Role of a LivingSubject (player) as a recipient of healthcare services from a healthcareProvider (scoper) (Fig. 13.6).

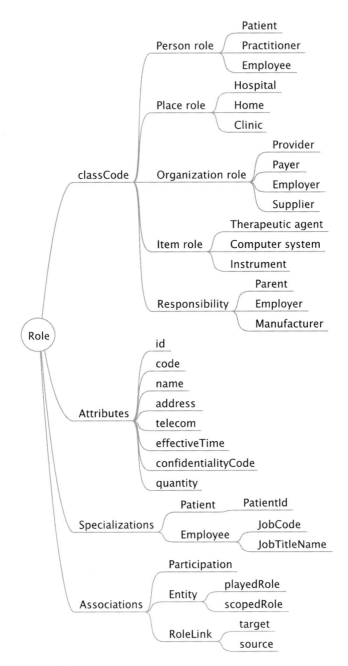

Fig. 13.6 HL7 RIM – Role

Association Classes

This simple backbone structure of Act, Role and Entity is sufficiently flexible to cover almost anything you may want to say. However, we also need explicit connectors between each of these classes.

ActRelationship is a relationship between two Acts, used to link Acts together, from a source Act to a target Act. There are various types of link, including composition, documentation, fulfillment etc. Every ActRelationship has a source and a target to which it points. An Act can have any number of ActRelationships, which may be organised as a hierarchy.

ActRelationship/typeCode describes the type of association between Acts:

- composition comprises (COMP) entries
- discharge summary documents (DOC) a hospital visit
- test report fulfills (FLFS) a test request
- discharge summary refers (REF) to a referral
- final report replaces (REPL) a preliminary report.

ActRelationship has two additional structural attributes, which are not always present but have important effects. The inversionInd attribute indicates that the direction of the relationship is inverted. The contextConductionInd attribute indicates whether context is conveyed through the relationship.

Participation defines the involvement of an Entity, in a particular Role, functions during the scope of an Act. Participants take part in Acts as either actors or targets in the Act. Actors do things, while targets are essentially passive. Participation is specific to a single Act. When the Act stops the Participation ceases. A particular Role can participate in an Act in many ways. Thus, a person in the role of surgeon may participate as primary surgeon or as assistant surgeon. Participation/typeCode describes the type of association between an Act and each participating Role:

- performer (PERF) such as surgeons, observers, practitioners
- subject (SUBJ) such as the patient
- location (LOC)
- author (AUT)
- informant (INF)
- responsible party (RESP)
- information recipient (IRCP).

Participation/contextControlCode is used to indicate how this participation changes the context.

RoleLink is a relationship between two Roles. It provides a simple way of linking Roles together such as between jobs in an organisation chart, family members or between members of a medical team.

Entity may have only two associations with Role: **player** and **scoper**. Entities may either play a Role directly, or may provide the scope for a Role. For example, Dr Smith plays the Role of doctor, but this Role may be scoped by the organization she works for, such as St Mary's Hospital. Similarly, the scope link may also be used to note the manufacturer of a medicine. An Entity may perform any number of Roles, but each instance of Role is only played by a single Entity.

V3 Data Types

In v3, data types have a similar role as in Version 2, providing fine detail. These are also the basis of the international standard ISO 21090:2011, Health Informatics - Harmonized data types for Information Interchange (see Fig. 13.7).

Basic Data Types

Basic data types are shown in Table 13.4.

Instance Identifier

The Instance Identifier (II) data type has two main flavors: UUIDs and OID-based identifier.

Universally Unique Identifiers (UUID) are software-generated identifiers, created on the fly to identify information artefacts uniquely. UUIDs are usually used when the identifier in question is generated by a software application without human assistance.

UUIDs are 16-byte (128 bit) numbers. The number of theoretically possible is large (more than 10 followed by 37 zeroes). The standard way of displaying a UUID is as 32 hexadecimal digits, displayed in 5 groups separated by hyphens in the form 8-4-4-4-12, such as:

```
550e8400-e29b-41d4-a716-446655440000
```

Object Identifier (OID) The second type of identifier is that held in some type of register. Here we use an identifier for the register itself and each item, which is registered is allocated an identifier that is unique within that register. The convention is HL7 is to use an OID (object identifier) to identify the register itself. The combination of an OID and a value is intended to be globally unique.

An OID is a node in a hierarchical tree structure, with the left-most number representing the root and the right-most number representing a leaf. Each branch under the root corresponds to an assigning authority. Each of these assigning authorities may, in turn, designate its own set of assigning authorities that work under its auspices, and so on down the line. Eventually, one of these authorities assigns a unique

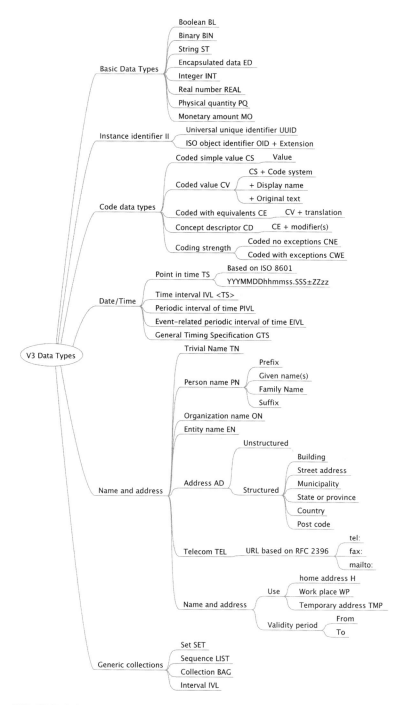

Fig. 13.7 HL7 v3 data types

Table 13.4 Basic v3 data types

BL	Boolean has only two possible values *true* or *false*
BIN	Binary
ST	Character String – unformatted text string
ED	Encapsulated Data – text data that is primarily intended to be read by a human. ED can include format information.
INT	Integer number – any positive or negative integer or zero
REAL	Real number
PQ	Physical Quantity – a measure quantity with units. Units should be coded using the Unified Code for Units of Measure (UCUM)
MO	Money – a currency amount

(to it as an assigning authority) number that corresponds to a leaf node on the tree. The leaf may represent an assigning authority (in which case the root OID identifies the authority), or an instance of an object. An assigning authority owns a namespace, consisting of its sub-tree. While most owners of an OID will "design" their namespace sub-tree in some meaningful way, there is no generally applicable way to infer any meaning on the parts of an OID.

HL7 has its own OID 2.16.840.1.113883 (iso.country.us.organization.hl7) and maintains an OID registry with around 3000 nodes. One way to obtain an OID in the UK is to use the company registration number. All companies registered in England and Wales may append their company registration number to the 1.2.826.0 root to obtain an OID that is unique to the company without further formality or charge. The hierarchy is:

$$\text{Top of OID tree}$$
$$1 - \text{ISO assigned OIDs}$$
$$1.2 - \text{ISO member body}$$
$$1.2.826 - \text{Great Britain} \left(\text{GB / UK} \right)$$
$$1.2.826.0 - \text{UK National registration}$$

So, for example, 1.2.826.0.1.9116995 means R-Outcomes Ltd.

Code Data Types

HL7 Version 3 has four code data types (CS, CV, CO, CE and CD) listed in increasing order of complexity:

CS (Coded Simple) is just a simple `code` value, optionally accompanied by a `displayName`, which is an aid to human interpretation. CS does not include any coding scheme identifier and is used only for codes that are defined by HL7 itself, such as structural attributes, realm, language and status codes.

CV (Coded Value) has a `code` value and a `codeSystem` identifier to identify an externally defined coding scheme, that is, any code not defined by HL7 itself. The `codeSystem` may have a `codeSystemName` as an aid to human interpretation. It may also have an optional `displayName` and/or an `<originalText>` element containing the text that was originally written. CV can be qualified as being CNE (coded no exceptions) or CWE (coded with exceptions).

CO (Coded Ordinal) has the property that codes are ordered, such as the stages of a disease (such as cancer stages, Stage I, Stage II, Stage III etc.)

CE (Coded with Equivalents) may be though of as an extension of CV, which allows a term to be coded in more than one way. It includes the original `code` (such as a local code used in the sending system) to be sent along with a `<translation>` element using the type of code required by the receiving system, which may have coarser granularity.

CD (Concept Descriptor) is the most complex code data type, providing the functionality of CE as well as `<qualifier>` elements to enable post-coordinated expressions to be exchanged. `<qualifier>` elements are made up of name-value pairs, where `<name>` is the type of relationship and `<value>` is the value of the qualifier. For example, the term 'compression fracture of neck of femur' can be represented as a post-coordinated SNOMED CT expression using compositional grammar as follows:

```
71620000|fracture of femur|: 116676008|associated morphology|=219
47006|compression fracture|,363698007|finding site|=29627003 |structure of neck of femur|
```

This expression can be represented in using the CD data type as:

```
<code code="71620000" codeSystem="2.16.840.1.113883.6.96"
    codeSystemName="SNOMED CT"
    displayName="fracture of femur">          OBJ
  <qualifier>                                    ATR
                                                 VAL
    <name code="363698007" displayName="finding site"
        codeSystem="2.16.840.1.113883.6.96" />
    <value code="29627003" displayName="structure of neck of
    femur"/>
  </qualifier>
  <qualifier>
    <name code="116676008" displayName="associated morphology"/>
    <value code="21947006" displayName="compression fracture"/>
  </qualifier>
</code>
```

Dates and Times

Dates and times are represented by a hierarchy of data types:

TS Point in Time
IVL <TS> Interval of Time
PIVL Periodic interval of time
EIVL Event-related periodic interval of time
GTS General Timing Specification

The time format is similar to that used in Version 2, based on ISO 8601. For example, YYYYMMDDhhmmss.SSS±ZZzz, where

YYYY is the year (always include the century)
MM is the month
DD is day in the month
hh is the hour (24 h clock)
mm is the minute in the hour
ss is second in the minute
.SSS is fraction of a second,
± is direction of offset from Universal Coordinated Time (UTC)
ZZ is hours offset from UTC
zz is minutes offset from UTC.

Most practical needs are met by TS, which may specify either a date or date/time. Interval of Time may be expressed with start and end dates/times, or as a duration or as an open range, with only the start or end date specified. The more complex specifications were developed to meet the potential requirements of complex medication regimes.

Name and Address

The data types used in v3 for names and addresses are similar to those used in Version 2. Each name or address can be structured or unstructured and may include codes to specify its type and use and a date range for validity dates.

Types of name and address include:

TN Trivial name (unstructured).
PN Person name in a sequence of name parts such as family name, given name(s), prefix and suffix, together with name use (legal, maiden name, former name, alias).
ON Organisation name.
EN Entity name (any thing).
AD Postal Address as a sequence of address parts, such as house, street, city, postal code, country and the address use (home, temporary etc.).

TEL Telecommunication Address is specified as a Universal Resource Locator (URL), which covers telephone numbers (voice, fax or mobile) e-mail addresses and web pages.

Generic Collections

Multiple repeats can be specified in four ways:

SET is an unordered collection of items without any repeats.
LIST is an ordered collection of items in a defined sequence (repeats are allowed).
BAG is an unordered collection of items (repeats allowed).
IVL is a range of values such as time or physical quantity.

Special Fields

All classes in the RIM are regarded as being specializations of the Infrastructure Root, which has four optional fields, which can be used in any RIM class or clone to support special communications needs. These special fields are:

nullFlavor When valued in an instance, this attribute signals that the class instance is null, and that the remainder of the information for this class and its properties will not be communicated. The value of this attribute specifies the flavor of null that is intended. HL7 v3 has no less than eleven different ways of saying that a value is unknown, but only the following are widely used: no information (NI), not applicable (NA), other (OTH), and unknown (UNK).

realmCode signals the imposition of geographical-specific constraints. The value of this attribute identifies the realm in question.

typeId identifies the type of HL7-specified message type or CMET, to which this message or part of a message conforms. Any fragment of an HL7 document conforms to one type only.

templateId This attribute signals the imposition of a set of template-defined constraints to an instance. Each template is an identified set of business rules. The value of templateId provides a unique identifier for the template in question. Any fragment of a document can be constrained by any number of templates, so multiple templates can be applied to any class.

Use of the RIM

The classes, attributes, state-machines and relationships in the RIM are used to derive domain-specific information models that are then transformed through a series of refinements to yield a static model of the information content of an HL7 standard.

The HL7 v3 standard development process, known as the HL7 Development Framework (HDF) defines the rules governing the derivation of domain information models from the RIM and the refinement of those models into HL7 standard specifications. The rules require that all information structures in derived models be traceable back to the RIM and that their semantic and related business rules do not conflict with those specified in the RIM.

The abstract style of the RIM and the ability to extend the RIM through vocabulary specifications make the RIM applicable to any conceivable healthcare information interchange scenario. Universal applicability makes the RIM particularly useful for an organization like HL7 that has to consider the needs of a large and diverse membership. The style of the RIM makes it very stable, which is another important characteristic for HL7.

The HL7 standards development process calls for the creation of domain specific models derived from the RIM and the incremental refinement of those models into design models that are specific to the problem area. These problem-area-specific design models narrow the abstractness of the RIM and include constraints on attribute values and class relationships that are use case specific.

In summary, the RIM has six backbone classes: Act, Act Relationship, Participation, Role, Role Link and Entity. The meaning of each class is determined by one or more structural attributes, such as classCode and moodCode. Each class has a predefined set of possible attributes and may have specializations, which provide additional attributes for specialised classes. Each attribute has a data type.

References

1. HL7. HL7 Standards – HL7 Version 3. www.hl7.org
2. Russler DC, Schadow G, Mead C, Snyder T, Quade LM, McDonald CJ. Influences of the Unified Service Action Model on the HL7 Reference Information Model. Proc AMIA Symp 1999:930. American Medical Informatics Association.

Chapter 14
Constrained Information Models

Abstract HL7 v3 works by constraining the RIM for specific use cases. Several types of constrained models are defined (DMIM, RMIM, HMD, MT and CMET). These all use common types of constraint, including omission, cloning, multiplicities, constrained data types and code binding. HL7 has developed a special graphical notation for specifying constrained information models. The clinical statement pattern is a common pattern used for clinical information in various profiles. Implementation Technology Specifications (ITS) describe how information is expressed as XML on the wire.

Keywords Constrained information models • DMIM • RMIM • HMD • Message type • CMET • Cloning • Multiplicities • Constrained data types • Code binding • Clinical statement pattern • Implementation Technology Specification (ITS) • XML • Documentation

Types of Model

A central idea of the HL7 V3 approach is to limit optionality by constraining or refining a general model for the specific use case being considered. This idea of constraining a general model to create an agreed subset and interpretation of the specification is widespread in the standards world. Constrained specifications are called profiles.

Many standards have multiple optional aspects and if different suppliers do not implement the same subset they will fail to interoperate. The use of profiles is a way to enforce a particular interpretation to ensure interoperability.

Constrained information models create a tree-like hierarchy of possible models. At the root of HL7 V3 lies the RIM. Everything else is a constraint on the RIM.

The following types of constrained model are recognised within HL7 V3, starting with the broadest, proceeding to the narrowest (Fig. 14.1).

DMIM   Domain Message Information Model
RMIM   Refined Message Information Model
HMD   Hierarchical Message Description

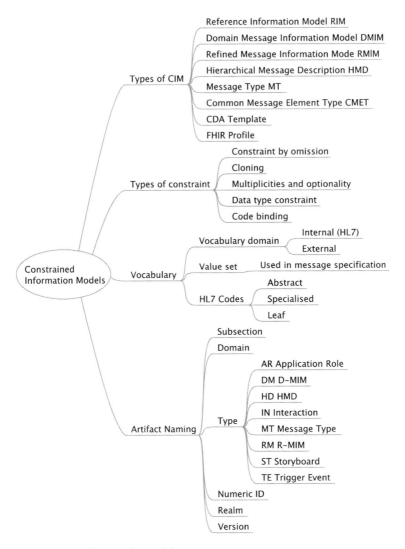

Fig. 14.1 Constrained information models

MT   Message Type
CMET   Common Message Element Type

DMIM (Domain Message Information Model) is a general model of a domain in
HL7 notation from which a related family of message specifications can be derived.
DMIMs have been defined for many subject areas. A DMIM may be created top-
down from domain experience or bottom-up as a superset of messages in a domain.
Once created a DMIM can be used a reference from which further messages may be
defined. DMIMs do not have a hierarchical structure and cannot be serialized. A
DMIM cannot be implemented as it is but needs to be further constrained as RMIMs.

The primary purpose of a DMIM is to provide a common point of reference to ensure compatibility between all artefacts, such as RMIMs in the same domain. Not all projects use DMIMs.

RMIM (Refined Message Information Model) is the most widely used constrained information model and may be thought of as a diagram of a message specification. RMIMs and DMIMs use the same notation. One important difference is that an RMIM has only one point of entry and can be expressed in a serialized format. Serialization is essential if a message is to be transmitted as a string of bits over a wire.

HMD (Hierarchical Message Description) is effectively an RMIM expressed in a tabular format. HMDs and RMIMs can contain the same information, but most people find that graphical RMIMs are easier to use and understand.

Message Type (MT) is a specific specification of a message, which can be used in a data interchange. Any one RMIM or HMD can be further constrained to create a set of closely related message types, which are then exchanged as a linear string of XML and validated using an XML schema.

Common Message Element Types (CMET) are reusable modules, which can be used in multiple messages, rather like a program sub-routine. Using CMETs can speed up the process of developing messages and increase consistency between different specifications.

Each CMET has two parts. The CMET reference is a special class, which can be added to an RMIM. When a CMET is referenced, or used in another diagram, it is shown with a special notation, a box with dashed edges, which contains the name of the CMET, its artefact id, classCode and level of attribution. This box is color-coded in a manner consistent with its root class. Each CMET has a unique artefact identifier (beginning with COCT_), which is the primary link between each CMET reference and its content.

The CMET content itself is defined as a small RMIM, which is stored in the CMET library, which is designed for common use by any HL7 committee. Relevant CMETs are included automatically in messages when they are constructed.

Each CMET has a single entry point, which is the point at which it is attached to any containing message, which references it. CMETs do not have exit points, which means they have to be at the terminal or leaf point in the hierarchical structure of a message.

CDA Templates and FHIR Profiles are also types of constrained model, described further in Chap. 15/CDA Templates and Chap. 21/Profiling Resources.

Types of Constraint

The RIM, DMIMs and RMIMs can be constrained by omission, cloning, multiplicity, optionality, data type constraint and code binding

The simplest form of constraint is by **omission**. Classes or attributes with classes are simply left out. All classes and all attributes (apart from structural attributes) in the RIM are optional, so you only use the ones you need.

The same RIM class can be re-used many times in different ways in DMIMs and RMIMs. This process is referred to as **cloning** and the classes selected for use in constrained models are referred to as clones. The idea is that you take a clone of a class from the RIM and constrain that clone in the constrained information model. Cloning limits the number of classes that need to be defined in the RIM, leading to a small stable RIM. The name of each cloned class in an RMIM is derived from its structural attributes. For example, a test request is represented in the HL7 V3 RIM as an observation request, so its structural attributes are classCode=OBS (observation) and moodCode=RQO (request or order).

The next form of constraint is to constrain **multiplicities** in terms of repeatability and optionality. Most associations and attributes in the RIM are optional and allow any number of repeats. These can be constrained by making such multiplicities non-repeatable mandatory (1..1) if you need one and only one; or non-repeatable optional (0..1) if you have any you can only have one.

In HL7 Version 3 specifications, the correct verb form for indicating a requirement is SHALL. The verb form for indicating a recommendation is SHOULD. The verb form for an option is MAY. Terminology used in standardization does not recognize the term 'must' and SHALL is always used to indicate a mandatory aspect on which there is no option. The negatives are SHALL NOT, SHOULD NOT, and MAY NOT.

The next type of constraint involves **constraining data types**. The HL7 V3 data types have been designed with a hierarchical structure. For example there are four code data types: CS (code simple), CV (coded value), CE (code with equivalents) and CD (concept descriptor) in increasing order of complexity. A more complex data type, such as CD can be constrained to a simple data type such as CV. Similarly the data type GTS (General Time Specification) can be constrained to IVL<TS>(Time Interval) or to TS (Timestamp). Data types can be further constrained to create data type flavors. For example the TS data type could be constrained to a date (TS.date) or year (TS.year).

The final type of constraint involves **code binding** – specifying what code value sets shall be used. The coding strength of a code may also be restricted to CNE (Coded No Exceptions) or may be specified as CWE (Coded With Exceptions). This may all sound quite complex but is simpler and more intuitive than it sounds. The simple rule is that you only specify what you need, leave out everything else or make it a simple as possible.

Vocabulary and Value Sets

The HL7 V3 standards talk about *vocabulary domains* and *value sets* and it is important to understand the difference between them.

A **value set** is the set of codes that may be used to populate a specific attribute in a message instance. The message designer usually specifies value sets. A value set may be a single code only, for example to specify a structural attribute, a subset of an HL7 defined code, or all or part of an externally defined coding system.

A **vocabulary domain** is the set of codes available to the message designer for a specific attribute. For example, the vocabulary domain for `Act.moodCode` is the set of all `moodCode` values defined and maintained by HL7.

Message users and implementers are concerned with value sets, while message designers need to think about vocabulary domains and select appropriate value sets from these.

The concept of vocabulary domains is most applicable to HL7's own internally defined vocabulary tables, which are quite extensive. These must be used for structural attributes and are widely used within data types. Each HL7-defined concept normally has a mnemonic code which is the code value used, a print name which explains its meaning, a concept ID used for internal reference, a level and type. Mnemonic codes have to be unique within a particular coding scheme. These tables have a hierarchical structure, with each concept being allocated a level, so a level 2 concept is the child of the preceding level 1 concept and so on. The code type may be:

- Abstract (A) which does not have a code but does contain child concepts
- Specialised (S) which has a code and contains child concepts
- Leaf (L) which has a code but no child concepts.

Artefact Names

HL7 V3 artefacts are identified using a common naming scheme, which is at first sight a bit complex. The format is

```
SSDD_AAnnnnnnRRVV
```

The first four characters identify the subsections and domains.

COCT   Common Message Elements

COMT   Common Message Content
FIAB   Accounting & Billing
FICR   Claims & Reimbursement
MCAI   Message Act Infrastructure
MCCI   Message Control Infrastructure

MFMI   Master File Management Infrastructure
POLB   Laboratory
PORX   Pharmacy
PRPA   Patient Administration
PRPM   Personnel Management
PRSC   Scheduling
QUQI   Query Infrastructure
RCMR   Medical Records

This is followed by an underscore character "_" and then the artefact type, identified with a two-character acronym.

AR   Application Role
DM   D-MIM (Domain Message Information Model)
DO   Domain
EX   Example
HD   HMD (Hierarchical Message Descriptor)
IN   Interaction
MT   Message Type
NC   Narrative Content
RM   R-MIM (Refined Message Implementation Model)
ST   Storyboard
ST   Storyboard Narrative
TE   Trigger Event

The artefact type is followed by a six digit identifier allocated by the committer responsible. The final characters are a 2-character Realm Code, where the identifying which international affiliate of HL7 is responsible for this. The default is UV (Universal) followed by a version number in the range (00–99).

For example: PRPA_RM001234UV00 may be interpreted as Patient Administration RMIM, with identifier 001234, used universally, revision 00. It is worth taking the trouble to memorize the main acronyms.

A Simple Example

Figure 14.2 shows a simple RMIM for an investigation report.

Every RMIM has an entry point, which states its name, Demo Report in this case, identifier and any descriptive notes that the author has provided.

The entry point or focal class, pointed at by the arrow is ObservationEvent. This is the default name for any act with classCode=OBS (observation) and moodCode=EVN (event). This has three other attributes: a unique identifier id (such as a UUID), code that states the type of report and effectiveTime, which refers to the date/time of the observation.

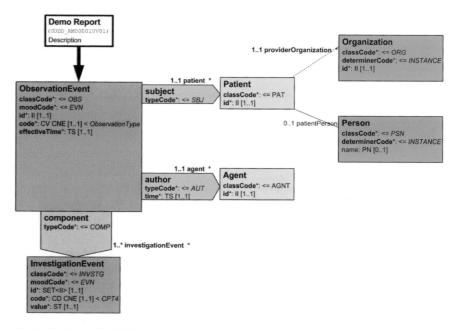

Fig. 14.2 Simple RMIM for an investigation report

This report contains one or more InvestigationEvent (classCode=INVSTG, moodCode=EVN), each of which has an id (such as a UUID or a line number), a code such as a CPT4 code to indicate what it is, and a value, which is a simple text string (ST).

The report has two participations: subject and author. The way to read the Participations is that the ObservationEvent has subject Patient and has author Agent.

The subject is a Patient, with an id, such as a hospital number. The Patient is scoped by an Organization, which has an agreed identifier (id). The combination of the Organization/id and the Patient/id should be globally unique.

The patient has an optional name, in the Person class (Entity). The playing association (patientPerson) between Person and Patient is indicated as [0..1], and is not in bold font, indicating that this is not mandatory. Similarly the name attribute in Person is not in bold font and is annotated as [0..1].

The author is an Agent, which could be a clinician, technician or a machine. The Agent has a unique identifier (id).

In this RMIM all elements are mandatory (and therefore required), which is why they are all written in bold font and suffixed with the "*" indicator.

R-MIM Notation

HL7 uses a special graphical notation for specifying RMIMs and DMIMs.

Each Act is represented as a red rectangle
Role as a yellow rectangle
Entity as a green rectangle
ActRelationship is usually shown as pink (salmon) arrow-shaped pentagon
Participation as a cyan (light blue) pentagon
RoleLink as a light yellow pentagon.

Each of the arrow-shaped pentagons has a source for the relationship and a target. The direction of the arrow indicates the meaning of the association, but this is not always the way that the diagram should be navigated. The direction of navigation, the way you read the diagram, is indicated by the location of the multiplicity shown just outside the class. This may sound confusing, but the important thing to remember is that the direction of the arrows is not always the way that the diagram should be read.

ActRelationship and RoleLink may be recursive, that is, each may point back to itself. This is indicated by a "pig's ear" box with a notched out corner which fits around one corner the Act or Role.

Each attribute uses exactly the same attribute name as is in the RIM; they cannot be changed. The attributes selected for use in RMIMs are formed by constraining or limiting the attributes as defined in the RIM. This allows checking and validation and is the key reason why the RIM may not be changed.

The attribute name in an RMIM diagram may be in **bold** print. This indicates that this attribute is mandatory, it must always be present, null values are not allowed. This is a responsibility of the sender Application Role.

The attribute name may have a star '*' next to it. This indicates that this attribute is required to be present in messages. If data is not available a 'null' value may be sent.

The multiplicity or cardinality of the attribute is denoted within square brackets [ ] to indicate how many times this attribute may be repeated. [0..1] indicates zero or one; [1..1] indicates exactly one. '*' indicates no upper limit, so [0..*] indicates zero to many.

The attribute's data type is specified after the attribute name, separated by a colon ':'. The specified data type must be either the same as or a valid constraint on the RIM data type for that attribute.

If the data type is a code, then the coding strength may be denoted by adding either CNE (coded no exceptions) or CWE (coded with exceptions) after the data type designator.

The value set or vocabulary domain to be used with each attribute is specified after either an '<=' or '=' symbol. '<=' indicates that the value may be taken from a vocabulary domain or the code specified or any of its descendants in a hierarchy. The equals sign indicates that the value should be as specified. The domain specification must be either a domain name defined in the vocabulary tables, or a single code value from the appropriate domain.

A string in quotes (e.g. "string") indicates a default value for this attribute.

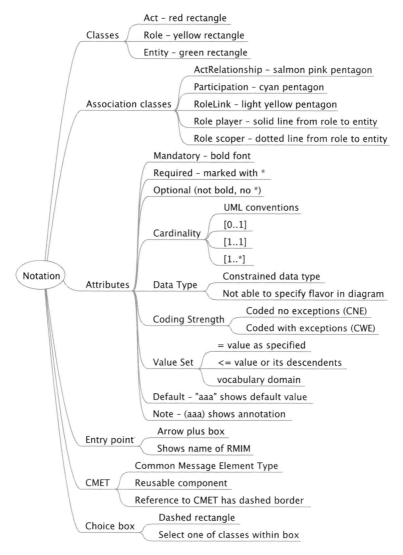

Fig. 14.3 HL7 V3 diagram notation

Finally, a brief description of attributes may be included, enclosed within parentheses, for example (description).

If the attribute information extends beyond one line, then second and subsequent lines are indented.

Choice Box is used in HL7 RMIMs to show alternative options. Each of the options is shown in a box with a dashed line border, from which a single choice is made. Associations may be made either to a specific class within the choice box, or to the outside border of the choice box, in which case that association applies irrespective of choice is selected (Fig. 14.3).

Tooling

RMIMs are built using a special tool-set developed by HL7. The original tools, based on Microsoft Access and Visio are in the process of being replaced by a new generation of tools, which use a slightly modified notation.

The basis of these tools is a set of inter-related XML schema, known as Model Interchange Format (MIF). MIF defines the primary artefacts that can be developed or exchanged as a result of HL7 V3 standards development and implementation.

Templates

HL7 templates are used to constrain and verify conformance to profiled HL7 Version 3 Refined Message Information Models (RMIMs). A template is an expression of a set of constraints on the RIM, which is used to apply additional constraints to a portion of an instance of data expressed in terms of some other Static Model. Templates are used to further define and refine these existing models within a narrower and more focused scope.

Each template is identified with a `templateId`, a globally unique identifier. Templates are used widely in CDA (see Chap. 9).

Clinical Statement Pattern

The HL7 Version 3 Clinical Statement is as a common pattern, which is used for the development of all types of clinical messages. For example, it is used in CDA Release 2 Level 3, for the exchange of complete electronic patient records between GPs, and for highly structured messages such as prescriptions and test reports.

HL7 defines a Clinical Statement as:

> An expression of a discrete item of clinical (or clinically related) information that is recorded because of its relevance to the care of a patient. Clinical information is fractal in nature and therefore the extent and detail conveyed in a single statement may vary.

Any clinical statement may have a number of participants, including subject, author, location, performer, participant and informer.

At the center of the clinical statement pattern is a choice box (ActChoice). A clinical statement to have any of the following specializations:

Observation covers a very broad range of statements relating to history, examination, tests, diagnosis and prognosis. Depending on the value of the moodCode, an observation can be an actual observation (mood=Event), a requested observation (mood=Request) or a goal set for a future observation (mood=Goal). Observation Events are usually reported using code-value pairs, where the code represents what is being observed and value represents the result. Observations may have child observations. For example a blood count may consist of sub-observations. The Observation class may also be linked to a specimen and to normal range values.

Procedure may refer to a specimen(s) or images and is used for all invasive procedures including surgical procedures and imaging. Procedures can have associated observations. The moodCode is used to distinguish between those that have happened (mood=Event) and those that are planned or proposed. Procedures in HL7 are clinical and exclude administrative events such as admissions, clinic appointments, which are encounters (below).

Encounter which covers most administrative procedures involving an interaction between a patient and a healthcare provider for the purpose of providing a healthcare service. Encounter includes admissions, discharges, transfers of care, appointment scheduling and waiting list management.

Substance Administration may refer to products such as medication mainly used for prescribing and administration of drugs. Depending on the value of the moodCode it can be used for requesting, recommending or administration of medicines. Both substance administration and supply (below) are associated with a product, material or substance.

Supply is mainly used for dispensing drugs or other medical supplies. This can support precise identification of the actual product supplied, such as manufacturer, batch or serial number, using the HL7 Common Product Model.

Organizer is a specialization of the act class designed to support grouping information into clusters or batteries. For example, the components of a full blood count is typically ordered and reported together.

Act is a generic class, which is used if none of the above apply; it is rarely used.

Several types of associations between clinical statements are provided such as containment, cause and effect, problem linkage.

Relationships Between Entries

The Clinical Statement pattern allows for a rich set of relationships between entries, to reflect the structure of clinical information and links between different items. The main relationships are direct, with a source and target, containment and association.

Examples of the types of ActRelationships frequently found in clinical statements include:

CAUS shows that the source caused the target, such as substance administration (e.g. penicillin) caused an observation (e.g. a rash), or observation (e.g. diabetes mellitus is the cause of kidney disease).

COMP is used to show that the target is a component of the source (e.g. hemoglobin measurement is a component of a full blood count).

GEVL (evaluates (goal)) links an observation (intent or actual) to a goal to indicate that the observation evaluates the goal (for instance, a source observation of *walking distance* evaluates a target goal of *adequate walking distance*).

MFST (is manifestation of) is used to say that the source is a manifestation of the target (for instance, source *hives* is a manifestation of target *penicillin allergy*).

RSON (has reason) shows the reason or rationale for a service (for instance, source *treadmill test* has reason *chest pain*).

SAS (starts after start) shows that the source Act starts after the start of the target Act (for instance, source *diaphoresis* starts after the start of target *chest pain*).

SPRT (has support) shows that the target provides supporting evidence of the source (for instance, source *possible lung tumor* has support target *mass seen on chest X-ray*).

HL7 Development Framework

HL7 Development Framework (HDF) describes the methodology for developing HL7 V3 standards [1]. The HDF is written for HL7 members who are developing standards within HL7 committees. However much of what it says is of universal relevance. The HDF adopts a project-oriented approach, based on a product life cycle with the following stages.

The **Project Initiation Process** (PIP) includes initiation, planning, and approval sub-stages, including the development of a detailed Project Scope Statement (PSS) and plan. The project plan identifies the business case and objectives, participants including sponsor committee, project leader, contributors and early implementers and a time schedule.

Domain Analysis Process (DAP) includes analysis and requirements documentation, including the development of a **Domain Analysis Model** (DAM), which includes:

- Business context including documentation using storyboards and identification of relevant actors and interactions

- Use case analysis documenting use cases and actors
- Process model, documented using activity diagrams
- Information model, documented using classes and attributes
- Business rules including trigger events
- Glossary

Specification Design Process (SDP) is the core of the process. It involves mapping the requirements as set out in the Domain Analysis Model to the HL7 RIM, data types and vocabulary to specify the message structures, value sets and dynamic processes.

Profiles

A profile is a set of information used to document system requirements or capabilities from an information exchange perspective and is expressed in terms of constraints, extensions, or other alterations to a referenced standard or another profile. Profiles of HL7 Version 3 are derived from a Version 3 specification, as balloted either by HL7 or by one of its affiliates.

The categories and use of profiles include annotation, constraint, localization implementable and conformance profiles.

Annotation Profiles document the standard exactly but with more information to further explain the base document to educate prospective users and/or implementers.

Constraint Profiles may contain unchanged and constrained elements, reducing the optionality and cardinality of the base specification (i.e., the HL7 V3 standard) in order to make the specification more exact.

Localization Profiles meet the same objectives as a constraint profile, with the addition of some additional elements (extensions). HL7 Version 3 allows localization of some parts of the standard but not others. In particular, HL7 does not allow anyone, apart from HL7 itself through a formal process, to change or modify the RIM or any of the Data Types. Localization can make full use of the constraint mechanisms and make certain changes to RMIMs, Data Types, Message Types, CMETs and Vocabularies.

Implementable Profiles are the most constrained constraint profiles and eliminate all optionality in the base specification (the HL7 V3 standard) in order to make the specification exact and approach plug-and-play interoperability. Optionality is eliminated when the conformance indicator for every attribute and association is either `Required` or `Not Permitted` and every vocabulary domain is bound to a value set.

A **conformance profile** indicates the set of interactions that a computer system (or application role) supports. It implies a commitment to fulfill all of the responsibilities of the interactions specified and to implement faithfully the artefacts that constitute the interactions and any further constraints or extensions. Conformance statements set out a computer system's conformance claim to a set of interactions.

Implementation Technology Specification (ITS)

The XML implementation technology specification describes how individual instances of message types shall be rendered in XML for serial transmission over the network and the structure of schemas used to validate each instance. Note that the HL7 generated XML schemas are not able to test all of the constraints defined in a HL7 message definition.

The generation of schemas and message representation is done automatically. Those not directly involved in that process do not need to understand the technical details. The key points are as follows:

- One XML element is defined to correspond to each attribute or association in the RMIM, with the exception of structural attributes, which are expressed using XML attributes.
- Each data type has a defined XML representation. The 'restriction base' feature in an XML schema is used extensively to define how data types are implemented.
- Schema files for CMETs are supplied separately and then used by each message schema as required.
- V3 data types and data type refinement use the W3C schema restriction element. Additional standard schema sections support RIM classes and the HL7-defined vocabulary definitions. These schema sections can be selectively combined with a specific message schema through the include function in the XML schema standard.
- HL7 messages share the same XML namespace. Message version information is conveyed as attributes within the message rather than by changes to the namespace identifier (Fig. 14.4).

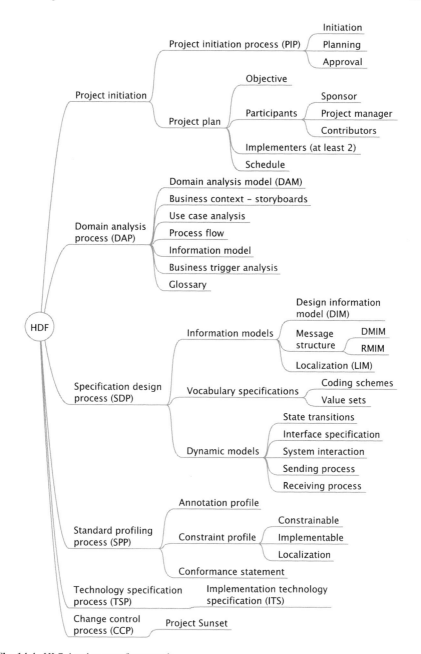

Fig. 14.4 HL7 development framework

Documentation

HL7 V3 documentation is voluminous. The full set of HL7 V3 standards is published annually and can be downloaded from the HL7 web site (www.hl7.org).
Foundation documents are the basis of the standard.

Note to Readers (7300 words)

Core Principles and Properties of HL7 Version 3 Models (17,200 words)

Refinement, Constraint and Localisation (14,600 words)
Reference Information Model (42,000 words)

Data Types – Abstract Specification (82,300 words)

XML Implementation Technology Specification – Data Types (43,600 words)
Vocabulary (6000 words)

HL7 Common Terminology Services (26,200 words)

Using SNOMED CT in HL7 Version 3: Implementation Guide (36,700 words)

HL7 Development Framework – HDF (21,300 words)
Specification and Use of Reusable Constraint Templates (17,500 words)
Glossary (24,900 words)

In addition HL7 has produced a wide range of domain specific standards covering:

Accounting and Billing
Blood, Tissue and Organ
Care Provision
Claims and Reimbursement
Clinical Decision Support
Clinical Document Architecture (CDA)
Clinical Genomics
Clinical Statement
Common Message Element Types (CMET)
Immunization
Laboratory
Master File/Registry
Medical Records
Medication
Message Control
Observations
Orders
Patient Administration
Personnel Management
Pharmacy

Public Health Reporting
Query
Regulated Products
Regulated Studies
Scheduling
Shared Messages
Specimen
Therapeutic Devices
Transmission

Reference

1. HL7 Development Framework. Version 1.3, 2009

Chapter 15
CDA – Clinical Document Architecture

Abstract CDA (clinical document architecture) is the most widely adopted implementation of HL7 v3. It is used for exchanging information in the form of documents. CDA has three levels: level 1 is a single human-readable document, level 2 can include multiple documents and level 3 can include structured information. Each CDA document has a common header and a variable body part. Templates are used to constrain the generic CDA model. The continuity of care document (CCD) is used to summarize a patients' health record. Consolidated CDA (C-CDA) is a harmonized set of templates for a variety of clinical documents.

Keywords CDA • Document paradigm • CDA levels • CDA releases • Header • Patient • Author • Steward • Relationships • Body • Section • Clinical statement • Template • Continuity of care document (CCD) • Consolidated CDA (C-CDA)

The Document Paradigm

Clinical Document Architecture (CDA) is the most widely adopted application of HL7 V3. The CDA paradigm takes the document metaphor seriously. It is illuminating to compare the differences between the database and document metaphors.

Databases are organised for rapid search and retrieval and are updated by transactions. The database structure is designed by a computer professional, updated by various people, who may or may not know each other, and is accessed by others using queries. The person who updates the database has little or no control over who, if anyone, will ever read the data, or for what purpose. The person who uses the database may know nothing about who entered the data and in what context. The lack of context makes it hard to evaluate whether or not you can rely on it.

In contrast, a document, electronic or paper, is organized as a stand-alone artefact to convey human understanding. Each document has a set of metadata stating the contextual details of who created it, for whom, when, where and about what subject. The author determines the content of the document, attests its veracity and can be held responsible for any errors. If readers have doubts about how to interpret it, they can contact the author requesting elaboration [1,2].

© Springer-Verlag London 2016   283
T. Benson, G. Grieve, *Principles of Health Interoperability*,
Health Information Technology Standards, DOI 10.1007/978-3-319-30370-3_15

We are all familiar with documents such as letters and invoices, although we seldom notice just how much information is contained in each. Even a simple letter contains a lot of fixed information that has little to do with the actual message content. This serves to provide context and verification data; it allows us to trust that it is what it purports to be, and lets us crosscheck its validity if we are suspicious.

On a typical letter from a company, there are usually three types of data: fixed, variable and case-specific. Fixed data on the letterhead includes the company name, logo, address, phone, fax, email and web addresses; it may also include the company's registered name and address, registration number and place of registration, tax identifiers, bank information and lists of directors or partners. Variable data is always present on a particular class of document, but may vary in every instance. On a letter, this may include the date, a reference, the recipient name and address and the author's name, position and signature. Case-specific data is the variable content of the letter, which contains the core of the message.

All documents share a number of properties, which are not shared with databases or transactions used to update database records. These properties include: persistence, stewardship, potential for authentication, wholeness and human readability (Fig. 15.1).

Persistence is a feature of documents. Every document has a life cycle; it is created, used and eventually destroyed (perhaps many years later). While it exists, it remains a single coherent whole. On the other hand, information in a relational database may be distributed across the rows of many tables. Different people may be authorized to update the different tables and after numerous updates, it may be impossible to recreate the information as it was originally, without sophisticated rollback processes.

Stewardship is another document property. At any time some person or organisation is responsible for looking after it. It is usually clear who is responsible for filing, copying, forwarding or destroying a document. Organizations invariably keep copies of all documents they send out. Again, this is not true of a database, where different rules may apply to different types of data.

Documents have the **potential for authentication**. They can be signed. Authentication is much simpler with documents than with database records. It is relatively easy to maintain an audit trail for the whole life cycle. Each document may be signed, physically or electronically. Validity can be attested in ways that are difficult to replicate with data base records. Only authenticated documents are likely to be of value in medico-legal disputes.

Wholeness each document is complete and whole in itself, including context information, such as who created it, when, where and for what purpose. This makes it easier for others to use it outside the immediate purpose for which it was created. Without strong evidence on the original context, it can be hazardous to place meaning on any statement.

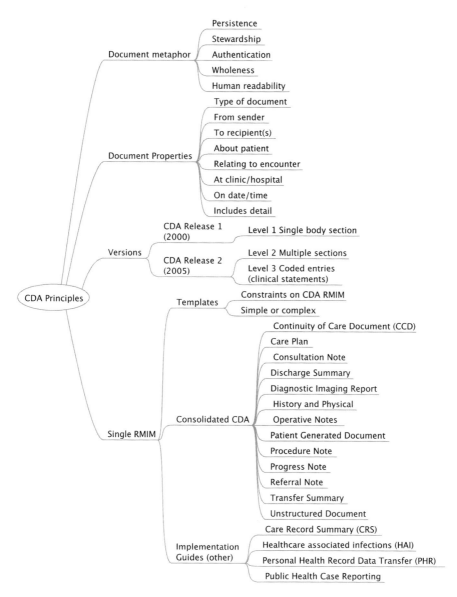

Fig. 15.1 CDA principles

Documents are **human-readable**. The meaning understood by a human reader is paramount, even when there is coded machine-readable information within the same statement. Human-readable messages have a long-term value (medical records may need to be preserved for 100 years or more), whereas machine-readable data depends on specific technology, which may not be available many years in the future. For example, few modern computers can read floppy disks or magnetic tapes that were ubiquitous a couple of decades ago.

The need for long-term human-readable persistence was one of the motivations that led to the development of XML and its predecessor SGML. XML is a simplification of SGML, which was used in highly complex documents, such as the technical documentation for aircraft and military equipment, which need to be rendered in a variety of different formats. The analogy between complex instruction manuals and medical records was not lost. Both are voluminous and difficult to navigate. A key feature of XML is that very instance of an XML file is referred to as a document, which is perhaps one of the reasons why people began to think seriously about the document metaphor.

CDA History

In 1997, the year that XML became an official standard, a group of people with an interest in both HL7 and SGML/XML met at the Kona Mansion in New Hampshire, where they set out a route map for using XML in medical documents, along with the emerging HL7 Version 3 reference model. This became known as the Kona Proposal, which set out a plan for three levels of document structure, which became known as Clinical Document Architecture (CDA).

The three levels are:

CDA Level 1 has a header and a human-readable body. The header contains basic meta-data, primarily intended to support information retrieval, while the body is a human-readable narrative or image. For example, the body can be a PDF document, a jpeg image or a text document, possibly containing simple formatting markup.

CDA Level 2 has the same header as Level 1, but allows the body to be either an unstructured blob (enabling compatibility with Level 1) or comprise any number of sections, which may be nested. Each section contains a narrative block, which in a form that is suitable for rendering in human-readable form.

CDA Level 3 allows each section to include structured machine-processable data in addition to a narrative block. Since its release, CDA Level 3 has become popular although it is more complex than Levels 1 and 2, because it offers the benefits of both human-readable and machine-processed documents. Machine-processed data is encoded using the HL7 V3 Clinical Statement pattern.

The CDA standard has evolved in a series of Releases:

CDA Release 1 published in 2000, is a simple standard, describing a header and body [3]. The header is based on the HL7 V3 RIM; while the body supports a variety of human-readable non-XML formats such as text or images. It specifies the structure of CDA Level 1 and Level 2 and is published as ANSI/HL7 CDA R1-2000.

CDA Release 2 published in 2005, is more complex and both the header and the body are based on the HL7 V3 RIM, allowing fine granularity of structured data [4,5]. The body may be non-XML (providing backward compatibility to Release 1)

Table 15.1 Relationships between CDA Release 1 and CDA Release 2

	CDA Release One (R1)	CDA Release Two (R2)
Date	2000	2005
Level 1	CDA R1 Level 1	CDA R2 Level 1
Level 2	CDA R1 Level 2	CDA R2 Level 2
Level 3	Not available	CDA R2 Level 3

or may be organised into one or more sections, which may have structured entries. It is published as ANSI/HL7 CDA-R2 2005.

When referring to any CDA implementation it is important to state clearly both the release and level that applies, such as CDA R2 Level 3 (Table 15.1).

One of the attractive features of CDA is that it lets you start simply, with Level 1 or 2, and then evolve over time. The lower levels of CDA provide low technical barriers to adoption, while providing a migration route towards more complex structured coded records. CDA can be deployed easily to enable web-based access to patient data in human-readable format. Health is a long-term business and records and documents need to be kept safely and be accessible many years into the future.

All levels validate against the same generic CDA XML schema. There is just one CDA schema for all clinical documents. Different types of clinical document all use the same base schema. Templates on the generic CDA schema provide additional validation and constraints.

CDA is at the core of many standards-based health information exchange architectures internationally. A key to its widespread acceptance is the "A" for architecture in CDA, which promotes reusability across a sufficiently wide range of documents to cover clinical information sharing, public health, quality reporting, and clinical trials.

Institutions like Mayo Clinic and the NHS in England have committed to CDA because it provides a single architectural foundation for their clinical information requirements that can be sustained over generations of application development. CDA lets you share information now without sacrificing scalability or reuse in the future.

All CDA documents have a Header and a Body.

Header

The CDA header is common to all three levels of CDA. The primary purpose of the header is to provide structured metadata about the document itself, which can be used in document registers and databases to classify, find and retrieve documents. These metadata include information about what the document is, who created it, when, where and for what purpose (see Fig. 15.2)

The root class of all CDA documents is an HL7 V3 Act called `ClinicalDocument`.

The following example shows a fragment of a header instance. Each line is discussed in detail below.

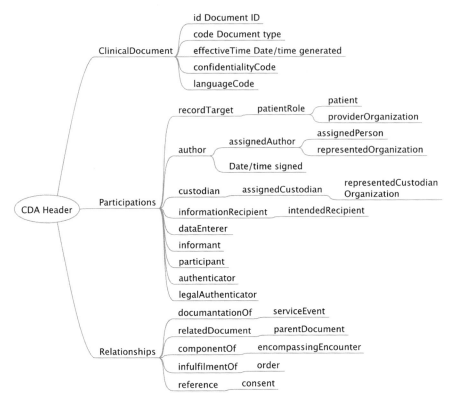

Fig. 15.2 HL7 CDA header

```
<ClinicalDocument xmlns:="urn:hl7-org:v3">
        <typeIdextension="POCD_HD000040" root="2.16.840.1.113883.1.3"/>
        <templateId root="2.16.840.1.113883.10.20.3"/>
        <templateId root="2.16.840.1.113883.10.20.16.2"/>
        <id root="db734647-fc99-424c-a864-7e3cda82e703"/>
        <code code="11488-4" codeSystem="2.16.840.1.113883.6.1"
            codeSystemName="LOINC"
            displayName="Consultation note"/>
        <title>Consultation note</title>
        <effectiveTime value="20091029224411"/>
        <confidentialityCode code="N" codeSystem="2.16.840.1.113883.
        5.25"/>
        <languageCode code="en-GB"/>
        ...
</ClinicalDocument>
```

<ClinicalDocument> is the root XML element of a CDA document. The
namespace for a CDA R2 document is xmlns:cda= "urn:hl7-org:v3".

The <typeId> on the next line is a technology neutral reference to the CDA
Release 2 specification in which POCD_HD000040 is the HL7 V3 artefact identi-

fier for the CDA R2 hierarchical description and `root="2.16.840.1.113883.1.3"` is the OID for HL7 registered models.

The `<templateId>` lines reference one or more templates that have been assigned a unique OID identifier. Use of the templateId indicates that the CDA instance conforms to the constraints specified by each templateId referenced as well as to the standard CDA schema. In the above example:

`<templateId root="2.16.840.1.113883.10.20.3"/>` indicates conformance with HL7 CDA general header constraints.

`<templateId root="2.16.840.1.113883.10.20.16.2"/>` indicates conformance with the HL7 Discharge Summary DSTU (draft summary for trial use).

`<id>` represents the unique identifier (UID) of a clinical document instance. The `<id>` element uniquely distinguishes this document from all other documents. This allows documents to move among systems without risk of ID collision. Often, the document instance `<id>` is implemented as a UUID, created on the fly, that does not require an extension, e.g.

```
<id root="db734647-fc99-424c-a864-7e3cda82e703"/>
```

The `<code>` element in the document header specifies the kind of document that is being created. The value set is usually drawn from an externally specified code scheme, which places no limit on the potential number of different types of CDA document that could be used.

```
<code code="11488-4" codeSystem="2.16.840.1.113883.6.1"
codeSystemName="LOINC" displayName="Consultation note"/>
```

However the `<code>` element is used for two separate jobs, which require different levels of detail. First it provides the most detailed coded specification of what the document is, which encourages fine detailed granularity (e.g. final discharge summary). The second more important role is to facilitate information retrieval, which tends to encourage less restrictive codes (e.g. clinical correspondance). It is always important to find and review all applicable records and not miss any important information and so information retrieval considerations point to using less restrictive codes. One potential solution is to use some form of hierarchical coding scheme but such schemes seldom provide a satisfactory way of doing more than one job. The CDA standard recommends LOINC (see Chap. 14) as the preferred vocabulary for `<code>`, but LOINC is not hierarchical.

The `<title>` element is a human-readable name for the document. This should always be included in the human-readable rendering of a CDA document.

`<effectiveTime>` signifies the document creation time, when the document first came into being. Dates and times are coded as local times using the HL7 adoption of ISO8601 (YYYYMMDDhhmmss±ZZzz).

```
<effectiveTime value="20091029224411-0500"/>
```

`<confidentialityCode>` is another mandatory attribute in the root class. It defines the overall confidentiality status of the document and has a default value of N for normal.

Example - normal confidentiality

```
<confidentialityCode code="N" displayName="Normal"
codeSystem="2.16.840.1.113883.5.25" codeSystemName="Confiden
tiality"/>
```

`<languageCode>` is represented in the form nn, or nn-CC. The nn portion is an ISO-639-1 language code in lower case; the CC portion, if present, is an ISO-3166 country code in upper case, e.g. en-GB.

```
<languageCode code="en-GB"/>
```

Every clinical document has at least three participations: patient, author and steward.

Patient

The `<recordTarget>` represents the patient who is the subject of the document. All CDA documents refer to a patient. The data required usually includes patient ID, name (given and family names), gender and date of birth. The `<providerOrga-nization>` is normally the organization responsible for issuing the patient ID.

```
<recordTarget>
        <patientRole>
                <id extension="12345" root="2.16.840.1.113883.3.933"/>
                <patient>
                        <name>
                                <given>Henry</given>
                                <family>Levin</family>
                        </name>
                        <administrativeGenderCode code="M"
                        codeSystem="2.16.840.1.113883.5.1"/>
                        <birthTime value="19320924"/>
                </patient>
                <providerOrganization>
                        <id extension="M345" root="2.16.840.1.113883.3
                        .933"/>
                </providerOrganization>
        </patientRole>
</recordTarget>
```

Author

The <author> element represents the human or machine that authored the document. In CDA, <author> contains a mandatory child element <time>, which is
the date or date/time that the document was originally written.

  <assignedAuthor> represents the author's role and typically includes the
author ID. The author could also be a machine or device, rather than a person, in
which case the ID may be based on the serial number.

  <assignedPerson> typically includes the author's name.

  <representedOrganization> identifies the author's institution or
department.

```
<author>
    <time value="20000407"/>
    <assignedAuthor>
        <id extension="KP00017"root="2.16.840.1.113883.3.933"/>
        <assignedPerson>
            <name>
                <given>Robert</given>
                <family>Dolin</family>
                <suffix>MD</suffix>
            </name>
        </assignedPerson>
        <representedOrganization>
            <id extension="M345" root="2.16.840.1.113883.3.933"/>
        </representedOrganization>
    </assignedAuthor>
</author>
```

Steward

The steward <custodian> is the organisation that keeps a permanent copy of the
record. The custodian may be the document originator, a health information
exchange, or other responsible party.

```
<custodian>
    <assignedCustodian>
        <representedCustodianOrganization>
            <id extension="M345" root="2.16.840.1.113883.3.933"/>
            <name>Good Health Clinic</name>
        </representedCustodianOrganization>
    </assignedCustodian>
</custodian>
```

Other Participants

Other optional participations may also be specified including:

`<informationRecipient>` is the intended recipient of the information.
`<dataEnterer>` is typically a transcriber.
`<informant>` is the source of the information.
`<participant>` is any other participant.
`<legalAuthenticator>` is the person who takes legal responsibility for signing a document on behalf of an organization. `<authenticator>` is any other signatory. CDA documents can support electronic signatures.

Relationships

A clinical document may have relationships with events or documents via ActRelationships. These may include information about related parent documents, specific services performed, the patient encounter, orders related to the document and patient consents.

Parent Document CDA documents can be linked using the `relatedDocument` ActRelationship to the `parentDocument`. A CDA document may replace, append or transform a parent document. Only one related document can be referenced.

Service Event CDA documents can be `documentationOf` a `serviceEvent`, such as an operation, and may include start and end times and details of the performer(s). This is used to record additional details of the context of the event, including details of the times and visit details, the responsible doctor, other participants, and location details.

```
<documentationOf>
    <serviceEvent classCode="PCPR">
        <code code="xxx" codeSystem="xxx" codeSystemName="xxx"
        displayName="xxx"/>
        <effectiveTime>
            <low value="20110501"/>
            <high value="20110501"/>
        </effectiveTime>
        <performer typeCode="PRF">
            <functionCode code="PCP"
            codeSystem="2.16.840.1.113883.5.88"/>
            <assignedEntity>
                <id extension="xxxx" root="xxxx"/>
                <code code="xxxx" codeSystem="xxxx"
```

```
                       codeSystemName="xxxx" displayName="xxxx"/>
                       <assignedPerson>
                             <name>
                                 <prefix>xxxx</prefix>
                                 <given>xxxx</given>
                                 <family>xxxx</family>
                                 <suffix>xxxx</suffix>
                             </name>
                       </assignedPerson>
                   </assignedEntity>
               </performer>
           </serviceEvent>
</documentationOf>
```

Encounter CDA documents may be thought of as a componentOf an encompassingEncounter, which may have location at a healthCareFacility and a responsibleParty, such as the specialist in charge and encouterParticipants who may admit, discharge or refer the patient.

Order CDA documents may be produced inFulfillmentOf one or more orders.

Consent CDA documents may include a reference to authorization of consent.

Body

Every CDA document has one header and one body part (see Fig. 15.3).

The body is either a <nonXMLBody> or a <structuredBody>. <nonXMLBody> is present to provide upward compatibility with CDA Level 1, and may contain any type of human-readable data including text (txt, rtf, html or pdf) or image (gif, jpeg, png, tiff or g3fax). Data encoded using XML may not be put in the <nonXMLBody>.

```
<component>
     <nonXMLBody mimeType="text/plain">
           <text> Text goes here </text>
     </nonXMLBody>
</component>
```

<structuredBody> is used for XML-encoded data. It is the root node for one or more <section>.

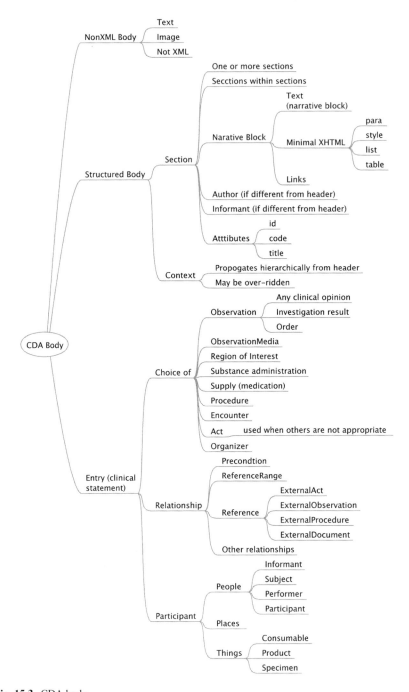

Fig. 15.3 CDA body

```
<component>
    <structuredBody>
        <section>
            <code code="xxxx" codeSystem="xxxx"
            codeSystemName="xxxx" displayName="xxxx"/>/>
            <title>xxxx</title>
            <text>
            xxxx<content styleCode="xxxx">xxxx</content>xxxx
            </text>
        </section>
    </structuredBody>
</component>
```

Section

Each <section> contains a human-readable narrative block, called <section/text>. This narrative block is one of the key components of CDA and contains the human-readable content of the <section>. One of the responsibilities of the originator of any CDA document is to ensure that the narrative block of each <section> accurately conveys the meaning of that <section> in a way that can be rendered appropriately for human readability.

<section/text> can include special XML mark-up, which is similar to but simpler than XHTML. However, relatively few documents use mark-up within narrative blocks other than <paragraph> and
 (line break), although this might change.

The original vision of CDA was for each <section> to comprise either a whole document content or at least a significant chunk, such as a composition or section as defined in EN 13606. However, many implementations of CDA, such as the NHS Care Record Service have chosen to implement Sections with rather fine granularity, so each Section is no more than a single line or entry.

Fine-grained sections allow sections to be filtered, sorted and rendered in different ways, chronologically, or by author, or by record type, for example, to display all allergies, diagnoses or medication records. Sections may contain sub-sections within them, although this is not common, because it adds to the processing complexity. Sections usually share the same context data as that found in the header, although this can be over-ridden for each specific section, although this is not common.

Clinical Statement

Each section can contain any number of entries, which are Clinical Statements, in a structured computer processable form. The Clinical Statement Pattern is described in Chap. 14.

Any clinical statement may inherit context information from the CDA header or context information may be defined within the clinical statement, in which case it over-rides the default inherited data. Examples of participations, which can be applied to any clinical statement, are `subject`, `author`, `performer`, `informant`, `location` and `participant`.

CDA Templates

CDA templates are used to specify how CDA is to be used for particular purposes and specific use cases.

A CDA template is an expression of a set of constraints on CDA, which apply additional constraints to a portion of an instance of data. Templates are used in a variety of different ways.

Narrative this can be used to reference an implementation guide or pattern e.g. *A valid legal authenticator must be provided.*

Schematron assertions e.g. *legalAuthenticator and not legalAuthenticator[@ nullFlavor]*

Static Model (RMIM) Publish a new static model making `legalAuthenticator` mandatory

The CCD specification (see next section) provides an exemplar of how the `templateId` can be used to reference a template or implementation guide that has been assigned a unique identifier. The following example shows how to formally assert the use of this implementation guide. Use of the `templateId` indicates that the CDA instance not only conforms to the CDA specification, but in addition, conforms to constraints specified in this implementation guide.

```
<ClinicalDocument xmlns:='urn:hl7-org:v3'>
    <typeId extension="POCD_HD000040"
        root='2.16.840.1.113883.1.3'/>
    <templateId root='2.16.840.1.113883.10.20.1'/>
    ...
</ClinicalDocument>
```

In addition to assigning a template identifier to the overall implementation guide, template identifiers can be assigned to other patterns, such as document sections and specific clinical statements within document sections. Using the templateId to reference one of these patterns indicates that the CDA instance conforms to the constraints specified in that pattern.

```
<Section>
    <templateId root='2.16.840.1.113883.10.20.1.14'/>
    …

    <Observation classCode="OBS" moodCode="EVN">
        <templateId root='2.16.840.1.113883.10.20.1.32'/>
        …

    </Observation>
</Section>
```

Templates allow constraints to be applied to all or any part of a CDA document including the roles (author or patient details), sections (such as business headings) and entries (such as clinical statements), to say exactly how each is to be specified with a narrower and more focused scope. CDA templates may be simple or quite complex.

Multiple CDA templates can constrain the same portion of a CDA document specification. A CDA Profile is a set of templates that correspond to a particular document type. A template list provides a set of templates which provide choices for the user.

CDA templates have a templateId and may be stored in a repository. The templateId is one of the hidden attributes of the HL7 RIM, which can be used in all RIM classes. The templateId is used to indicate which template is being used; it is useful in document validation, software and human-readable specifications. Validators use templateId to check that a document complies with the rules specified in the template; computer software uses templateId to indicate how this part of a document should be used. Humans use the templateId to reference how each part of the specification is to be used. A templateId may be an OID, a UUID or a locally specified identifier.

Each template has a set of metadata to describe the purpose and use of the template. The metadata includes a globally unique identifier, a name, description, version, an identifier of the model from which it is derived, the RIM version and publication details. The use of standard metadata allows templates to be stored in repositories, which can be queried and the templates shared.

Currently most CDA template constraints have been implemented in Schematron, and are used primarily for validating CDA document instances.

Continuity of Care Document (CCD)

CCD (Continuity of Care Document) maps the functionality of the ASTM Continuity of Care Record (CCR) – also referred to as ASTM E2369-05 [6] – into HL7 V3 CDA format.

Both CCD and CCR can be used to summarize the most relevant facts about a patient's healthcare, covering one or more encounters. They provide a means for one practitioner, system, or setting to aggregate all of the pertinent data about a patient and forward it to another practitioner, system or setting to support the continuity of care. The primary use case is to provide a snapshot in time containing the pertinent clinical, demographic, and administrative data for a specific patient.

CCD sets out a set of constraints on CDA, using templates. Although the stated purpose of CCD is to communicate clinical summaries, it is increasingly being used as a framework for developing other types of message.

One way of looking at CCD is to consider it as a set of templates, because all parts are optional and it is practical to mix and match the ones you need. All sections of a CCD document are optional and may be combined together in any way. CCD is the semantic equivalent of a CCR – both are in XML and both adhere to ANSI-based specifications. Implementers may choose either one or the other standard as the primary data format, but cannot mix them. CCD has been endorsed by HIMSS and HITSP as the recommended standard for exchange of electronic exchange of components of health information [7] (Fig. 15.4).

Each instance of a CCD document is identified by a universally unique identifier (UUID) generated by the originating system. The use of UUID is mandated as the simplest way to ensure that each document generated is unique and will be understood to be unique by any receiving system.

A separate originator document ID may be provided, which is a human readable identifier for a document, used by the originator. It is not guaranteed to be universally unique, although no such identifier should be used which is known to contain any duplicates. It should be used only in combination with an identified patient and a specific date and time of document creation.

The date/time created represents the exact clock time that the document is created and must include a time zone offset. It is not the time that the document is sent.

Document type is the type or title of the document, e.g. Discharge Summary. Ideally coded, but may be just text. CCD documents should also have a title.

Each CCD refers to a single patient or subject, who is the person seeking to receive, receiving, or having received care. It can only refer to one patient. Examples are a treated patient, a client of a physiotherapist, each particular member of a target population for screening, each particular member of a group of people with diabetes attending a session of medical education. Each patient has one or more identifiers. For example, patients may have different identifiers in different units. Patients may also have a full set of demographic details (name, date of birth, sex, address etc.)

A CCD always has at least one source. Multiple sources may be specified when it is useful to specify the person(s), organization and or system responsible for generating the document.

There may be any number of intended recipients (and copy recipients) of any document. This is optional, because some documents do not have an explicit recipient. The recipient may be any party (person or organization), including the patient.

Each CCD document may have one primary purpose, which is the reason that a clinical document is generated, such as patient admission, transfer, consult/referral, or inpatient discharge. It may be associated with an indication (text or code) and a relevant date/time. Each document may also reference prior documents.

Each CCD document contains several items of technical metadata. Metadata is information about the document, which is used to support electronic processing and information retrieval.

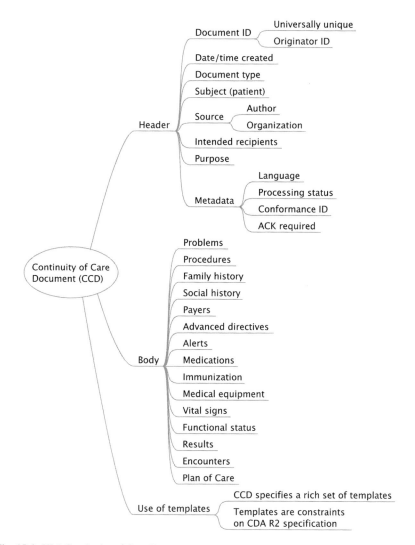

Fig. 15.4 HL7 Continuity of Care Document

Language code in the form nn, or nn-CC (e.g en or en-US). The nn portion shall be a legal ISO-639-1 language code in lower case. The CC portion, if present, shall be an ISO-3166 country code in upper case.

Processing status indicates whether a document is being used in production, testing or training.

Conformance ID is a unique identifier, which identifies the specific version of the clinical document to which conformance is claimed, such as an XML Schema or Schematron.

ACKrequired specifies the circumstances under which acknowledgement of receipt and or processing is required (always, on success or on error).

CCD Body

The CCD Body contains sections corresponding to the main sections of the CCR.

The *Problems* section provides a problem list of current and historical clinical problems.

The *Procedures* section includes surgical, diagnostic, or therapeutic procedures or treatments pertinent to the patient.

Family history describes relevant family history.

Social history includes administrative data such as marital status, race, ethnicity and religious affiliation as well as information about the patient's occupation, lifestyle, social, environmental history and health risk factors.

For each *payer*, all the pertinent data needed to contact, bill to, and collect from that payer should be included as well as authorization details.

Advance directives document the existence of living wills, healthcare proxies, and resuscitation status.

Alerts describe allergies, adverse reactions and alerts related to current or past medical history.

The *medications* section lists the patient's current medications and medication history.

Immunization lists current immunization status and immunization history.

Medical equipment includes both durable medical equipment and implanted devices.

Vital signs may include the most recent, maximum and/or minimum, or both, baseline, or relevant trends.

Functional status contains information on the "normal functioning" of the patient at the time the record is created and provides an extensive list of examples. Deviation from normal and limitations and improvements should be included here.

The *results section* contains the results of observations, including abnormal values or relevant trends, generated by laboratories, imaging procedures, and other procedures.

Encounter lists healthcare encounters pertinent to the patient's current health status or historical health history.

Plan of care contains active, incomplete, or pending orders, appointments, referrals, procedures, services, or any other pending event of clinical significance to the current and ongoing care of the patient. The plan of care section also contains information regarding goals and clinical reminders.

Consolidated CDA

Consolidated CDA (C-CDA) is the recommended standard for exchanging key clinical information within the Meaningful Use Stage 2 (MU2), in the context of the 2014 CEHRT (certified electronic health record technology) requirements. The referenced release is *HL7 Implementation Guide for CDA Release 2: Integrating Health Enterprise (IHE) Health Story Consolidation, Draft Standard for Trial Use (DSTU) Release 1.1 - US Realm* (CCDA R1.1) [8]. The standard is also accompanied by a Companion Guide [9]. Consolidated CDA contains several different types of commonly used CDA documents:

- Continuity of Care Document (CCD)
- Care Plan
- Consultation Note
- Discharge Summary
- Diagnostic Imaging Report (DIR)
- History and Physical Note (H&P)
- Operative Note
- Patient Generated Document
- Procedure Note
- Progress Note
- Referral Note
- Transfer Summary
- Unstructured Document (not used for MU2 requirements)

A key objective of Consolidated CDA has been to bring together in one place all of the CDA templates, schema and other guidance needed for MU2 compliance, to provide a single source of truth, using common templates across different documents, to minimise ambiguity and simplify implementation.

References

1. Benson T. The message is the medium. Health Serv J. 1997;107:(5538; IT update): 4–5. http://abiesuk.blogspot.co.uk/2007/02/all-at-sea.html
2. Spronk R. HL7 version 3: message or CDA document? Ringholm Whitepaper, version 1.2, 2007. http://www.ringholm.de/docs/04200_en.htm
3. Dolin RH, Alschuler L, Beebe C, Biron PV, Boyer SL, Essin D, Kimber E, et al. The HL7 clinical document architecture. J Am Med Inform Assoc. 2001;8:552–69.
4. Dolin RH, Alschuler L, Boyer S, Beebe C, Behlen FM, Biron PV, Shabo A. HL7 clinical document architecture release 2. J Am Med Inform Assoc. 2006;13:30–9.
5. Boone KW. The CDA™ book. London: Springer 2011.
6. ASTM. Specification for Continuity of Care Record, E2369-05; 2006
7. HL7 Implementation Guide: CDA Release 2 – Continuity of Care Document (CCD).
8. Health Level Seven (HL7). Implementation Guide for CDA Release 2: Integrating Health Enterprise (IHE) 9. Health Story Consolidation, Draft Standard for Trial Use (DSTU) Release 1.1 - US Realm, July 2012.
9. Companion Guide to HL7 Consolidated CDA for Meaningful Use Stage 2. Standards and Interoperability Framework, Office of the National Coordinator for Health IT, US Health and Human Services 2012.

Chapter 16
HL7 Dynamic Model

Abstract The dynamic model specifies the interoperability message flow between two parties, including the interactions, trigger events that lead to them, application roles, message types, interaction sequence, and message wrappers. The implementation technology specification (ITS) describes the wire format.

Keywords Dynamic model • Interaction • Trigger event • Application role • Message type • Interaction sequence • Message wrapper • Acknowledgement • Implementation technology specification (ITS)

The following aspects of the Dynamic Model have to be specified (see Fig. 16.1):

- Interactions specify the message content the sender and receivers
- Trigger events determine when certain messages can be transmitted
- Application roles (sender and receiver) represent groupings of functionality that an application can do. This includes receiver responsibilities
- Message type(s)
- Interaction sequence
- Message wrappers
- Acknowledgements

Interaction

An interaction is the smallest unit (atomic) of communication that can stand on its own. It is a one-way transfer of information and ties together HL7's static models of payload content and the dynamic model of information flow and system behaviour.

Formally, an interaction is a unique association between a specific message type, a particular trigger event that initiates or triggers the transfer, and the application roles that send and receive the message type.

In HL7 Version 3, each interaction is described in a table with its name and artefact ID, together with the sending and receiving Application roles, the trigger event,

© Springer-Verlag London 2016

T. Benson, G. Grieve, *Principles of Health Interoperability*,
Health Information Technology Standards, DOI 10.1007/978-3-319-30370-3_16

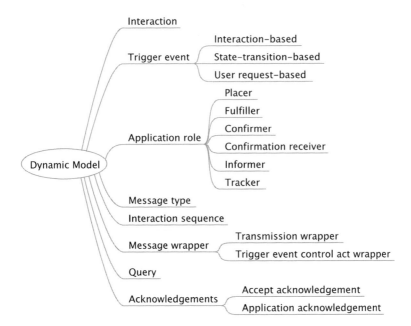

Fig. 16.1 HL7 dynamic model

the message type, the trigger event type and the wrapper types and their artefact identifiers.

Trigger Event

A trigger event is an explicit set of stated conditions, which can be recognized by a computer system, that initiates an interaction. A trigger event may be the result of human action, such as a mouse click, a state transition for an information object (such as the successful completion of a business transaction), an exception condition (such as an error), or be specific to a point in time (e.g. midnight).

The context for each trigger event is specified in use cases, and storyboards, which form part of the requirements specification.

Trigger events may be classified as being interaction based, state-transition based or user request based.

Interaction Based Trigger Events can be based on another interaction. For example, the response to a query (which is an interaction) is an interaction based trigger event.

State-Transition Based Trigger Events result from a state transition as depicted in the state transition model for a particular message interaction. For example, the change in status of an order from 'request' to 'fulfillment' is a common state transition. In practice, state-transition based trigger events are frequently encountered.

User Request Based Trigger Events may be based on a user request, such as clicking a mouse. The term 'environmental' is also used. For example, the trigger event that prompts a system to send all accumulated data to a tracking system every 12 h is considered environmental as no human user is involved.

Application Role

An Application role is a collection of communication responsibilities intended to be implemented as a group. Communication responsibilities are identified as the interactions that the system is able to send or receive. Application roles may be specializations of other existing application roles, inheriting the responsibilities of its parent, with additional or more specialized responsibilities added, or they may be the merging together of other application roles acting as components.

From the application role definitions, the reader can identify the purpose for information flow between two healthcare applications and the roles that those healthcare applications play in that exchange.

The application role description sets out what one application does, with respect to information exchange. It lists all of the interactions, sent or received, consequent to one particular trigger event. It is silent about the application functionality behind it – and how this is achieved.

Application roles have responsibilities, which are restricted to sending messages (interactions). Any other responsibilities and actions are outside the HL7 model. The sender role has the responsibility to send a message in response to a trigger event, and the receiver role may have responsibilities to initiate further transactions such as an acknowledgement, error report, response to query etc. These are referred to as **receiver responsibilities**.

The application role is a key element in specifying conformance and for contractual arrangements between users and service providers. It is the intent of HL7 that healthcare systems be able to declare conformance to the HL7 specification by creating an implementation profile that identifies the application roles supported by that implementation. Conformance to an application role means supporting each of the interactions specified.

Typically, one application role supports several interactions. For instance, a query is meaningless unless it includes a response, so the application role for the query questioner requires at least two interactions (query and response) to be supported, and similarly for the query answerer.

The names given to application roles provide one of best ways of finding the transaction sets, already defined, which meet a particular requirement. The naming convention is to state the subject of the interaction (e.g. Residential Address) followed by the application role category.

HL7 uses the following generic terms for application roles:

- Placer: An application that is capable of notifying another application about a significant event, and expects the receiver to take action. For example, a clinical system places and order for a laboratory test.
- Fulfiller: An application that is capable of receiving a request from a Placer application. For example, a laboratory system is to fulfill a an order
- Confirmer: An application that is capable of accepting a request from a Fulfiller application. For example, the laboratory system confirms receipt of order.
- Confirmation receiver: A role implemented by a placer indicating what types of confirmations it accepts. For example the clinical system receives confirmation from laboratory.
- Informer: An application that is capable of notifying another application about a significant event, but does not expect any action on the part of the receiver. Paired with tracker. For example a patient admission system informs laboratory that patient has been admitted.
- Tracker: An application that is capable of receiving information about a significant event, but is not expected by the receiver to perform any action. For example laboratory tracks patients who have been admitted.

In theory, application roles should be helpful to the reader in understanding the business roles and functionality provided by a set of interactions. However, the use of abstract terms, such as manager, tracker, placer and filler, makes this less useful than it might be.

Message Type

A message type is the most precise specification of a message, with explicit constraints about what data elements are sent and what values each data element may have. These constraints should be as tight as possible to minimize any chance of ambiguity.

Message types are derived by the intersection of specific interactions, application roles, and trigger events. The same message type may be associated with any number of application roles and be used in response to many different trigger events. However, an interaction can only ever have one trigger event and one message type.

Interaction Sequence

The precise flow of messages may be represented using a UML sequence diagram, which shows the application roles and the flow of message types between them in sequential order.

Message Wrapper

Whenever domain content (as a payload) is transmitted in the form of messages they use message wrappers, analogous to a letter's envelope. HL7 defines two types of wrapper: a transmission wrapper and a trigger event control act wrapper. Each HL7 Version 3 message typically consists of a transmission wrapper, a trigger event control act wrapper and the domain content.

The **transmission wrapper** includes a unique reference ID for each message instance sent, the precise date and time the message was created and the identity of the sending and receiving systems.

The **trigger event control act wrapper** sits inside the outer transmission wrapper and may include details of a previous interaction, which has triggered this interaction. Different variants of the Trigger Event Control Act wrapper are used for asynchronous messaging and for queries, where the response needs to be coupled with the query (Fig. 16.2).

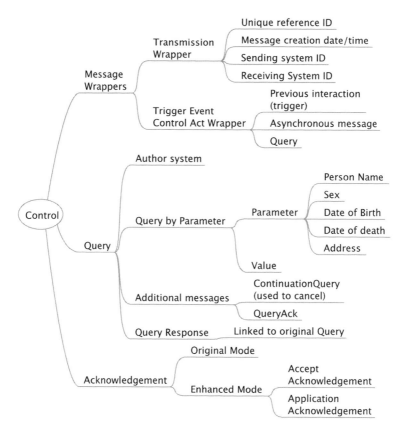

Fig. 16.2 Message communications control

Query

Queries are used to interrogate databases, such as to retrieve patient identification details from a patient master index.

The standard query message is an extension of the control act, using query by parameter. For a simple patient demographics query, the parameters could be patient name, sex, date of birth (and/or death) and address.

The query response is linked to the original query message.

Acknowledgement

Most HL7 transactions involve two or more messages: an originating message and an acknowledgement, in one of two modes – original mode or enhanced mode.

In **original mode acknowledgement** there are just two messages, the first, originating message comes from the sending system and the second, an acknowledgement is sent by the receiver saying whether it was able to process the originating message. Original mode acknowledgement is more straightforward to implement, especially for simple point-to-point interfaces.

Enhanced mode acknowledgement is more complex, but is suited to a multi-hop environment that uses an intermediary such as an interface engine between the sender and the final recipient. In enhanced mode acknowledgement, two separate acknowledgements are sent.

The first, the **accept acknowledgement** is a message indicating whether the receiving system, which could be an interface engine, was able to take custody of the sender's message, but does not indicate whether it was able to process the information contained within it

The second, the **application acknowledgement** indicates whether the final receiving application was able to process the sender's message successfully, indicating end-to-end completion of the transaction.

Safety

Safety is paramount in healthcare. Examples of safety procedures include:

- Acknowledgements sent at both transport level (message received) and application level (message processed)
- Explicit validation by both sender and receiver systems
- Use of automatic patient matching, with fallback of manual matching if not entirely unambiguous

- Routing messages to alternative recipient if not actioned within a specified time (for example if a named recipient is on leave)
- Messages are not removed from a task list until all actions specified have been performed
- If any user edits a message the original is kept unchanged (deletionless messages)
- A full audit trail is maintained.

Chapter 17
Sharing Documents and IHE XDS

Abstract IHE XDS (cross-enterprise document sharing) can be used to share documents between different organizations using a common portal. The XDS registry holds document metadata, which can be searched to retrieve appropriate documents stored in XDS repositories. The local specifications for metadata are defined in an affinity domain. The metadata includes information about each document, the patient, author, event and technical data. A number of related specifications have been defined as extensions.

Keywords XDS • Source • Repository • Registry • Consumer • Patient Identity Source • Metadata • Affinity Domain • PIX (patient identifier cross-referencing) • PDQ (patient demographics query) • Submission set • Direct project • HISP (health information service provider) • DSUB (document metadata subscription) • ATNA (audit trail and node authentication) • BPPC (basic patient privacy consent)

Sharing Documents

When patient records are fragmented across multiple care providers it is hard for anyone to grasp the whole picture.

> Approximately 75 % of Medicare spending pays for care for beneficiaries who have five or more chronic conditions and see an average of 14 different physicians each year. [1]

Fragmentation of the patient's health record leads to errors, duplication of work and waste. The case for sharing records across provider organizations is strong, although there is a danger of information overload [2].

Integrating the Health Enterprise (IHE) Cross-Enterprise Document Sharing (XDS) enables healthcare documents to be shared over a wide area network, between hospitals, primary care providers and social services. It is part of the IHE IT Infrastructure Technical Framework Integration Profiles (ITI-TF-1) [3].

Rather than having one big database at the center, IHE XDS offers a distributed collaborative approach to sharing clinical documents held by different healthcare organisations. It is based on standardized metadata.

© Springer-Verlag London 2016

T. Benson, G. Grieve, *Principles of Health Interoperability*,
Health Information Technology Standards, DOI 10.1007/978-3-319-30370-3_17

Metadata is information about an information item that is used in search to find it later. It describes the content (what), time (when), people (who) and locations (where) applicable to any item. Clinical and patient portals rely on common metadata to find the information being sought. Unless every source system supplies metadata in the same way, there is a risk that information from some sources cannot be found. This is a clinical safety risk.

XDS-based portals include a central registry with one or more repositories, which may be physically or logically separate. A book library provides an analogy. The library has shelves of books and a central index. Index cards in the central index contain limited information about each book and points to its shelf address. Similarly, the XDS registry contains metadata used to index each item in the repository and a URL pointer to its location.

Common metadata is essential to retrieve and share data that originates from multiple heterogeneous sources, as found in health and care services [4]. Sharing may be across the whole web of care, including health, social and voluntary sector services, family and carers. It may also cover secondary uses in audit, management, commissioning and research.

Metadata Standards

The Dublin Core [5] is a metadata standard that is widely used in libraries for indexing text documents. The UK e-Government Metadata Standard (e-GMS) [6] set out to standardize metadata across government, building on the Dublin Core foundations. The focus on traditional paper documents makes it not suitable for health portals.

The ebXML registry standard [7] provides a broad specification for business portals. It is the foundation on which XDS has been built [8].

HL7 CDA (Clinical Document Architecture) contains a common document header [9], which can be thought of as metadata. However, IHE XDS and CDA are not tightly aligned.

The optionality provided in XDS and CDA standards allows implementers wide scope for local variation. All elements are optional and so implementers need to specify a local Affinity Domain (XDS) and Templates (CDA). Guidance is needed on a core set of items required.

Documents and Statements

Health portals can contain items at two distinct levels of granularity, which we refer to as clinical documents and clinical statements [10]. These have rather different uses and granularity, although the metadata needed to enable retrieval are similar.

Both share common properties of persistence, coherence, wholeness, human readability, stewardship and potential for authentication.

A **Clinical Document** is a discrete electronic composition about an identified patient to be read or used by a human. Examples of Clinical Documents include pdf, images, audio and video. A clinical document may include structured information within the body of the document (e.g. in CDA Level 3), but such information is usually difficult to process.

A **Clinical Statement** is the smallest meaningful category of stand-alone medical information. Think of it as a single notes line, having an independent existence, which can be selected and combined with others in different ways to generate new ways of looking at data. Examples include diagnosis, medication entry, procedure and test result, such as a blood pressure reading. Clinical statements are designed to be easy to process. They can be selected and reordered on the fly to compile lists or charts from multiple sources, sorted by type, date, author or source.

The requirements set out below cover both clinical documents and clinical statements. We refer to these collectively as items. A set of metadata for an item is made up of elements.

Requirements

There are six core requirements for any item being indexed:

1. Metadata about an item contains information about what it is, who created it about whom, when and where.
2. A registry contains metadata on items coming from multiple source systems. The metadata needs to be standardised.
3. Metadata must be complete, because if one item contains a value for a metadata element but a similar item containing similar item does not, then search will find the former but not the latter. This is a clinical safety issue. It is safer to specify a small set of mandatory elements, than a larger optional set.
4. All metadata elements should be required (not optional) and multiple instances not supported. If more than one, then additional metadata records need to be created. There is nothing to stop the same information item having multiple metadata records.
5. Metadata is computer-processed. It needs to use unambiguous identifiers, codes and dates. Unique identifiers are needed for all items, people, places and organisations.
6. Metadata is derived from multiple source systems, so it should avoid the limitations imposed by any specific implementation platform and be platform-independent.

Metadata is limited by what can be provided by source systems. We cannot assume that anyone will ever actually retrieve any specific item, or that items found in a portal are complete or up-to-date, although each item should be accurate at the time it is created.

Metadata is used in search mainly outside the originating organisation, not to support internal business transactions. It does not include transaction-specific information, although such information is frequently found within the items being referenced.

Standardised metadata, covering both clinical documents and clinical statements, is low-hanging fruit, which can enable information sharing between all those who help care for patients. Lack of consensus around stringent requirements for common metadata has delayed the progress of clinical and patient portals and increased the costs of deployment.

How XDS Works

The XDS Document Registry stores metadata, which is used to retrieve documents, while the actual documents are stored in any number of XDS Document Repositories. The Registry and Repositories are logically and physically separate.

The XDS Registry contains standardized metadata describing the content of each item just as the library's card index contains metadata about each book. The XDS Repository can contain any type of electronic content much like the library shelves contain any type of book or printed material. Every item in the Repository has corresponding metadata in the Registry just as every book has a library card in the index.

XDS enables user applications to retrieve documents, such as letters, results, images and folders, from one or more repositories in a quick and consistent way. Each document is retrieved in its original form, which can include structured data.

XDS has five actors: Document Source, Document Repository, Document Registry, Document Consumer and Patient Identity Source (Fig. 17.1).

These actors are contained within a single XDS **Affinity Domain** (XAD), which establishes the rules and conventions about the type of clinical documents, metadata codes, security constraints and other policies that shall be used. One of the tasks in setting up an affinity domain is to specify the metadata code sets that are allowed

The **Document Source** submits documents, for example HL7 CDA XML documents, to a local Document Repository (each organization may maintain its own Repository) with metadata about each. An Integrated Document Source/Repository combines the functionality of the Document Source and the Document Repository. Where an integrated source/repository is not used, the 'Provide and Register Document Set' transaction is used to provide both the document(s) and the corresponding metadata to the Document Repository.

The **Document Repository** provides a persistent store for each document and uses the 'Register Document Set' transaction to submit standardised metadata to the Document Registry. It assigns a uniqueId to each document for subsequent retrieval

Fig. 17.1 XDS actors and information flow

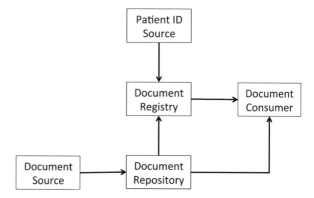

by a Document Consumer. Each care provider only registers information it is willing to share with others. These documents remain in the local Repository.

The **Document Registry** provides an index to all registered documents in the system about each patient using the patient identifier as the key. One Registry can index documents in any number of repositories. The central Registry supports queries and maintains a link back to the Repository where each document is stored. This provides a way to find, select and retrieve documents irrespective of where they are actually stored. To maintain security, the Registry has no access to the content of any document, but relies on standardised metadata to retrieve relevant items.

The **Document Consumer** is a user application that submits queries to the Registry to locate documents that meet the specified query criteria. The Document Registry returns a list of metadata, which includes the location and identifier of each corresponding document in one or more Document Repositories.

In 'Registry Stored Query' transaction, the definition of the query is stored on the Registry. To invoke the query, an identifier associated with the query is transmitted along with parameters defined by the query.

The consumer then retrieves the document set using the 'Retrieve Document Set' transaction from the relevant Document Repositories, for display in a browser.

The **Patient Identity Source** provides a unique identifier for each patient in the affinity domain. A server supporting IHE PIX/PDQ may be used to provide the Patient Identity Source. PIX (Patient Identity Cross-Referencing) provides cross-referencing of patient identifiers from multiple patient identifier domains, while PDQ (Patient Demographics Query) is used to retrieve patient demographic details.

The 'Patient Identity Feed' transaction is used to convey the patient identifier and corroborating demographic data, captured when the patient's identity is established, modified or merged, in order to populate the Registry with appropriate patient identifiers.

The main flows between each of these components are shown in Table 17.1.

Trans-actions are numbered and typically prefixed with ITI, e.g., transaction number 18, to query for documents for a particular patient is referred to as ITI-18.

Table 17.1 XDS Transactions

From	To	Name	Reference ID
Document Source	Document Repository	Provide & Register Document Set-b	ITI-41
Document Repository	Document Registry	Register Document Set-b	ITI-42
Patient Identity Source	Document Registry	Patient Identity Feed (HL7 v2)	ITI-8
		(HL7 v3)	ITI-44
Document Consumer	Document Registry	Registry Stored Query	ITI-18
Document Consumer	Document Repository	Retrieve Document Set	ITI-43

The XDS document Registry is a profile of the ebXML Registry standard (ISO 15000:2004 parts 3 and 4),[6] which defines the requirements for the information model for the ebXML registry. The ebXML registry describes objects that reside in a repository for storage and safekeeping. The information model does not deal with the actual content of the repository, but represents metadata about the content stored in the repository. The registry information model provides a high-level schema for the ebXML registry. The registry information model defines what types of objects are stored and how stored objects are organized.

Documents are exchanged using SOAP and HTTP, while SQL is used for information retrieval. Various document formats may be used, including HL7 CDA Release 2, DICOM and PDF. The format of the metadata in XDS Registry is largely based on HL7 v2.

XDS Metadata

An XDS document is the smallest unit of information that can be registered. The metadata about each document is defined in the Affinity Domain and specifies what information is stored in the Registry (Fig. 17.2).

The XDS metadata attributes fall into five groups:

Document data
Patient data
Author data
Event data
Technical data

Rules on multiplicities are:

[1..1] required exactly one
[1..*] required one or more
[0..1] optional, zero or one
[0..*] optional, any number

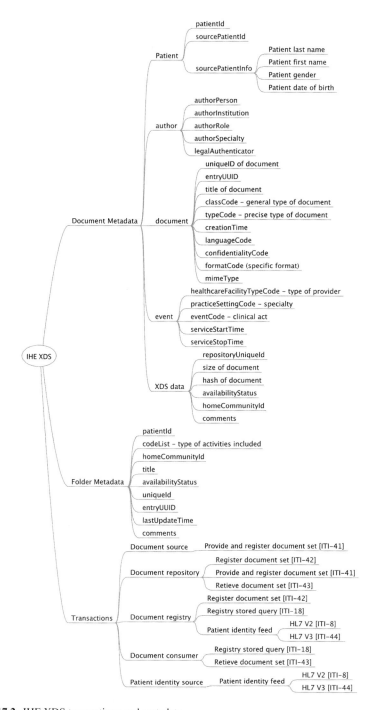

Fig. 17.2 IHE XDS transactions and metadata

Coded Attributes

When establishing an XDS affinity domain, one of the tasks is to agree value sets for the following attributes of documents: class, type, healthcare facility, practice setting, confidentiality, format and events(s).

All information needed to register a document has to be supplied by the document source or by the repository. Registration is usually an automated process so it is not practical to find or add additional data needed at the time of registration. For this reason, the amount of metadata specified for any Affinity Domain is limited to what can be reliably provided by source systems.

Document Data

Document data is metadata about the document. Some document data, such as classCode, typeCode, uniqueId, confidentialityCode, languageCode and title (if used) must be submitted by the document source. The repository also generates other metadata, such as entryUUID and availabilityStatus.

classCode required [1..1] The classCode specifies the particular kind of document being registered (e.g. prescription, discharge summary, and laboratory report). The Affinity Domain needs to specify a set of classCode values using a coding scheme such as LOINC or SNOMED CT. IHE recommend that this value set have a coarse level of granularity (10–100 entries).

In clinical statements Class may be used to distinguish between events that are planned and those that have actually taken place, between those about the patient and about others (family history), or to indicate negative findings. Type is used to say what it is at a fine level of detail. Note that in negative findings, subsumption works upside down – knowing someone does not have liver cancer does not imply that they have neither cancer nor a liver problem.

typeCode required [1..1] The typeCode specifies the precise kind of document. This describes the type of item in some detail, but should exclude information about the clinical specialty or mode of care (which are recorded as Specialty and CareType respectively). The Affinity Domain needs to specify a value set for typeCode with a fine level of granularity using a coding scheme, such as LOINC or SNOMED CT.

For clinical documents the typeCode is usually a specialization of the classCode. The typeCode typically corresponds to the title of the document. In a clinical statement, Type is used to give the heading of what is being described, but not the content.

uniqueId required [1..1] The uniqueId is a globally unique identifier assigned by the document creator to the document. This may be a UUID, which is generated on the fly.

entryUUID required [1..1] This is a universally unique identifier that is used in the Registry. It is not to be used as an external reference outside the Registry (uniqueId is used for that purpose). UUIDs are formatted according to RFC4122 formatted in hexadecimal notation. An example of a properly formatted UUID is:

urn:uuid:10b545ea-725c-446d-9b95-8aeb444eddf3

confidentialityCode required [1..*] This code or codes specify the level of confidentiality of the XDS Document. The XDS Affinity Domain specifies these codes (e.g. normal, sensitive). Normally only one code is used. Requires both a coding scheme and code value.

availabilityStatus required [1..1] is assigned by Registry. Each XDS Document shall have availability status of either: *approved* – meaning available for patient care, or *deprecated* meaning obsolete. This attribute is set to *approved* as part of the submission of new XDS Documents. It may be changed to *deprecated* under the primary responsibility of the Document Source.

languageCode required [1..1] specifies the human language of character data in the document. The values are language identifiers as described by the IETF RFC 3066 such as en-US or en-GB.

title optional [0..1] represents the title of the document. Clinical documents often do not have a unique title, and the display name of the document typeCode is adequate to inform the user of the content.

Patient Data

XDS is patient-centric. The Registry is keyed on the patientId and can normally only access one patient at a time.

patientId required [1..1] Each patient has a single unique ID within the XDS Affinity Domain (patientId). This is composed from an Assigning Authority ID, typically an OID, and the patient ID issued by that authority. The Registry only supports a single Assigning Authority. Usually the affinity domain maintains its own Enterprise Master Patient Index (EMPI) for this purpose.

Some commentators have proposed that an externally maintained ID scheme such as the NHS number be used for patientId. However, every entry in an XDS system must have exactly one patientId, and we cannot usually be certain that an externally maintained identifier will be known for every subject.

Mapping from other ID schemes is typically performed using PIX/PDQ. PIX (Patient Identifier Cross-referencing) and PDQ (Patient Demographics Query) are closely related profiles that build on the EMPI to support federated patient identification management. The PIX server cross-references patient identifiers from one or more external patient identifier domains where patients have identifiers, and maps these local identifiers to the single identifier (patientId) used within the XDS system. PIX feeds are usually HL7 v2 ADT messages. PDQ provides a method for local applications to submit queries to retrieve the common XDS patientId.

sourcePatientId required [1..1] represents the patient's medical record identifier known to the document source. It contains an authority domain Id (an OID) and a patient's Id in that domain. The sourcePatientId is not updated once the document is registered (just as the content and metadata itself are not be updated without replacing the previous document). It is used primarily as an audit/checking mechanism.

One issue is that many health organizations support multiple patient Ids. For example they use a local hospital number as well as a national number (such as the NHS number in England). In some cases they also use departmental patient numbers (such as in A + E and pathology departments).

sourcePatientInfo optional [0..1] This contains demographics information of the patient known to the Document Source and may include:

- patient first name
- last name
- sex
- birth date

Author Data

author required if known [0..*] represents the humans and/or machines that authored the document. In practice a decision may be made to restrict the optionality to a single required author [1..1]; author does not have a simple value but is a container for sub-attributes authorPerson, authorInstitution, authorRole and authorSpecialty.

authorPerson required if known [0..1] represents the person or machine that authored the document. The document author may be the patient. Within a single author there is normally one authorPerson. The decision may be made not to allow machines to "author" documents, but to require the machine supervisor to be named. It may be decided that each document requires one author [1..1].

authorInstitution required if known [0..*] represents a specific healthcare facility under which the human and/or machines authored the document. An issue here is whether to record both the site (such as a specific hospital) and the responsible organisation, if appropriate.

authorRole required if known [0..*] represents the role of the author with respect to the patient when the document was created. In practice a decision may be made to allow only one authorRole [0..1].

authorSpecialty required if known [0..*] represents a specific specialty within a healthcare facility under which the human and/or machines authored the document. In practice a decision may be made to only allow one authorSpecialty [0..1].

legalAuthenticator optional [0..1] represents a participant who has legally authenticated or attested the document within the authorInstitution. Legal authentication implies that a document has been signed manually or electronically.

Event Data

creationTime required [1..1] represents the time the author created the document in the document source.

healthcareFacilityTypeCode required [1..1] This code represents the broad type of organizational setting of the clinical encounter during which the documented act occurred. This phrase is ambiguous for some documents such as laboratory and radiology reports. For example, the healthcare facility for a laboratory report could be (a) when the test was ordered, (b) when the specimen was collected, or (c) when the test was done. This has to be specified. It is often useful to use this field to specify the type of organization that triggered the document in the first instance, such as hospital, clinic, domiciliary, out of hours, community, nursing home, maternity, GP surgery. This provides an orthogonal dimension to typeCode.

practiceSettingCode required [1..1] This code specifies the clinical specialty where the act that resulted in the document was performed. As with healthcareFacilityTypeCode, the practiceSettingCode of a laboratory report could be the clinical specialty that ordered the test or the specialty that reported it. It is usually more informative to specify the requesting specialty.

eventCode optional [0..*] This list of codes represents important clinical acts such as surgical procedures being documented. If one or more eventCodes are included, they must not conflict with the values of the classCode, typeCode or practiceSettingCode.

serviceStartTime required if known [0..1] represents the clinically relevant start time of the service being documented. It is not necessarily when the document was produced or approved. For a laboratory report this could be when the specimen was taken. Encounter time is not coded in XDS metadata but may be coded in documents managed by XDS. Other times such as document creation or approval are recorded if needed within the document.

serviceStopTime required if known [0..1] represents the end time the service being documented took place. Must be later than start time and should not be a future time.

Technical Data

homeCommunityId required [1..1] is a globally unique identifier for a community (Affinity Domain), an OID. Each community has one Registry. This is required for cross-community access (XCA).

repositoryUniqueId added by Repository [1..1] This unique identifier for the Document Repository may be used to identify and connect to the specific Document Repository where the document is stored once its metadata has been retrieved from a Document Registry. This identifier is usually an OID, which should be registered in an appropriate OID Registry.

formatCode required [1..1] uniquely specifies the format of the document. Along with the typeCode, it should provide sufficient information to allow any potential XDS Document Consumer to know if it will be able to process the document. It is recommended that an OID be used.

size computed by Repository [1..1] is the size of the document stored by the XDS Document Repository. This value is computed by the Document Repository and included in the Register Documents Set Transaction

hash computed by Repository [1..1] is the hash key of the XDS Document. This value is computed by the document Repository and used by the document Registry to detect improper resubmission of XDS documents.

mimeType required [1..1] is the MIME type of the document in the Repository. The options supported include:

* text/plain
* text/xml

comments optional [0..*] Optional free text comments may be associated with the document metadata for human readability. The use of comments within metadata is specific to the affinity domain. Use of comments is not recommended.

Each metadata attribute described above is an attribute of the XDSDocumentEntry object. The attribute name is defined with a prefix of XDSDocumentEntry when referenced by other objects, for example XDSDocumentEntry.patientId.

Submission Sets and Folders

In addition to individual documents discussed above, XDS may also be used with Submission Sets and Folders. Documents are typically submitted in Submission Sets. Each submission set may include author (person, role, specialty and institution), title, submission time, comments, availability status, and coded elements for type of clinical activity and identifiers for the patient, source and submission set identifiers. These attributes are similar to those for individual documents discussed above. Folders provide a way of grouping documents in a directory-like structure.

XDS Extensions

XDS is the core specification of a constellation of related IHE specifications and profiles (Fig. 17.3). IHE defines profiles, and the transactions associated with those profiles.

The basic XDS.b profile has been refined to support special requirements for DICOM images (XDS-I), HL7 CDA medical summaries (XDS-MS), and structured laboratory reports (XDS-Lab).

Other related IHE profiles cover point-to-point transmission (XDM and XDR), patient identity management (PIX/PDQ), information retrieval (DSUB and MPQ) and a number of security profiles.

Point to Point Transmission

Cross enterprise document sharing using reliable messaging (XDR) and cross enterprise document sharing using media (XDM) are sister profiles of XDS, which share many of the same attributes, but do not require the full XDS infrastructure.

XDM – Cross-Enterprise Document Media Interchange XDM facilitates person-to-person interchange using media such as CD, USB memory or as zipped email attachments.

XDR – Cross-Enterprise Document Reliable Interchange XDR provides a reliable and automatic point-to-point transfer of documents and metadata for one patient between EHR systems. It may be used with direct TCP connections or off-line over SMTP. XDR uses the same metadata as XDS and can be used to feed XDS submissions or XDS Query/Retrieve can be used to feed XDR transmission. The XDR metadata allows a receiving system to associate the message with the appropriate patient and clinicians for automated processing.

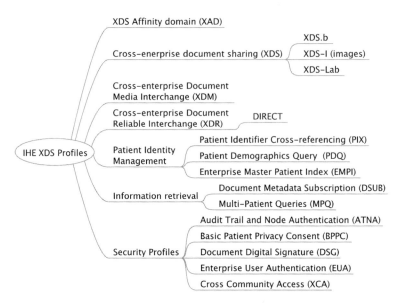

Fig. 17.3 IHE XDS profiles

The **Direct Project** sponsored by the ONC (Office of the National Coordinator) specifies a simple, secure, scalable, standards-based way for senders (health organizations, physicians and patients) to transmit authenticated encrypted health information directly to known trusted recipients over the Internet.

Direct provides a simple alternative to paper and fax for communication of health information among providers and patients, in a way that is more secure and includes encryption and non-repudiation. The focus is on the technical standards and services needed to push content from a sender to a receiver in a secure way. Direct is silent about the structure, format and terminology used in the message content.

The Direct project has introduced the concept of a **Health Information Service Provider** (HISP) at both sender and receiver ends. The communication pathway has three logical stages:

- From sender to sender's HISP, which locates the receiver's HISP address from a routing information directory.
- From sender's HISP to receiver's HISP.
- From receiver's HISP to receiver.

HISP may be a separate business or technical entity. Using HISP moves responsibility for managing the trust model on behalf of users from the user to a different level, which makes it easier to implement in complex organisations.

The content sent is packaged using MIME or XDM. The confidentiality and integrity of the content is handled through S/MIME encryption and digital signatures. The authenticity of the sender and receiver is established with X.509 digital certificates. Message routing is handled through SMTP.

Information Retrieval

DSUB – Document Metadata Subscription enables notification of documents arriving in an XDS Registry, which enables an event-driven publish/subscribe model of data exchange. For example a subscriber could request to receive notifications when a particular type of document for a particular patient is registered in XDS. DSUB is a subscription and notification mechanism. It allows for the matching of metadata during the publication of a new document for a given patient, and results in the delivery of a notification that a new document instance has been registered. That document may then be retrieved using XDS. There are important security, access control and audit issues when using a publish/subscribe model for protected health information. The Multi-Patient Queries (MPQ) profile defines a way to query information about multiple patients.

MPQ – Multi-Patient Queries profile defines a mechanism to enable aggregated queries to an XDS Document Registry based on criteria needed by areas related to data analysis, such as quality accreditation of healthcare providers, clinical research trial data collection and population health monitoring.

Security Profiles

Security of protected health information is paramount and any distributed system is more open to attack than one with a secure perimeter. IHE has developed a set of profiles to address these security concerns including ATNA (Audit Trail and Node Authentication), BPPC (Basic Patient Privacy Consent), DSG (Document Digital Signature), EUA (Enterprise User Authentication) and XUA (Cross-enterprise User Assertion)

ATNA – Audit Trail and Node Authentication provides a centralized audit trail and node-to-node authentication to create a secured domain. It assumes that users are authenticated by local systems before allowing network access. Enterprise User Authentication (EUA) is one option. Remote nodes are authenticated using digital certificates (X.509). Communications between nodes may be restricted to other secure nodes in that domain. The audit trail logs all security-related operations. This would allow a security officer in an institution to audit activities, assess compliance with security policies, detect instances of non-compliant behaviour and facilitate detection of improper creation access modification and deletion of protected health information.

BPPC – Basic Patient Privacy Consent profile provides mechanisms to record patient privacy consent and to enforce it. BPPC sets up a set of consent policies for an Affinity Domain, which patients can choose from. These may be based on opt-out or opt-in. Each policy is identified using an OID and implemented using role-based access control (RBAC) and XDS confidentialityCodes. Patient consents are recorded HL7 CDA documents which are human readable and machine processable and should be digitally signed using DSG. In BPPC, the XDS consumer enforces confidentiality.

DSG – Document Digital Signature provides a means of attesting that a document is a true copy and origin to ensure integrity, non-repudiation and accountability. A digital signature is a separate XDS document and uses X.509 certificates (see Chap. 5).

XUA – Cross-Enterprise User Authentication provides the means to communicate claims about the identity of an authenticated person.

XCA – Cross Community Access XDS Registries can work together to offer a unified service using XCA, much like multiple libraries can participate in a cooperative network and offer a unified service. Multiple registries may be federated together to appear as a single virtual Registry/Repository enabling seamless information integration and sharing and local autonomy over data. Federated information management relies on SAML, the federated identity management standard.

References

1. Congress of the United States, Congressional Budget Office. Budget Options Vol I, Healthcare. December 2008, page 77.
2. Duftschmid G, Rinner C, Kohler M, Huebner-Bloder G, Saboor S, Ammenwerth E. The EHR-ARCHE project: satisfying clinical information needs in a Shared Electronic Health Record System based on IHE XDS and archetypes. Int J Med Inform. 2013;82(12):1195–207.
3. IHE IT Infrastructure (ITI) Technical Framework, Volume 1 (ITI TF-1) Integration Profiles Revision 12.0 September 2015, Section 10 Cross-Enterprise Document Sharing (XDS.b): 89–126
4. Benson T. Metadata requirements for portals. In: Cornet R et al., editors. Digital healthcare empowering Europeans: proceedings of MIE2015, Studies in health technology and informatics, vol. 210. Amsterdam: Ios Press; 2015. p. 577–81.
5. Weibel S, Baker T. The Dublin Core Metadata Element Set. RFC5013, IETF; 2007.
6. e-GMF. e-Government metadata framework v3.1 2006 http://bit.ly/14y4q37
7. Electronic business eXtensible Markup Language (ebXML) – Part 3: Registry information model specification (ebRIM) ISO/TS 15000-3:2004.
8. IHE. Infrastructure Technical Framework: Cross-Enterprise Clinical Documents Sharing (XDS). IHE ITI Technical Committee, Version 3 (2006).
9. Dolin RH, Alshuler L, Boyer S, Beebe C, Behlen FM, Biron PV, Shabo A. HL7 clinical document architecture, release 2. J Am Med Inform Assoc. 2006;13(1):30–9.
10. Benson T, Conley E, Harrison A, Taylor I. Sintero server – simplifying INTEROperability for distributed collaborative healthcare. IHIC 2011, Orlando. http://www.hl7.org/events/ihic2011/papers/Benson%20IHIC2011Sintero20110426-Final.doc

Part IV
Fast Healthcare Interoperability Resources (FHIR)

Chapter 18
Principles of FHIR

Abstract FHIR was developed after implementation experience with the other HL7 standards, and built to copy the web, particularly RESTful interfaces. FHIR is organised around the concept of "Resources" and defines many types of resources that describe the healthcare space. All resources have references to other resources, extensions, and a human readable XHTML display. FHIR has an open license, a focus on implementation, and a formal maturity process linked to implementation outcome.

Keywords Web • Free • Open • License • FHIR • RESTful • API • Resource • URL • Resource types • Data types • References • Extensions • XHTML • Narrative • Manifesto • Implementers • Common scenarios

Origins

By mid-2009, HL7 had invested many years of development into the v3 RIM-based standard. As described in Chap. 13, the v3/RIM specifications were designed to address one of the principal limitations of the V2 specifications – the ad-hoc and inconsistent nature of the information exchanged using it.

The v3/RIM specifications were comprehensive in scope, rigorous in detail, and consistent in application. In these regards, the effort was an unqualified success, and the requirements gathering and analysis, methodology and tooling together set a new standard for healthcare information exchange that was the successful underpinning for a number of large national programs. In particular, CDA, which is part of the v3/RIM specifications, has been widely adopted for information exchange between loosely coupled healthcare systems around the world.

However, in spite of its success in these terms, by 2008–2009 it was clear that there were several issues that brought v3 and the HL7 organisation itself to a crossroads.

© Springer-Verlag London 2016   329
T. Benson, G. Grieve, *Principles of Health Interoperability*,
Health Information Technology Standards, DOI 10.1007/978-3-319-30370-3_18

Consistency

The adage quoted earlier in this book – *if you ve seen one v2 interface, you've seen one v2 interface* – turned out, rather than being about v2, to be much to do with the nature of the problem being solved. Across the healthcare system, there is little consensus about what information should be used, how it should be represented, and when it should be exchanged. While v3 excelled at producing consistent definitions, the consequence of this was that it was actually harder to use it inconsistently, and v3's rigorous consistency turned out to be a false advantage.

Complexity

The v3/RIM specifications are a set of very wide and deep specifications, with many layers. In order to implement successfully, the specification designers and the implementers must read and understand many layers of documents. Further, the custom tooling stack that HL7 had to produce to string all the layers together requires its own expertise, and the end outcome is engineering artefacts that do not align with what else is commonly used in the industry. Truly successful implementations of v3 or CDA – and there are some – require an extensive custom stack of libraries and templates.

Conformance Testing

One of the main drivers for developing v3 was to enable more meaningful conformance testing, but the combination of the inherent inconsistency in the exchanged information, and the introduction of the abstraction that the RIM language represents meant that in spite of the significant increase in cost, the incremental improvement in the ability for conformance testing to deliver useful outcomes did not justify the increase in cost required.

In practice, a successful implementation required both the intimate involvement of dedicated experts in v3, (who either started or ended as key authors of the specification itself), and the resources of a national program – that is, billion dollar projects.

While the v3/RIM program had met its own goals, it was unfortunately clear that it had not met HL7's wider goals of making interoperability cheaper and easier (or, alternatively, of producing standards that could compete in the market, since healthcare interoperability standards is a market of its own).

In response to the growing awareness of this issue, HL7 created a 'Fresh Look' task force that was asked to examine the best ways HL7 could create interoperability solutions, with no pre-conditions on what those solutions might be. The FHIR project came out of this work, and from two other sets of considerations. The first was around considering the strengths and weaknesses of the existing standards, and the second was a search for exemplars of interoperability done well.

As an exemplar, the very first draft of FHIR (then known as RFH, *Resources for Health*) was based on a typical example of a RESTful API.

The rest of this chapter describes the basic principles of the FHIR specification – a mix of technical choices and management philosophies that combine to make it an exciting new approach to healthcare interoperability, and one that has generated wide implementation interest.

API Based Approaches

Interoperability specifications can be grouped into one of several different approaches:

- **Messages**: define a series of fixed information that can be exchanged between applications when a specific event occurs.
- **Services:** define a set of functional operations that one system exposes for others to use, along with the expectations of behaviour around these.
- **Documents:** define a set of fixed information packages that can be exchanged or stored for later use.

In the end, in order to have working interoperability, systems have to agree what is exchanged, when it is exchanged, and why. So all these approaches need to end at the same place – how they differ is about which parts of the overall solution are standardised, and which are left to the discretion of the implementers. For instance, a messaging system will typically bind the information descriptions to a particular set of technologies, so the messages can be exchanged, but leave it to system implementers to decide what kind of service to offer, whereas a service based approach will define the services offered, and the information that is exchanged (to some degree) while leaving the technology bindings to the implementer.

During the mid-2000s *Service Orientated Architecture* (SOA) became a popular way to manage collections of disparate systems, especially within large enterprises. Typically, SOA meant building a collection of services that could be connected by an enterprise service bus, based on a combination of web services using the SOAP standard. In response to this, HL7 teamed up with OMG to develop service based specifications, in the *Healthcare Services Specification Project*. This project defined a number of useful services (http://hssp.wikispaces.com/specs), but never gathered the critical mass to transform interoperability, for a variety of reasons, perhaps most importantly that it did not solve the complexity problem.

All along, there has been another form of interoperability, called an Application Programming Interface (API). An API is a set of services that a programming library offers that can be used by another program to achieve its own goals. These are the interoperability services which operating systems are built on. Historically, these were limited by technology to providing services within a single process on one computer, but later this was extended to provide for invoking operations across the network.

In the service / message / document categorization, an API is a service interface, even though many people do not regard it as such. Typically, an API is bound to a particular technology, and is focused on a providing a concrete definition of a set of capabilities that a providing system defines, and the definition tries as much as possible not to dictate how it is used, as use case specific definitions tend to lead to fragile outcomes.

Gradually, the technology supporting programming interfaces became more capable of supporting distributed operations, and the difference between services/ SOA and API started to blur. Eventually, this led to the definition of REST.

RESTful Interfaces

The notion of 'REST' (*Representational State Transfer*) was first defined by Roy Fielding in 2000 as a set of design constraints, methods, and architectures that leads to scalable, reliable, easy to use interfaces [1].

Since then, a community has developed that adopts many of the base principles of REST. Many companies have published significant web APIs that are implemented using the RESTful paradigm:

- Google
- Apple
- Facebook
- Twitter
- Basecamp
- etc.

FHIR is a RESTful specification – FHIR aspires to follow RESTful principles as much as possible. This is a recognition of the influence that RESTful interfaces have had on the industry – they have shown that it is possible to build large integration-based ecosystems quickly, that developers can easily integrate with RESTful services, and that these services scale very well in a technical sense.

Here are the basic REST principles:

1. **Uniform Interface**: Individual resources are identified using URLs, and can be represented in multiple different ways (e.g. XML, JSON). Clients manipulate the resource through the representations using self-descriptive messages. Hypermedia (hyperlinks) and hypertext act as the engine for state transfer.
2. **Stateless Interactions:** None of the client's context is stored on the server side between requests, so all of the information necessary to service the request is contained in the URL, headers, or body.

3. **Cacheable:** Responses can be cached, and responses must define themselves as cacheable or not.
4. **Client and Server** are separated from each other so the client is not concerned with the data storage while the server is not concerned with the user interface.
5. **Layered System:** At any time, a client cannot tell if it is connected to the end server or to an intermediate. Intermediaries can help enforce the security policies, enable load-balancing, etc.

A controversy associated with REST is just exactly what makes an interface conformant to these basic principles. For this reason, many interfaces describe themselves as 'REST-ful' to avoid the question of whether they conform to "REST". FHIR is a RESTful interface; the specification does not adopt REST as a religion. Instead, the religion of FHIR is to do what works in practice, preferably what other RESTful interfaces such as those provided by companies like Twitter, Google, and Facebook do.

There are two key differences between FHIR and the kind of RESTful interfaces that typical cloud providers offer:

1. Most Cloud-based RESTful services are provided by a single provider to meet a specific business purpose on a single server. On the other hand, FHIR is a general specification for exchanging data between multiple parties. A consequence of this is that the FHIR specification is broader and looser than normal RESTful APIs. This theme is pursued further below (see Chap. 21).
2. Healthcare information will never be limited to exchange across RESTful APIs – there are many reasons to use other kinds of exchange, and the FHIR specification extends to cater for messaging and document approaches as well.

FHIR is a RESTful specification in the sense that RESTful approaches are generally the option of choice, and all of the defined content is required to work within a RESTful paradigm. However, it is possible to define service-based specifications that would augment or substitute parts of the RESTful interface.

Repository Specification/Overview

As a RESTful specification, FHIR is organised around the concept of a repository, which is a list of resources of a particular kind. This is explicitly represented in the URLs defined by FHIR. Here is a typical FHIR URL:

```
http://server.example.com/fhir/Patient/23455
```

Structurally, this URL has three parts (Table 18.1):

```
[base-address]/[Type]/[id]
```

Most of the functionality of a FHIR interface is provided by these three services (Table 18.2).

Table 18.1 FHIR URL structure

base-address	Identifies a FHIR System service. That is, a server that that makes information available in conformance with the FHIR specification
Type	Identifies a FHIR Type service that manages a collection of resources that all have the same type. The type must be one of those defined in the FHIR specification
Id	Identifies a FHIR Instance service that manages an instance of a resource within the collection

Table 18.2 FHIR base services

Instance Service	Allows a client to retrieve the current content of a resource, to update the content to a new state, to delete the resource, or to see its modification history
Type Service	Allows a client to search through the existing resources that already exist, create a new instance of a resource, or get the history of all changes to resource of the type
System Service	Allows a client to determine what functionality is provided by the system, perform batches and transactions across multiple resource types, and get the history of all changes to all resources

Note that servers choose which of these functions to make available for which types based on the use cases that they support. The only function that servers must provide is to return a conformance statement that allows a client to determine what functionality it provides. A server that returns a conformance statement that says that no other functionality is provided is a fully conformant server (although not very useful).

The FHIR API provides a record centric approach to data exchange. Instead of asking the server to perform some operation, the client tells the server what the contents of the record should be. These services are often called CRUD services, since the client can Create, Read, Update and Delete records (though, of course, in healthcare, very few records can be truly deleted at all).

In addition to these base services that apply to all the types of resources defined by FHIR, FHIR allows for the additional special services that provide specific functionality beyond simple CRUD services. FHIR defines a number of useful services itself, such as asking a server whether it considers a resource is valid, linking two patients, or retrieving all the records associated with a particular patient. Servers can define their own addition services.

Resource Types

Most of the significant content of the FHIR specification is in the definition of the particular types of resource. FHIR defines around 100 types of resource, representing widely different types of content. For each resource type, the specification defines:

- The scope and intent of the definition of the resource, along with additional background information about how to use it well.
- The specific data content of the resource, using the common definition framework, and a set of common data types (see Chap. 20).
- Additional terminology and content rules that all resources must meet in order to be valid.
- A set of common search parameters that can be used to find particular resources, and to join across them.
- Mappings with other specifications.
- Additional specific services for the resource type.

The actual resources defined in the FHIR specification are categorized as follows (Table 18.3):

The list of resources is gradually growing as new use cases are included. Note that there is no formal support in the specification for organisations other than HL7 to define their own resources, and most of the library implementations (see Chap. 22) assume that this does not happen.

Figure 18.1 helps to explain what kind of things are resources, and what are not.

The general intent is that level 1 corresponds to FHIR data types, and levels 2 and 3 (Attribution and Core Business) correspond to resources in the FHIR specification. Level 4 (Specialty Business) is delivered as profiles on the core business resources (see Chap. 21) and Level 5 (Care Delivery) corresponds to packages or implementation guides that use the profiles and base resources.

An alternative way to visualize the FHIR specification is shown in Fig. 18.2.

The FHIR specification defines a series of domain resources that deal with healthcare data exchange. Around those domain resources the FHIR specification provides:

- An infrastructure for exchanging resources (the RESTful API mentioned above, and other exchange infrastructure not described in this book).

Table 18.3 FHIR resource types

Conformance	Resources describe how a system does or should work	Conformance, StructureDefinition, ValueSet
Infrastructure	Resources defined as part of the API to provide API related services, or basic IT infrastructure	Bundle, List, AuditEvent
Administration	Resources to manage the administrative side of healthcare – who the participants are, where they are or should be, and managing workflow	Patient, Encounter, Appointment, Order / OrderResponse
Clinical	Clinical summaries, record keeping and planning	Observation, Condition, Care Plan, AllergyIntolerance
Financial	Resources that support the financial services associated with the provision of healthcare	Claim, Coverage, ExplanationOfBenefit

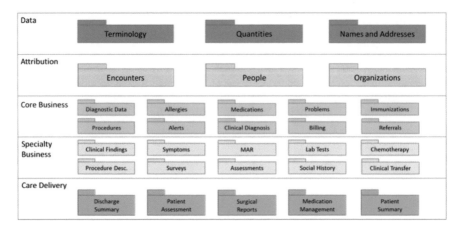

Fig. 18.1 Conceptual information architecture for health

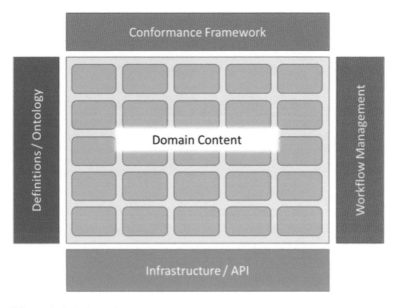

Fig. 18.2 Logical FHIR architecture

- A definitional/ontology layer that provides narrative descriptions of the content, mappings to other specification, and a computable set of definitions (not further described in this book).
- A conformance framework (see Chap. 21).
- A set of resources for managing workflows – requests to perform actions etc. (still yet to be developed).

References Between Resources

One key feature of resources is that they refer to other resources. For instance:

- Clinical resources have a reference to a Patient resource to identify the subject of the clinical content.
- A Condition can reference an Observation as evidence for the diagnosis.
- A Composition is a wrapper around a set of references to the resources that are part of the composition.

The references between the resources build up a web of information about a set of patients, providers, institutions, diseases etc. only limited by the scope of the institutions that maintain the coherent record set. References are represented using a URL that maybe either absolute or relative.

A relative reference:

```
<context>
<reference value="Patient/123456" />
</context>
```

An absolute reference:

```
<context>
<reference value="http://server.domain/path/Patient/123456" />
</context>
```

Relative references are interpreted relative to the base-address of the server providing the information. Absolute references have the advantage of not being subject to interpretation and possible confusion as to the correct base as content is transported around a system. However, relative references have the advantage of being stable, as groups of resources are ported from address to address, or where the server providing them is made available on multiple different addresses.

Servers are allowed to insist that references between resources must be valid (analogous to foreign key constraints in SQL), but are not required to by the base specification. Whether references must be valid depends on the purpose of the server; must production clinical record keeping systems (e.g. EHRs) would be expected to require referential integrity, but middleware servers usually would not.

Links between resources are always done by reference, rather than just inserting the content directly into the resource. This is done to create technical stability – both the representation of the resources, and to ensure that applications are always following references when resolving links. However, there is one circumstance that has to be treated a little specially.

Sometimes, when creating a set of resources from a secondary data feed – e.g. from an HL7 v2 message, implementers find themselves creating resources where there is no identity (e.g. no primary key), and no possibility of determining one. A typical example is where an application sends an HL7 v2 ORU message and provides just a name for the principal interpreter of the laboratory report (OBR-32):

```
|^Smith&John|
```

There is no way to create a properly identified Practitioner resource from this; a secondary data processor is unable to determine which John Smith this actually is. Since such secondary data use is common, FHIR has to cater for this, and it does so by allowing for 'contained resources'. A contained resource is placed in the header of the resource that defines its identity:

```
<DiagnosticReport xmlns:="http://hl7.org/fhir">
  <text>…(snip).. </text>
  <contained>
    <Practitioner>
      <id value="p1"/>
      <name>
        <family value="Smith"/>
        <given value="John"/>
      </name>
    </ Practitioner >
  </contained>
    <!-- other content -->
    <performer>
      <reference value="#p1" />
    </ performer >
    <!-- other content -->
</DiagnosticReport>
```

Containing a resource inside another resource means the information in the resource is owned by the container, and can never be accessed independently from it. For this reason, it is a practice that should be avoided as much as possible, as it creates difficulties handling the content. Note: the difficulties are inevitable in this case, since the missing information – e.g. the identifiers – was lost from the source. But using contained resources should be avoided wherever possible to avoid losing information.

Extensibility

When developing any healthcare exchange standard, there is a fundamental challenge: the business of providing clinical care is wildly variable around the world. The process flows, and even the ways people think about the problems they are trying to solve and describe, vary wildly. Further, there is no central standards body that sets standards for the provision of healthcare process around the world – indeed, for cultural and political reasons, different countries solve the problems of healthcare radically differently. Even within countries, standards, regulations and funding policies with regard to forcing common practices have limited success.

This is the context within which FHIR is developed and implemented. And because of this, HL7 can not produce really tight, easy to use specifications. Instead, they are full of flexibility to support variable business practices and different ways of understanding and describing the same things. The idea is that countries and projects etc. will then take the specification and add their own rules about how it is used based on whatever agreement they are able to get from the eco-system in which the exchange is going to occur. This means layers, and complexity. There is no way around this.

Generally, a standard can try and manage this in one of two different ways:

- Define every data element that anyone is ever going to use, and then let the particular implementations exclude things that they do not use.
- Define a basic set of data, and let particular implementations add extra stuff when they need it.

Both approaches have their problems. The first is going to produce a huge, comprehensive specification, and it will take lots of time (and $$$) to produce and use it. Many people will prefer not to use it at all due to its unwieldy size. The second approach will mean that every implementation will add their own extra stuff, and none of them will be able to talk to each other using the extra stuff – and it is going to matter.

In v2, HL7 took the second approach, and implementers or projects (and even jurisdictions) are allowed to define 'Z-segments', which allow them to add any additional data to messages, or even to define Z-events that have entirely custom messages. Without the flexibility offered by Z-Segments, v2 wouldn't really be a workable standard (that is not to say that every message includes a Z-segment. Not at all – just that if this mechanism was not available, it would be very difficult to commit to HL7 v2 in practice).

But Z-segments are a notorious problem as well – use of Z-segments is very often ill disciplined, and poorly documented, so that you often work with messages where you have to guess what the content means – not a good place to be in healthcare. Also, it is hard to get vendors to exchange content because they use Z-segments differently.

In order to avoid the well understood problems of Z-segments in v2, HL7 decided to go with option #1 for v3: model everything known in the base models. This turned out to have the predictable problems described above: lots of time and money to produce and implement the specifications. Beyond this, there was still the option for implementers to add additional content in other XML namespaces, which is regularly used in practice. This is actually worse than Z-segments, because many schema-driven implementations simply can not handle these alternative namespace extensions at all.

This was, then, a central question for FHIR – how would this problem be handled? The principal design goals were two-fold:

- Implementers needed to be able to use and extend the resources.
- Using extensions needed to be manageable, so that using them would not create operational disasters, or attract a stigma, as it does for v2.

Extensions in FHIR

Any element in FHIR resource can carry one or more *extension* elements, in addition to their normal content. A basic extension is a pair: a URL that identifies the extension, and a value. In XML, an extension looks like this:

```
<extension              url="http://hl7.org/fhir/StructureDefinition/
iso21090-en-qualifier">
  <valueCode value=" NB"/>
</extension>
```

The URL not only identifies the extension – any system can retrieve a formal definition of the extension from that URL. The information in the definition allows the system to display and/or process the data.

The value of the extension must be one of the base FHIR data types (see Chap. 20). This means that every implementation – including code generated from the FHIR schema – can read and write all extensions, without needing to access the definition of the extension.

Any implementer is allowed to define and publish their own extensions, but implementers are encouraged to use a central registry to find existing extensions rather than define their own. HL7 also encourages implementers to register their extensions with HL7 through their local affiliate (if outside USA) or HL7 itself, or even to ask HL7 to define them, but they do not have to. This allows implementers to choose how much governance they want to opt-in to.

This position does not please everyone, and it is not perfect, because while the price of choosing not to be completely interoperable is partly born by the implementers, it is only partly borne by them. Other implementers who consume the information, system purchasers, national programs – they bear part of the price too.

The FHIR specification encourages responsible use of the extensions facility by the following methods:

- Making it easy to find and register extensions to foster re-use (using the FHIR API).
- Providing strong social networks that spread expectations around the use and registration of extensions.
- Leveraging social media such as Stack Overflow to encourage implementers to consult with the community, and avoid the need for making their own extensions.

At the time this book went to press – shortly after DSTU two was published – these processes seemed to be working.

Importance of Human Display

One of the basic features of FHIR is that every single resource includes a human displayable form, called the narrative. Constructing this is the responsibility of the system or person that authors the resource, and it can be used by any system to display the contents of the resource to a human, whether or not the system understands the data content of the resource completely (or at all). This is an idea taken from the CDA specification (see Chap. 15).

Technically, the narrative is a limited form of HTML. It is allowed to contain text formatting, lists, tables, images, and styles, but not any form of active content such as forms, scripts, objects or use of local storage. Here is an example of a patient resource with a simple table that summarizes the patient's important properties:

```
<Patient xmlns:="http://hl7.org/fhir">
  <text>
    <status value="generated"/>
    <div xmlns:="http://www.w3.org/1999/xhtml">
      <table>
        <tbody>
          <tr>
            <td>Name</td>
            <td>Peter James Chalmers (&quot;Jim&quot;)</td>
          </tr>
          <tr>
            <td>Address</td>
            <td>534 Erewhon, Pleasantville, Vic, 3999</td>
          </tr>
          <tr>
            <td>Contacts</td>
            <td>Home: unknown. Work: (03) 5555 6473</td>
          </tr>
          <tr>
            <td>Id</td>
            <td>MRN: 12345 (Acme Healthcare)</td>
          </tr>
        </tbody>
      </table>
    </div>
  </text>
  <!-- content -->
</Patient>
```

Note that the narrative has a flag called *status* to inform consumers whether the narrative contains information not found in the data content; this may influence when it is appropriate to use.

It is important to understand what the narrative is intended to achieve, and what it does not do.

The first function of the narrative is as a fall back, a safety net. It provides a floor to ensure that there is always a way for an application to present the content of the resource to a user if there is any reason to think that the system may not fully understand it. For example, a system might display a list of diagnostic tests in a summary table for a clinical user. This summary table would be built from data taken from many resources. The system might choose to add an option 'see original' to the interface, so that a user could see the original narrative and confirm for themselves that the system had processed the content correctly.

Obviously it would be preferable for the system designers to be sure in advance that there was no possibility of misunderstanding the content in resources, but in practice this is much more difficult than it sounds. This is particularly true over time, as changing business arrangements lead to the creation of new integrations faster than the basic system design can adapt. This is made more likely by the flexibility with regard to extensibility that FHIR offers.

In addition, the narrative can be used by general processing software to present the contents to human users in a wide variety of situations. This empowers simple software to provide useful functionality without having to become expert at the processing the contents of the resources.

The idea of integrating narrative and data goes deeper than merely a fallback for safety or convenience, however; at its heart, clinical information is a combination of a narrative story, and supporting data:

> The medical record is not data. It contains data, as do many forms of writing, but it is not data, nor is it simply a repository into which data are poured. Although its raw material is information—some of which, importantly, can only be expressed with words and not with numbers—a finished medical record is information that has been transformed by the knowledge, skill, and experience of the physician, motivated by the healing impulse, into an understanding of human experience that makes the care of the patient possible [2].

Including narrative in every resource makes this duality a core feature of the specification.

It is still expected that the specialist software that supports direct clinical processes will use the data from the resources: these are the systems that need to understand the data fully, that construct resources and build the narrative themselves. There is no intent that the narrative should support these use cases – or that its presence should make them more difficult to implement.

For this reason, the narrative does not need to be particularly efficient or contain a beautiful presentation of the content of the resource, though it is allowed to. Because it is primarily a clinical safety back up, this means that there only needs to be one presentation, not multiple different presentations for different contexts.

The presence of narrative in the resource does mean that information is duplicated in the resource. There is, then, a possibility that there will be disagreement between the narrative and the supporting data. If this happens, it creates uncertainty about the correct content of the resource which is unresolvable without consulting the source system for the resource.

Implementers frequently initially create interfaces where the narrative and the data disagree with each other. Almost always these represent implementation errors in one or the other, but if only data was present, this would not be detected. One of the benefits of the narrative is that end-users can participate in this quality checking, instead of being held hostage to the errors. One feature that the basic reference platforms (see Chap. 22) offer is the ability to generate basic narrative automatically from the provided data – this can help users check that the data and provided narrative are consistent with each other. In practice, the presence of the narrative acts as a quality check on the data.

Narrative is not actually required to be present in every resource, and so some implementers decide that their narrow context of use means that they do not benefit from it, and so they will not create it. Typically, this is possible where there is a small bounded community that exchange content amongst themselves, with a high degree of governance and a solid testing framework that is used to ensure safe exchange of data. In this context, it is easy to dismiss the value offered by the narrative, and focus only on the price that it carries (constructing it, testing it, carrying it around and storing it). The problem with deciding that this means narrative is not required is that narrow bounded communities rarely stay simple like this; at some stage in the future it is likely that that the boundaries will lapse, and resources will flow both inwards and outwards. For this reason, implementers are strongly recommended to always include a basic narrative for clinical safety purposes (see additional notes in Chap. 22/Implementers Safety Checklist).

Relationships with Other Organisations

HL7 has built some strong working relationships with other organisations in order to collaborate on healthcare solution design over the past few decades. The most important organisations (see Chap. 6) are:

- DICOM
- IHE
- IHTSDO
- SMART Health

DICOM is a widely adopted standard for exchange of medical images (XRays, CT Scans, MRI etc). HL7 collaborates with the DICOM community to ensure that the interface between the imaging ecosystem and the wider healthcare system is seamless. In FHIR, the DICOM community works through the HL7 process to make the ImagingStudy and ImageObjectSelection resources available. These resources support exposing the availability of images from DICOM end-points to the wider EHR system.

IHE exists to deal with the problem that HL7 can not solve – the lack of agreement around use cases. IHE picks particular use cases that a portion of the community can agree to, and eliminates as much optionally as it can. IHE does not solve the actual problem of disagreement around use cases: narrower use cases allow for less flexibility and more likelihood of working 'out of the box' but are only applicable for a smaller population of potential stakeholders.

The FHIR community collaborates closely with IHE to make the resources AuditEvent, DocumentReference and DocumentManifest available that support exposing XDS repositories (see Chap. 17) over a RESTful interface, which is described as *Mobile Health Documents* (MHD). IHE also publishes profiles of FHIR resources and API functions for the same reason it profiles other HL7 standards.

IHTSDO HL7 collaborates with IHTSDO in the following ways:

- Ensure that it is clear how to use SNOMED-CT properly in HL7 specifications.
- HL7 has a SNOMED-CT extension namespace in which it can define content using the SNOMED-CT definitional framework (though it has not done so).
- Drive greater consistency between the HL7 and SNOMED-CT definitional frameworks.

In FHIR, IHTSDO and HL7 jointly own the page *Using SNOMED-CT with FHIR* [3]. In addition, HL7 and IHTSDO are collaborating to produce mappings between resource elements and their SNOMED-CT definition equivalents for the resources where this is appropriate.

SMART Health is a team run out of Boston Children's Hospital and Harvard Medical School that aims to create innovation in the health system by enabling apps to run against electronic health record (EHR) systems as simply as apps work on a smart phone.

The SMART team collaborates deeply with the FHIR project. SMART uses FHIR for their content model, and SMART provides a particular framework for integrating applications together using FHIR that is suitable for use in many contexts where FHIR is used. *Smart on FHIR* is discussed in Chap. 22.

The FHIR Manifesto

The FHIR team has adopted a manifesto, a declaration of the goals and priorities of the FHIR project:

- Focus on Implementers.
- Target support for common scenarios.
- Leverage cross-industry web technologies.
- Require human readability as base level of interoperability.
- Make content freely available.
- Support multiple paradigms & architectures.

Focus on Implementers The FHIR specification is written for implementers; it is not a theoretical exercise in standards development. Implementers are developers, systems analysts, informatics professionals – anyone involved in making systems work. The team works hard to prioritize what matters to this target audience, and tries to keep arcane theory and abstract layers of structure elsewhere (e.g. the FHIR wiki).

Target Support for Common Scenarios The ethos of the team is that the common things that everyone does should be easy, while the edge cases and difficult requirements should be more difficult. This is a goal, rather than an outcome that can be measured, and how to do this remains an ongoing debate throughout the FHIR project.

Leverage Cross-Industry Web Technologies Wherever possible, the intent of the FHIR project is to use web technologies. Some examples are XML, JSON, RESTful APIs, XHTML, KML, OAuth, SCIM, etc. In other cases, where requirements analysis or architectural constraints prevent using existing content, the FHIR project provides mappings to the relevant content.

Require Human Readability as Base Level of Interoperability The primary expression of this intent is that all the resources can – and should (as discussed above) – have a human readable narrative representation.

Make Content Freely Available The FHIR team wants to make content freely available. The specification itself is licensed under the most permissible license available: Creative Commons Public Domain (https://creativecommons.org/public-domain/). This license allows any form of re-use, and is one of the key reasons for the popularity of the FHIR specification. Software, machine re-interpretations of the specification, derived specifications, republication of portions of the specification – all of these are allowed by the license.

   Beyond this, the goal of the FHIR team is to make interoperability sufficiently cheap that it becomes a commodity, and that clinical and business users take for granted that the information they need is available when and where they want it.

Support Multiple Paradigms & Architectures The most visible part of the FHIR specification is the RESTful API (see Chap. 19). While this is a focus for many of the implementers, the set of use cases for data exchange in healthcare is wider than the problems that can be solved using RESTful. For a variety of architectural, legacy, and workflow reasons, it is not the only or even best solution for all problems. The FHIR team is committed to ensuring that the FHIR content can be shared by means of documents, messages, and services outside of just the RESTful API.

The FHIR Development Process and Maturity Levels

FHIR uses an agile specification process. This is derived from the Agile Programming community, who have a set of software development methods. The Agile Manifesto is based on twelve principles [4]:

1. Customer satisfaction by early and continuous delivery of valuable software.
2. Welcome changing requirements, even in late development.
3. Working software is delivered frequently (weeks rather than months).
4. Close, daily cooperation between business people and developers.
5. Projects are built around motivated individuals, who should be trusted.
6. Face-to-face conversation is the best form of communication (co-location).
7. Working software is the principal measure of progress.
8. Sustainable development, able to maintain a constant pace.
9. Continuous attention to technical excellence and good design.
10. Simplicity—the art of maximizing the amount of work not done—is essential.
11. Self-organizing teams.
12. Regular adaptation to changing circumstance.

The FHIR community is working on developing a specification equivalent of these principles, within the constraints imposed by the structures that community has to work with. A key part of the FHIR process is frequent delivery, changing requirements, and adaptation to changing circumstance. As a working standard rather than a software project, the pace of change is significantly slower than in typical agile software cycles.

Nevertheless, the rate of change in the FHIR specification, or even the fact that it is changing at all, is an ongoing challenge for any portion of the FHIR community that needs to commit to a particular integration solution. At some point, each project has to commit to a particular version, knowing that there will be ongoing changes, possibly even in response to the issues raised by that particular project.

Observing the FHIR specification over time, there is a clear emerging process around the introduction, development, and stabilization of parts of the standard. When new concepts and/or requirements are first introduced, they are very facile, and significant change is expected. In fact, it is guaranteed to happen as the community experiments with the ideas. Over time, the design will be refined and adjusted in the light of experience. As this process occurs, an increasing number of systems implement the working design in prototype and then production systems, and change starts to become increasingly costly. Note that this process can easily be visualized as a typical group dynamic of forming–storming–norming–performing (Tuckman's stages of group development) [5].

The challenge for implementers with this is two-fold:

- Different parts of the specification are running this same process on different timelines.
- It is hard to determine where in this process an artefact is, unless you are intimately involved with the full FHIR development process.

For this reason, the FHIR specification explicitly marks implementable artefacts with a FHIR Maturity Model (FMM) level. Implementation artefacts are things an implementation might conform to, and include the data types, resources, RESTful API, XML, JSON etc.

The FHIR Maturity Model is a formal set of criteria that can be used to determine where in the overall lifecycle a particular artefact is up to. The FMM levels are modelled on the well-known CMM grading system [6]. The FMM level can be used by implementers to judge how advanced – and therefore stable – a resource is. The following FMM levels are defined:

0. The resource or profile (artefact) has been published on the current build.
1. FMM0+the artefact produces no warnings during the build process and the responsible WG has indicated that they consider the artefact substantially complete and ready for implementation.
2. FMM1 + the artefact has been tested and successfully exchanged between at least three independently developed systems at a connectathon whose results have been reported to the FHIR Management Group.
3. FMM2+the artefact has been verified by the work group as meeting the DSTU Quality Guidelines and has been subject to a round of formal balloting with at least 10 implementer comments drawn from at least three organizations resulting in at least one substantive change.
4. FMM3+the artefact has been tested across its scope (see below), published in a formal publication (e.g. DSTU), and implemented in multiple prototype projects. As well, the responsible work group agrees the resource is sufficiently stable to require implementer consensus for subsequent non-backward compatible changes.
5. FMM4+the artefact has been published in two formal publication release cycles at FMM1+ (i.e. DSTU level) and has been implemented in at least five independent production systems in more than one country.

Implementers can use the assigned maturity level to determine where a particular artefact is in the overall process; as the level gets higher, the artefact becomes increasingly stable. At the higher levels, HL7 plans to have a formal consultation process with stakeholders that have registered as implementers of the particular artefact, though the exact details are still to be resolved.

Beyond FMM level 5 is *normative*, when a particular artefact is declared frozen, such that future changes will not break existing implementations. While a few FHIR artefacts have achieved level 5, there are no plans as yet to freeze anything to normative status.

The most important aspect of the FHIR maturity model is that the definition of success – of completion of the development of a resource – reaches far past *publication* and means that the FHIR community has to be focused on the implementation success of the specification. This is the most concrete and tangible manifestation of the FHIR project's focus on implementation.

References

1. Fielding RT. Architectural styles and the design of network-based software architectures. PhD Dissertation University of California, Irvine 2000 http://www.ics.uci.edu/~fielding/pubs/dissertation/rest_arch_style.htm
2. Foote RS. The challenge to the medical record. JAMA Intern Med. 2013;173(13):1171–2.
3. http://hl7.org/fhir/snomedct-html
4. Fowler M, Highsmith J. The agile manifesto. Software Development. 2001;9(8):28–35.
5. Tuckman BW. Developmental sequence in small groups. Psychological Bulletin. 1965;63(6):384.
6. Paulk M. Capability maturity model for software. Pittsburgh: Wiley; 1993.

Chapter 19
The FHIR RESTful API

Abstract The FHIR RESTful API provides a consistent set of HTTP services for understanding the system capabilities, finding (search) and managing resources (read, create, update, delete), and subscribing to content using a set of predictable URLs. FHIR also defines a framework for actions (execute) on the server, and a way to prevent version conflicts on the server.

Keywords API • Security • XML • JSON • System service • Conformance • Search • Joining • Type service • Instance service • Operations • Versions

This chapter assumes that readers are familiar with the HTTP and related specifications. Good short tutorials are available [1].

The FHIR RESTful API has the following parts:

- System Service
- Type Service
- Instance Service
- Operations
- Version tracking

Common Behavior

All parts of the API share some common behavior:

Security The FHIR API makes no rules on the kind of security that protects the operation. While there is certainly an expectation that in almost all production use, there will be some security, the variety of contexts in which the FHIR API is used means that there is no single security approach that meets all requirements. In fact, some non-clinical uses such as terminology distribution may not require any API level security at all. See Chap. 22 for some additional information.

Content-Type FHIR defines 2 content types for FHIR resources: application/xml + fhir, and application/json + fhir. Servers using the XML and JSON representations

© Springer-Verlag London 2016   349
T. Benson, G. Grieve, *Principles of Health Interoperability*,
Health Information Technology Standards, DOI 10.1007/978-3-319-30370-3_19

for resources (see below) should use these content types for the Accept and Content-Type headers. Servers are allowed to accept and/or return other content types. One common use of other types is to return HTML representations for casual use of browsers for debugging – a convenience for implementers.

HTTP Version Multiple HTTP versions are in currently use on the web: 1.0, 1.1, 2.0. FHIR is not bound to any version's specific features, and can be used with any of them. Note that the features of HTTP/2.0 are not generally needed for RESTful interfaces.

Error Handling For all the interactions or operations, except the batch operation, servers return an error response if the operation fails. An error response is any HTTP code of value 300 or greater. An HTTP response code between 200 and 299 indicates that the operation succeeded. Whenever the server returns an error, it should also return an OperationOutcome resource that contains a human readable HTML representation of the error (preferably in the language indicated by the client), and a structured error as an issue (or more than one) that provides detailed evidence of what the problem was.

System Service

The system service is found at the base address of the server. It provides the following functionality:

- Returns the system's conformance statement on request.
- Handles transactions and batches.
- Returns a system wide history – a list of changes on all resources.
- Performs system wide search operations.

Conformance Statement Clients can retrieve the system's conformance statement by performing an HTTP OPTIONS request on the base-address, or a GET on [base-address]/metadata (provided since not all clients can do the OPTIONS command). The response is a statement of how the system conforms to the FHIR specification by detailing all the functionality it provides, including:

- What resource types are supported.
- What interactions and operations are supported for each resource type.
- What search parameters are supported for each parameter.
- What security is required to use the system. Note that systems may choose to secure the conformance operation or not, or return a different conformance statement if the user is known or not.

In addition, the conformance statement can specify the applicable profiles, value sets etc. that the system supports, and provide a really detailed statement of the functionality provided by the server.

A growing number of tools use these conformance statements for functions such as:

- Generating client code.
- Creating search forms.
- Comparing systems for interoperability.

The Conformance sub-system is described in more detail in Chap. 21.

Batches and Transactions The system service can accept a Bundle resource that contains a set of operations to perform in a single HTTP operation. This operation is done by POSTing a Bundle Resource to base-address. There are two modes for this set of operations, Batch Mode and Transaction Mode.

In *Batch Mode*, the bundle contains a set of requests that the server performs sequentially, as if they were individual requests. The server takes the result of each request, and adds it to a single bundle, which is then sent back to the client as the response. This is a way for the client to reduce the effects of network latency. In a batch, each request is processed separately and sequentially, and their success or failure, along with details, it reported back to the client.

In *Transaction Mode*, the operations are processed as a single operation, and all of them fail or succeed together. In addition, the server is responsible for updating internal references between resources in the transaction to their final identities. This allows a client to submit a set of resources that refer to each other, and have the server sort out all the details. In order to ensure that servers are capable of performing these operations, there are additional rules about what the content of a transaction can be. This is one of the most complex parts of FHIR, and the specification should be consulted for full details of regarding how transactions work.

System Wide History The System service is able to return a complete list of changes to all resources on the system. Clients can invoke this by performing a GET on [base-address]/_history. The response is a Bundle resource with a list of all the changes on the system, in reverse order (most recent first). For each change, the server lists:

- The date at which the change occurred.
- Whether the change was insertion, deletion, or update.
- The contents of the final resource (if insertion or update).
- The version of the resource.

This is most commonly used for publication/subscription services, to allow a secondary server to replicate the contents of a primary server. Obviously this list can get very large, so clients can reduce the size of the list by using the lastUpdated parameter, where the value of the lastUpdated parameter is the time stamp on the return from the last query to the server.

```
GET [base-address]/_history?lastUpdated=2015-11-04T13:44:45
```

Implementation Notes:

- The results from _history can be paged (see discussion below under search).
- Transaction boundaries are not represented in the history. The subscribing client may have to deal with transient referential integrity issues.
- On the lastUpdated boundary, a client may receive the same event more than once.

System Search   The system service can also search across all resources, by a GET on [base-address]/_search with some parameters to specify the search:

```
GET [base-address]/_search?_text=diabetes
```

This will return all the resources that have 'diabetes' in their text, or a related word (at the discretion of the server). For this system wide search, only the search parameters that apply to all resources can be used for system wide search. For additional details about search, see below.

Type Service

The type service maintains a set of resources that all have the same type as described by the FHIR specification. The address of the type service is found at [base-address]/[Type] where Type is name of the type, such as /Patient. Note that case is significant, and /patient is not correct.

The type service provides the following functionality:

- Create a new instance of a resource.
- Get a history of changes to the resource type.
- Search through the resource collection.

Create a New Resource   To create a new resource, POST the resource to the Type Manager:

```
POST [base-address]/[Type]
```

If the resource is acceptable to the server, it assigns a new identity to the resource, stores it in the new location, and then returns the new location to the client so that it knows where the resource has been stored. How much checking to perform is at the discretion of the server.

Type Specific History   The Type service is able to return a complete list of changes to all resources of its type on the system. Clients can invoke this by performing a GET on [base-address]/[Type]/_history. The response is a Bundle resource with a list of all the changes on the system to this particular resource type. Other details are the same as for the System wide history. Note that the specification does not provide any ability to filter by the contents of the resources (e.g. By search parameters) because of a series of problems relating to resources falling out of scope by being deleted or the content changing.

Search The type manager allows for a client to search through the list of resources, and return only the subset of resources that meets some particular criteria. Search is the single most important interaction in that it provides a general purpose tool for navigating and joining the resources on the system, and it has received a lot of attention through the implementation process.

To search a resource type, the client performs a GET from the Type Manager:

```
GET [base-address]/Patient?gender=male
```

This searches through the list of patients and returns all the male patients. Multiple parameters can be provided:

```
GET [base-address]/Patient?name=peter&gender=male
```

Clients are not required to provide any parameters. For example, GET [base-address]/Patient is a request to return all the patients. However, many servers will refuse to perform these general open searches. Generally, servers implementing searches need some kind of strategy to limit the load impact that search has on the system.

In order to make search implementable on the server, the FHIR specification separates between the contents of a resource and the search parameter. Clients are not searching directly against the contents of a resource and cannot construct arbitrary searches using some kind of path-based notation (e.g. XPath) into the resource content. Instead, servers declare the search parameters that are available and inform the clients of what features in the resource these relate to. The servers can then index these parameters using whatever technique is appropriate for their implementation in order to provide high performance search facilities. While servers are able to define their own search parameters, the FHIR specification provides a set of common search parameters for clients to use, and most servers use these parameter names and implement most or all of these search parameters.

For each resource type, the specification provides a set of standard search parameters. The selection of these is based on user request through the development life cycle, with a particular focus on providing parameters for joining resources.

The specification defines different types for search parameters than for elements in a resource. The definition of these types is wholly concerned with how they apply as a search on a set of elements across multiple resources. Table 19.1 summarizes the various search parameter types.

The full search specification provides an exhaustive list of details about how each search parameter type works, and what data types they are applied to.

Once a set of resources has been selected, the server must return the set of matching resources. The search results may be very large, so the first obvious thing to do is to return the result as a set of pages, which the user driving the client can walk

Table 19.1 Search parameter types

number	A search parameter that refers to a numeric element
date	A search parameter that refers to a date/time element
string	A search parameter that refers to a simple string, e.g. a name part. Parameters are case-insensitive and accent-insensitive
token	A search parameter that refers to a coded element or identifier. Its value is either a string or a pair of namespace and value, separated by a "|", depending on the modifier used
quantity	A search parameter that refers to a quantity element
reference	A search parameter that refers to a resource reference
uri	A search parameter that refers to a URI element (e.g. an external reference)

Table 19.2 Other parameters

_sort	Specify that the return result is sorted by the nominated parameter. Multiple search parameters can be specified
_summary	Ask for a limited subset of the results to be returned to save bandwidth
_elements	A variant of summary where the client specifies which elements it wants returned

through one at a time. The way paging works in FHIR is collaborative between the client and the server: the client asks for a set of results and indicates what the page size should be. The server responds with the first page and a set of links to the other pages in the search. The server can add additional parameters of its own to the follow up URLs to help it maintain continuity as the client slowly walks through a set of resources selected from a base list subject to ongoing change. This means that the client can not jump right into the middle of a search.

Other parameters that the client can use to affect how the search results are returned are shown in Table 19.2.

Joining Using Search A particularly useful technique is to join across references between resources. There are two different ways to perform joins: *parameter chaining* and *include*.

Parameter chaining is allowed when the search parameter type is reference. The first way to use reference parameters is by value:

```
GET [base-address]/Observation?subject=Patient/345
```

This is a request to get all observations for a particular patient. However, a client can go a step further and select all observations for a set of patients:

```
GET [base-address]/Observation?subject:patient.gender=male
```

This is a request to get all observations for male patients. This in itself is not a particularly useful query, but chained parameters are an important building block for really useful queries. Here is a more useful example:

```
GET [base-address]/DiagnosticReport?subject=Patient/1231243&obser
vation.code= http://loinc.org|1234-5
```

This is a request to fetch any diagnostic reports for a particular patient that include a particular LOINC code as one of the diagnostic observations.

The other way to join across resources is to ask the server to return resources related to the matching resource. For example:

```
GET   [base-address]/Observation?code=1234-5&_include=Observation:
subject
```

This is a request to retrieve all the observations meeting a set of criteria and in addition to return all their subjects. The point of this is to save the client from simply retrieving them immediately in order to properly display the search results, which would have additional network latency costs. When paging applies, the server should return the included resources for the matching resources in each page as part of the page – else there is no reduction in network cost.

What Search Is Not Over the years, the search facilities included in FHIR have become rather powerful, though few servers implement all the features. However, search is still limited in scope to returning sets of resources. It is not a general purpose query language that allows the client to return arbitrary assemblies of data for presentation or analysis. This is as yet an unaddressed feature.

Instance Service

The instance service provides interactions to allow a single instance of a resource to be managed:

- Read
- Update
- Delete
- Get Version History

Read This is the simplest operation in the RESTful API: GET a resource by providing its identity:

```
GET [base-address]/[Type]/[id]
```

Where [id] is the logical id of the resource. As an example:

```
GET [base-address]/Patient/345
```

Note that the logical id – and all ids in FHIR – can be any combination of letters (A..Z, a..z) digits (0..9), the special characters '-' and '.', and can contain between 1 and 64 characters. Ids are case sensitive.

Update In order to change the content of a resource, a client PUTs the updated content of the resource to the address that is the resource's identity:

```
PUT [base-address]/Patient/345
<body: new patient record>
```

Servers may allow a client to put a resource to an as yet unoccupied location, or, in other words, to allow the client to dictate the identity rather than have the server assign it via a Create operation as described above. This is really useful in multiple contexts, especially where sets of resources are transferred, but does require that the server can trust the client(s) to assign properly unique identifiers. For the reasons, servers declare in their conformance statement whether they allow clients to do this.

Servers are expected to apply appropriate business rules before accepting an update – e.g. check that the resource is valid, that immutable elements have not changed, and that referential integrity is maintained. The exact rules vary from application to application.

Servers are not required to return exactly the submitted resource when it is read after being updated. Servers may chose to differ from the submitted resource because of business rules, or because of underlying information limitations (common in legacy systems). However, servers should make changes as minimal as possible to foster the creation of stable exchanges.

Servers may choose to enforce version integrity on updates – e.g. a client can only update the resource if it is based on the most recent version. For additional details, see Versioning below.

Delete A client asks a server to delete a resource by sending an HTTP DELETE operation to the resource identity. The server may then choose to delete the resource. Once deleted, a resource can no longer be 'read' (see above) or found via a search operation. Deleted resources can be brought back to life by updating them to some valid resource.

Note that in many healthcare applications, there is no way to delete existing records; instead, records must be retained and marked as no longer current by some means. For this reason, the Delete operation is frequently not supported at all.

Resource History Service This service provides access to historical versions of a resource. A client can request the full history of a particular resource:

```
GET [base-address]/Patient/345/_history
```

Or to access a particular version from the history:

```
GET [base-address]/Patient/345/_history/2
```

Note that the version id does not need to be a serially incrementing number. There is no way to update a previous version of a resource. For further information, see Versioning below.

Operations

All the interactions described so are part of the base FHIR specification, and apply to all resource types. In addition to these, FHIR defines a special type of interaction called an *operation*. A client invokes an operation to ask the server to execute some action. The outcome of the action might be to return a set of resources, with no lasting impact on the source system resources (other than audit trail entries), such as *return the entire patient record* or it may make a difference to several different resources – such as *reserve the next available non-urgent appointment for patient X*.

An operation is invoked by POSTING a set of parameters to an operation endpoint. Operation end-points have the pattern [manager-url]/$[name], where the manager-URL is one of the service managers defined above (system, type or instance), and the name is the one defined for the operation. So to ask for the value set expand operation (see Chap. 21/Terminology Service) to be invoked:

```
POST [base-address]/ValueSet/[id]/$expand
<body: parameters>
```

The parameters resource is a special resource that contains a list of name/value pairs.

When all the parameters have primitive type representations and the operation is a read-type operation, then the parameters can all be passed as URL parameters, and the operation can be invoked as a GET operation:

```
GET [base-address]/ValueSet/[id]/$expand?filter=text
```

The response from invoking an operation is either a list of parameters, or, if there is a single return parameter with the name *result*, then the resource itself is just returned directly (which may be a bundle containing other resources).

Operation Definitions In order to invoke an operation correctly, then, a client needs to know the following things:

- What the operation does
- Whether it is invoked at the system, type or instance level
- What the input parameters are
- What the output parameters are

The *OperationDefinition* resource contains all this information. Clients can inspect the server's conformance statement to see what operations are available, and then from there, access the definition of the operation that describes how it is invoked.

The specification itself describes a number of operations. Table 19.3 includes the important ones:

In addition to this list, several terminology related operations are defined – $expand, $validate-code, $translate, $lookup, and $closure. These are documented in Chap. 21.

As well as the operations defined in the specification itself, many implementation guides define their own operations, and servers can define their own. A typical FHIR service consists of a set of basic FHIR Interactions and with a few additional operations that provide a means to shift common operations from the client to the server.

Table 19.3 FHIR operations

$validate	Find out whether a server considers a particular resource to be valid or not
$process-message	Process a FHIR message (FHIR messaging is not described in this book)
$find	Find a current list e.g. a patient's current problem list (see Chap. 20)
$everything	Retrieve the entire set of resources related to a patient or just to an episode
$populate	Given a questionnaire, fill out as much of the answers as possible from stored data
$document	Given a composition, build a complete document (Chap. 20/Composition provides more information on documents)

Version Tracking

An important part of the API is managing versions and editorial contention. In order to maintain record integrity, it is important to prevent two users from editing a record at the same time, which would have the consequence that the user that completes their work last overwrites the changes from the other user, with the result that an update is lost. More generally, in systems with multiple applications in the ecosystem, communicating directly with each other, it is necessary to track information currency more carefully. In addition, the ability to look up past versions of a resource is important for review and audit purposes.

In FHIR, a resource is visualized as a logical entity that is actually represented by a series of sequential versions; one of these is the 'current' version. An update creates a new entry in the sequential version list, and sets that as the current one. Note that the current version must be the most recent – there is no way to roll changes back.

Deletes, when they are allowed, work similarly: the delete creates an entry in the version history marked *deleted*, and sets that to the logical current version. A full version history may include multiple delete/update cycles, though this would be very unusual in practice.

FHIR supports record versioning by marking each resources explicitly with two version related values: *meta.versionId* and *meta.lastUpdated*. On the RESTful API, these two elements are updated by the server whenever the content of the resource is changed by a client, or by an internal server process. The server ignores any client provided values for these elements, and replaces them with its own.

The versionId contains whatever internal version marker the server cares to assign, although it must be a valid id (1–64 characters, uppercase and lowercase a to z, digits, and '-' or '.'). The value does not have to be serially incrementing, or even ordered in any detectable way. The only rule is that it must be unique for the version of the resource for the given resource id.

The lastUpdated element is a marker for a human user of how old the information in the resource is. It is not used for version tracking, and does not need to be more precise than resolution to the nearest second.

This data is also represented in the HTTP headers when the resource is read (Table 19.4).

Table 19.4 Version information

meta.versionId	HTTP ETag header. The versionId is a weak ETag, so a versionId of 3141 would be represented as: `ETag: W/"3141"`
meta.lastUpdated	HTTP Last-Modified header. A lastUpdated of 2015-11-30T13:04:20Z would be represented as: `Last-Modified: Mon, 30 Nov 2015 12:04:20 GMT`

The FHIR specification encourages servers to provide full version support, since it is an important part of keeping records consistent and assuring that all changes are properly accounted for. But since many healthcare systems do not implement versioning internally, they can not provide it on their FHIR interface. For this reason, version tracking is not mandatory, though some implementation guides will require it.

Lost Updates can be prevented using a combination of the ETag and the If-Match header. When the client reads a resource, it gets the versionId in both the header and the resource:

```
HTTP 200 OK
Date: Sat, 09 Feb 2013 16:09:50 GMT
Content-Type: application/json+fhir
Last-Modified: Sat, 02 Feb 2013 12:02:47 GMT
ETag: W/"23"
{ "resourceType" : "Patient",
  "id" : "347",
  "meta" : {
    "versionId" : "23",
    "lastUpdated" : "2013-02-02T12:02:47Z"
}, etc.
```

When the client updates the resource, it submits the request with an If-Match header that quotes the ETag from the server:

```
PUT /Patient/347 HTTP/1.1
If-Match: W/"23"
```

If the version id given in the If-Match header does not match, the server returns a 409 Conflict status code instead of updating the resource. Some servers leave it to the client to decide whether it wants a version specific update, while others require it, and return a 412 Pre-condition failed status code if no If-Match header is found.

Reference

1. For example: Podita P. HTTP: the protocol every web developer must know- part 1. Envatotuts + 2013. http://code.tutsplus.com/tutorials/http-the-protocol-every-web-developer-must-know-part-1--net-31177

Chapter 20
FHIR Resources

Abstract Each resource has logical table and UML definitions, along with literal XML and JSON templates. All resources have a set of common data, and also a set of data elements that use common data types. The data types Decimal, Identifier, Coding, CodeableConcept, Quantity and Timing are the most likely to cause implementers trouble. Most implementers will work with Patient and the infrastructure resources Bundle, List, Composition, Provenance, AuditTrail and OperationOutcome.

Keywords Definitions • Logical • UML • XML • JSON • Identity • Metadata • Data types • Decimal • Identifier • Coding • CodeableConcept • Quantity • Timing • Patient • Bundle • List • Composition • Provenance • AuditTrail • OperationOutcome

FHIR resources are small reusable structures that are defined to work with the FHIR RESTful API described in Chap. 19, although they can also be used in other ways too. Generally, resources have the following features:

- Identity (a URL) by which they can be addressed.
- Type – for FHIR, this is one of the types of resource defined in the FHIR specification.
- Defined set of data, as described by the definition of the resource type.
- Version that changes if the contents of the resource change.

Resource Definitions

A resource contains a set of elements defined in a strict hierarchy. Elements either have child elements, or a primitive value. All elements can have extensions, and extensions either contain a value or other extensions.

Resource and Elements are defined as types. All the types defined by FHIR have a logical definition, which is represented as either as a logical table or as a UML diagram, plus XML and JSON representations. The XML and JSON representations are algorithmically derived from the logical definition, as explained below.

© Springer-Verlag London 2016   361
T. Benson, G. Grieve, *Principles of Health Interoperability,*
Health Information Technology Standards, DOI 10.1007/978-3-319-30370-3_20

Figure 20.1 presents the logical content of the type as a hierarchy of named elements, where each element has some flags that provide additional information about the element, its cardinality rules, the type(s) that can apply to the element, and some additional description and rules about the content. Table 20.1 lists the icons that can be used in logical models of resources.

For further discussion about choice elements, see below.

In addition, an element may have the flags shown in Table 20.2 associated with it:

A definition is also presented in UML (Fig. 20.2):

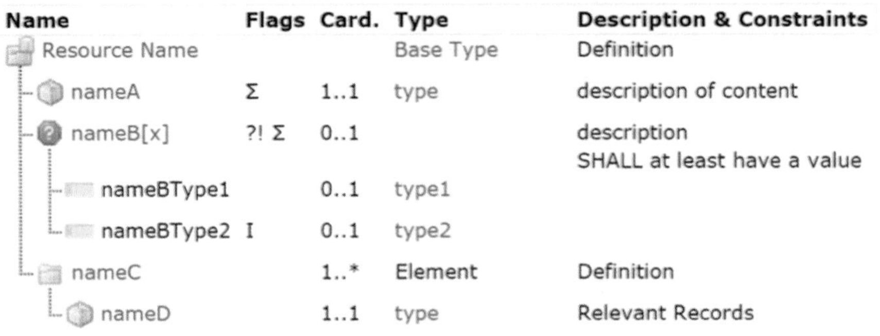

Name	Flags	Card.	Type	Description & Constraints
Resource Name			Base Type	Definition
nameA	Σ	1..1	type	description of content
nameB[x]	?! Σ	0..1		description SHALL at least have a value
nameBType1		0..1	type1	
nameBType2	I	0..1	type2	
nameC		1..*	Element	Definition
nameD		1..1	type	Relevant Records

Fig. 20.1 Logical definition example

Table 20.1 Icons in logical models of resources

	The base element for a resource, if the logical model is for a resource
	An element that is part of the resource and has elements within it that are defined in the same resource
	An *choice* element which can have one of a several different types
	An element that has a primitive data type (see below)
	An element that has a complex data type (see below)
	An element that contains a reference to another resource
	This element has the same content as another element defined within this resource (reuse of content is allowed within a resource)
	An extension
	A complex extension – one with nested extensions
	An extension that has a value and no nested extensions
	Introduction of a set of slices (see Chap. 21/Profiling Resources)

Table 20.2 Flags used in logical models of resources

?!	This element is a modifying element – see below
S	This element is an element that must be supported – see Chap. 21/Profiling
Σ	This element is an element that is part of the summary set – see search in Chap. 19/Type Service
I	This element defines or is affected by constraints – see below

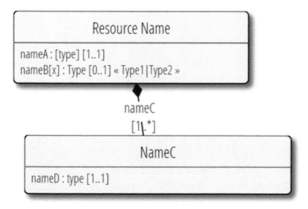

Fig. 20.2 UML representation of logical model of resource

```
<ResourceName xmlns="http://hl7.org/fhir">
  <nameA><!-- 🔒 1..1 type description of content  --><nameA>
  <nameB[x]><!-- 0..1 type1|type2 description  --></nameB[x]>
  <nameC> <!-- 1..* Definition -->
    <nameD><!-- 1..1 type>Relevant Records  --></nameD>
  </nameC>
<name>
```

Fig. 20.3 XML template for resource

The UML representation is concise, which means that the UML diagram can be more useful for visualizing the entire resource, but it also that less information is shown. In particular, the short description shown in the logical view is not present, and readers that only use UML can sometimes misunderstand elements.

XML To help implementers that use XML, a template form is also shown that describes exactly how a resource looks in XML (Fig. 20.3):

Implementers can copy this form into an XML editor, and then replace the element content with the relevant data types and correct data, and the outcome will be a completed resource.

JSON A similar template is also provided for JSON (Fig. 20.4):

For this example, the JSON template shows two nameA elements, to show the difference between primitive and complex data types.

JSON vs. XML Supporting both XML and JSON creates a cost for implementers, and some wonder whether this is worthwhile. The FHIR Community is split roughly 50:50 between using XML and JSON. Many implementers are locked into one or the other by their infrastructure, or by past architectural decisions. For these

```
{
  "resourceType" : "ResourceName",
  "nameA1" : "<[primitive]>", R! // description of content
  "nameA2" : { [Data Type] }, R! // description of content
  // nameB[x]: description. One of these 10:
  "nameBType1" : { Type1},
  "nameBType2" : { Type2},
  "nameC" : [{ // Definition R!
    "nameD" : { Reference(ResourceType) } // R!  Relevant Records
  }]
}
```

Fig. 20.4 JSON template for resource

reasons, the FHIR specification will continue to support both XML and JSON. The Implementation consequences of this are discussed in Chap. 22/Choosing between XML and JSON.

Choice Elements   Some elements in FHIR resources may have more than one data type, depending on the type of data that applies. A typical example is for the value of the Observation resource – what is observed may have multiple different types.

When an element can have multiple different types, the name of the element takes the form nnn[x], where 'nnn' is the base portion of the name, which is fixed, and [x] takes the name of the data type used.

This means that if the element name is value[x], and the valid data types include string and CodeableConcept, then when it appears in the instance (whether in XML or JSON), it will have the name "valueString" or "valueCodeableConcept" for those types respectively.

Elements that have a choice of data types never repeat (this follows inevitably from how elements are represented in JSON). Choice elements are like polymorphic elements in object-orientated languages and are often represented that way in FHIR implementations in such languages.

Modifier Elements   A few elements may be marked as 'modifier' elements (technically, in the element definition, isModifier = 'true'). This means that these are elements that may contain a value that changes the interpretation of the element that contains it. Typically, these are elements such as a 'status' that could have a value 'entered-in-error', or negating elements such as 'didNotHappen'.

The 'modifier' flag is set to alert implementers to the potential significance of the element. Implementers are not required to do anything in particular with a modifier element, but implementers need to be sure that their application will never ignore the value inappropriately.

Common Features of All Resources

All resources have a set of common base features:

- A logical ID
- Common Metadata
- Implicit Rules Tag
- Language

Logical ID

The first element in a resource is named *id*, and contains the id that is also found in the URL as described in Chap. 18/Instance Service. The logical id can be any combination of letters (A..Z, a..z) digits (0..9), the special characters dash (-) and stop (.), and can contain between 1 and 64 characters. The logical id is case sensitive.

On the RESTful specification, the logical id is duplicated between the URL and the resource. It is either fixed by the RESTful API, or about to be fixed (in the body of a POST to the type manager). It is useful to have the id in the resource as well for various implementation related reasons, and there are uses for the id in the resource in other implementation contexts.

Common Metadata

The next part of the content is the *meta* element that contains common metadata that describes the resource. It is important to note that the content of the meta is controlled by the context of use of the resource, and can change without being tracked as a history entry, or affecting the logical interpretation of the content of the resource. The common metadata contains:

- versionId / lastUpdated
- Profiles, security labels, Tags

VersionId/LastUpdated See Chap. 19/Version Tracking.

Profiles A series of URLs that claim that the content conforms to the structure definition found at the referenced URL. For a discussion of their use, see Chap. 21.

Security Labels Codes that are used to indicate particular conditions that apply to access control engines. This subject is further discussed in Chap. 22

Tags A set of arbitrary workflow tags that apply to this resource. These are intended to be used by clients for ad-hoc workflow support. An example of their use might be for a user to tag a set of interesting clinical cases with a code for easy recall later.

Implicit Rules Tag

The goal of the FHIR specification is to enable global interoperability. Any resource created by any application can be safely used by any other application anywhere in the world (though the two applications may not be able to support common work flows). This will become increasingly important as healthcare information is increasingly shared more widely by patients in their own context.

In order to do this, FHIR imposes a set of requirements around what is a valid resource. One of those requirements is that there cannot be implicit knowledge about the content or the context of the resource that needs to be known in order to correctly or safely process the resource. However, in the past, such arrangements are very common – in fact, almost the default in the context of v2 messages.

Some environments may not able to conform to the requirement that no implicit knowledge is needed to safely process a resource. The implicitRules element exists to make this allowable. An application creating a resource where implicit knowledge is required marks the resource with a URL referencing a location where the rules are described. Any application processing the data is required to check the implicit rules element, and refuse to process the content if it does not recognize the value.

An example of implicit rules is where a profile asserts a default value for an element. This is discussed further in Chap. 21/Profiling Resources.

Language

FHIR resources can be tagged with their primary language. The element contains the same content as xml:lang. Note that the data and/or narrative in the resource may contain content in a language other than the stated one, because a human wrote it, and the system is not aware of the language. The primary purpose of the language tag is for indexing. See the W3C discussion about the use of language tags for additional information [1].

Common Features of Most Resources

Other than a few technical resources that are defined purely for use in the technical infrastructure (Bundle, Parameters, Binary), all resources specialize the DomainResource Type. The DomainResource Type defines 3 additional properties:

- Contained resources (see Chap. 18/References between Resources)
- Narrative (See Chap. 18/The importance of Human Display)
- Extensions and modifier Extensions (See Chap. 18/Extensibility)

FHIR Data Types

The FHIR specification defines two kinds of data types: primitive data types, and complex data types. Primitive data types are the base data types from which everything else is built. Complex data types are re-usable assemblies of elements that represent common patterns encountered in more than one place in the specification.

Primitive Data Types The primitive data types (Table 20.3) are based on the types defined in XML schema, with some additional constraints to create a simpler implementation experience. Also, these types have an explicit representation in JSON.

Complex Data Types These data types (Table 20.4) represent common patterns encountered across healthcare data, and are defined after extensive experience in v2, v3/CDA, and other related healthcare specifications.

Most implementation problems and discussion relating to data types arise from decimal, CodeableConcept, Identifier, Quantity, and Timing. (Note: The use of HumanName is discussed below under patient)

Table 20.3 Primitive data types

FHIR name	Value domain
boolean	true I false
integer	A signed 32-bit integer (for larger values, use decimal)
string	A sequence of Unicode characters
decimal	Rational numbers that have a decimal representation
uri	A Uniform Resource Identifier (NB URIs are case sensitive)
base64Binary	A stream of bytes, base64 encoded
instant	An instant in time, known at least to the second and always includes a time zone. Note: used for precisely observed times (e.g. system logs etc.)
date	A date, or partial date (e.g. just year or year + month) as used in human communication. (NB There is no time zone in date)
dateTime	A date, date-time or partial date (e.g. just year or year + month) as used in human communication. If hours and minutes are specified, a time zone SHALL be populated
time	A time during the day, with no date specified
code	A value from a set of controlled strings defined elsewhere (see below)
oid	An OID represented as a URI, e.g. urn:oid:1.2.3.4.5
id	Any combination of 'A'..'Z', 'a'..'z', '0'..'9', '-' and '.', up to 64 characters
markdown	A string that may contain markdown syntax
unsignedInt	Any non-negative integer (e.g. $>=0$)
positiveInt	Any positive integer (e.g. > 0)

Table 20.4 Complex data types

Attachment	Additional data content defined in other formats. E.g. images or reports in a format such as PDF
Coding	A representation of a defined concept using a symbol from a defined "code system" (see below)
CodeableConcept	A value that is usually supplied by providing a reference to one or more terminologies or ontologies, but may also be defined by the provision of text
Identifier	A string of characters that uniquely identifies a single object or entity
Quantity	A measured amount (or an amount that can potentially be measured)
Range	A set of ordered Quantity values defined by a low and high limit
Ratio	Two Quantity values expressed as a numerator and a denominator used e.g. for titers and costings
Period	A time period defined by a start and end date/time
Timing	Specifies when an event occurs, for an event that may occur multiple times. Used for various treatment regimens including medications
HumanName	A name of a human with text, parts and usage information
Address	An address for postal delivery, and also used for finding a patient or person
ContactPoint	Details for all kinds of technology-mediated contact points, including telephone, email, etc.
SampledData	Data that comes from a series of measurements taken by a device, with upper and lower limits. There may be more than one dimension in the data e.g. ECG
Signature	An electronic representation of a signature and its supporting context e.g. XML DigSig, JWT or a picture of the signature
Annotation	A text note which also contains information about who made the statement and when it was made

Decimal

The precision of a decimal value has significance; 0.010 is different from 0.01, and implementations should be able to preserve the difference for display to human users, who may make use of the implicit precision. This means that the typical floating point primitive types provided in many implementations are not suitable for reading and writing decimal values. Note that many XML and JSON libraries will require special arrangements to support decimal values properly.

Implementations are not required to preserve significance when preforming calculations with decimal values, though they can choose to do so if they want. There is no absolute limit to the size of the decimal numbers, in either magnitude or significance, though large or highly precise numbers are very rare in healthcare (except for location co-ordinates, which may be very precise).

Identifier

An identifier is a string of characters that uniquely identifies a single object or entity, and does so globally. All implementers are used to working with identifiers that are unique within a given scope – these are ubiquitous in information systems (e.g. Primary Keys). However, in FHIR, since the focus is on sharing information across scope boundaries, all identifiers must be globally unique.

This is achieved by treating identifiers as a pair of values: a namespace, and the actual unique value within that namespace. The combination of these makes the identifier globally unique. The Identifier type shares this base approach with both the v2 CX data type, and the II data type in v3 and CDA. There is a key difference between the II datatype and the FHIR Identifier type, though: in the II datatype, there is root, and extension, and you only use extension if root is not unique; in the FHIR Identifier type, it is always the combination of namespace and key that conveys uniqueness.

Here is an example in JSON:

```
{
    "system" : "http://fhir.hospital.address/NamingSystem/mrn",
    "value" : "14564234"
}
```

The hospital MRN is 14564234. It is globally unique – that is, different from any other "14564234" because it is scoped by the identifier system http://fhir. hospital.address/NamingSystem/mrn. The system can be any URI that conveys uniqueness – that is, a web address, or a UUID or an OID (using the URI schemes urn:uuid: and urn:oid: respectively). Web URLs are preferred because implementers can resolve them to a definition that provides additional information about the identifier.

A few identifiers are inherently globally unique URLs (or UUIDs and OIDs that can be represented that way). In this case, the system is URIs themselves, which is marked by a reference to the internet standard that defines what a URI is:

```
{
    "system" : "urn:ietf:rfc:3986",
    "value" : "urn:oid:1.2.3.4.5.6"
}
```

Implementers should always take the time to choose the system value carefully. In many countries, implementers will be asked to register their identifier systems on a public registry, and use the URL the registry assigns to the identifier for the system URL – this means implementers can look up the URLs easily on a public maintained server.

In a few cases, the system is unknown. Typically, this happens when a POC testing device scans a patient wristband barcode, but does not know the system. In these cases: the system is simply omitted:

```
{
  "value" : "14564234"
}
```

The Institution gateway that process content from the POC testing devices is often able to populate the system.

Identifiers are often considered to have a 'type'. Common types for identifiers are:

- Institution Medical Record Number
- National Patient Identifier
- National Provider/Prescriber identifiers
- Account Number

Identifiers in FHIR have a type that is a coded value. This allows applications to extract an identifier from a list of identifiers, and populate the correct element in a user interface, or report, or other messaging format, without specifically knowing the actual identifier system. E.g. put the patient's 'MRN' in the first element top-left on the screen – this is more efficient than trying to maintain a look up table of all the system identifiers for every identifier that is an MRN, especially as the scope of information broadens, and as the patient gets involved. There is a set of codes pre-defined for common accepted types of identifiers.

Unfortunately, the type of an identifier is actually a difficult concept to describe – for some identifiers, there is only one of them (e.g. there is only one national provider identifier for any given country, and the name. scope, assignment policies etc. vary widely). For other identifiers, the 'type' is a well understood concept, but how to use it is far from clear. This applies to Medical Record Number – simple small institutions have just one, but once institutions aggregate – for efficiency! – patients start to get multiple MRN's and choosing the 'right' MRN for a particular task becomes very contextual. Finally, there are many ways to look at types, and there is no fixed list of codes – implementers can use alternative codes as they need.

So although FHIR offers an Identifier Type, implementers need to take care when using it, as special local arrangements are typically needed to make proper use of it.

Coding/CodeableConcept

The terminology data types – Coding and CodeableConcept – are perhaps the most difficult to use properly, as they sit at the boundary of the terminology and information models. Terminology concepts and practices are considered in Chaps. 7, 8, 9, 10, and 11, and many of the in principal issues to do with terminology use are dealt with in Chap. 11. In addition to the formal terminologies discussed on those chapters, these data types are used with code systems and enumerations defined informally by all sorts of standards, jurisdictions, and vendors.

Table 20.5 Elements in the coding data type

system	A URI that identifies the terminology system that provides the underlying definition of the code
version	The version of the underlying code system in use
code	An expression of meaning in a syntax defined by the system.
display	A human readable display for the code for applications that do not know the system

The FHIR data types have been designed to be as easy to use as possible after the benefit of decades of experience with the v2 and v3/CDA data types.

The foundation of the FHIR terminology data types is the Coding data type. This data type has 4 elements (See Table 20.5).

These 4 elements are the basic elements for referring to a terminology: system and code identify the concept, version may be important for interpreting the code, and display is a fallback for debugging and display.

System is a URI that identifies the system. Fixed URIs are published in the FHIR specification [2] for the common code systems. For other code systems, implementers must search the public registries and/or register their own system URLs. The system value should always be present – a code without a system is useless.

Version is a string of characters that identifies the version of the system being used. Many systems specify a particular identifier for a version (e.g. a URI for SNOMED CT, the published version number for LOINC). If no version is specified, a date may be used. The proper use of the version is one of the more controversial aspects of this data type. Some of these issues are discussed in Chap. 7/The Chocolate Teapot and in Chap. 11.

Code A code system defines a set of symbols ('codes') that can be used to indicate a particular meaning. The code must be valid as specified by the terminology system. Some systems define a grammar by which additional meaning can be built by combining codes (e.g. SNOMED CT, ICD-10, UCUM).

Display A human readable representation of the meaning of the code. This is a value as specified by the code system, and not the text that was coded (see below for that).

The Coding data type has one additional element called 'userSelected' that is set to true if the user explicitly chose the code from the code system, rather than more generally choosing a concept. This indicates if there is a code present that can reliably be assumed to represents the user's express meaning (see further discussion in Chap. 11)

The CodeableConcept data type builds on the Coding data type. It represents a single 'concept' by one or more Coding data types, and a text representation. The CodeableConcept data type exists to cater for several problems with using terminology:

• Coded representations are usually an approximation of the original intended meaning of the 'concept'

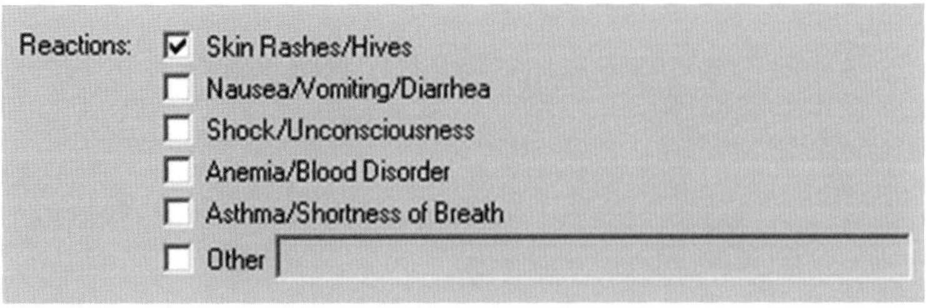

Reactions: ☑ Skin Rashes/Hives
            ☐ Nausea/Vomiting/Diarrhea
            ☐ Shock/Unconsciousness
            ☐ Anemia/Blood Disorder
            ☐ Asthma/Shortness of Breath
            ☐ Other

Fig. 20.5 Example of pick list and optional text entry

- Different code systems approximate the meaning differently
- Applications often are required to provide multiple different codes for different purposes
- Especially for clinical concepts, users are often allowed (or encouraged) to pick a coded representation of a concept, and are also allowed to choose to enter a text description (see Fig. 20.5)

A CodeableConcept must have at least one code, or a text. It can have both and, in fact, it is best practice to provide a text, even when codes are provided, because this can reliably be used as the source system's recommended display for the concept, instead of leaving receiving systems to pick the best display name.

Experience across v2, v3/CDA, and FHIR shows that implementers often do not take due care when using code data types; they do not invest in getting the right terminology system, that systems are registered correctly, in ensuring the that codes match the text, that they choose codes from the right value set (see Chap. 21/Value Sets). The single most common confusion in FHIR is between value set identifiers and code system identifiers, which is discussed in Chap. 21/Using Terminologies.

Quantity

The Quantity data type has 5 elements as shown in Table 20.6.

A quantity has a value and the units of the measure, along with a coded representation of the unit, and the comparator. See notes above about decimal for comments on the precision of the value.

The units are represented twice: in human and computable forms. The unit is whatever is appropriate for display to a human, whereas the system and code provide a computable form of the units and have the same rules as for Coding.system and Coding.code. Three common code systems are used for Quantity.system/code:

- UCUM (Universal Codes for Units of Measure) [3] – these are the best codes to use for physical units
- SNOMED CT – these are used for arbitrary units such as 'scoops', or 'tablets' – mostly these are encountered in dosage instructions

Table 20.6 Quantity data type elements

Element	Type	Description
value	decimal	Numerical value (with implicit precision) (see above)
comparator	code	How to understand the value (<|<=|>=|>)
unit	string	Unit representation
system	uri	System that defines coded unit form
code	code	Coded form of the unit

- ISO 4217 – currency codes, for encoding financial values

The comparator is the most difficult element to interpret correctly. It commonly arises in measured values in diagnostic settings, where the measurement method is calibrated to produce the best support for differential diagnosis, and values may be out of range. Typical examples are <0.5 mg/mL for C-reactive protein, or >30,000 U/L for a cancer marker.

Timing

In prescriptions and care plans, it is common to find instructions to perform a particular action – such as administer a medication or perform some exercise – at some regularly recurring pattern. Many variations on this theme exist, to specify how often something can be done, or how long it should be done for, or that it should be done relative to some other event (typically eating or sleeping). Sometimes, a list of specific times is provided. Further, some bounds may be set (e.g. – from tomorrow, do X for a week), or simply 'do this 10 times'. Finally, many institutions use some of code for common patterns, and a few codes are ubiquitous, such as BID for twice a day.

All of these ideas are represented in the Timing data type. Table 20.7 summarizes the core properties of the Timing data type:

Common codes are

- BID Two times a day at institution specified time
- TID Three times a day at institution specified time
- QID Four times a day at institution specified time
- AM Every morning at institution specified times.
- PM Every afternoon at institution specified times.
- QD Every Day at institution specified times
- QOD Every Other Day at institution specified times
- Q4H Every 4 h at institution specified times
- Q6H Every 6 h at institution specified times

There is a lot of complexity packed into the Timing data type – the problem is inherently complex. Most applications choose to offer users inputs tailored to a particular limited clinical problem rather than a general input form. For example,

Table 20.7 Properties of the timing data type

event	dateTime	One more times at which the event is to occur
repeat.bounds	Duration/Range/Period	A time period over which the schedule should be followed. Duration and Range are time periods not fixed to a particular stating date
repeat.count	Integer	A limited number of times to perform the action (though this is often implicit by limited medication package size)
repeat.duration	decimal + unit of time	How long the action lasts for (may be a range, from duration to durationMax)
repeat.frequency	Integer	How often the action should be done, understood as [frequency] times per [period] e.g. 2 times per day
repeat.period	Duration	
repeat.when	Code	Indicates an event that the action is connected to. When there is an event, period changes to be the offset from the event (e.g. 30 min before eating)
code	CodeableConcept	A coded value that represents a repeating pattern. Except for the common codes listed below, a full repeat pattern should also be provided for applications that do not know the codes

the kinds of rules for exercise schedules are only a subset of these capabilities, and most medications come with a set of limited schedules to pick based on the way a particular medication works.

Important Resources

Of the 96 resources included in FHIR DSTU2, some are worth detailed comment in this book, based on a combination of their ubiquity and some complexities around their use. Chapter 21 considers the conformance resources. Here we discuss Patient, Bundle, List, Composition, Provenance, Audit Trail and OperationOutcome.

Patient

Patient is the first resource most implementers use, as it is the simplest and most ubiquitous use case: manage a list of patients. Some parts of the Patient resource are simple and self-evident. For instance, a patient has a set of demographics elements:

- Names
- Addresses
- Telephone numbers and email addresses
- Date of birth
- Photo

These are the most commonly encountered and widely supported demographics elements. There are many other possible elements scattered across other standards, jurisdictions, and business processes; these can be added as extensions where they are necessary.

In addition, the Patient resource includes a list of contact people or organizations for the patient. This includes next of kin, but is also a wider set, and can include employers and formal guardians that are organizations, such as wards of the state. Names, addresses, and contact details can be provided for these parties.

A patient resource has a list of identifiers. There is a lot of scope for problems in the area of patient identifiers, and although this is a widely encountered problem, it is not really well understood. Most of the issues are underlying patient identification problems are discussed elsewhere in this book, though there are some FHIR specific issues to do with the intersection of identifiers and resource identity.

In a typical FHIR environment, there are multiple servers supporting the patient interface. Usually one of these is authoritative, but multiple systems are involved in maintaining the collective patient list and there are patient link, unlink and merge operations, often with differing perspectives on which system has final authority. To support these environments, the Patient resource allows for links between patients (Fig. 20.6).

The link element allows a patient record to be linked to multiple other patient records. Each link has a type, which indicates what the link means (Table 20.8).

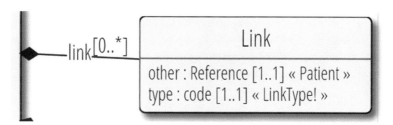

Fig. 20.6 UML representation of Link

Table 20.8 Patient link types

replace	The patient resource containing this link must no longer be used. The link points forward to another patient resource that must be used in lieu of the patient resource that contains this link
refer	The patient resource containing this link is in use and valid but not considered the main source of information about a patient. The link points forward to another patient resource that should be consulted to retrieve additional patient information
seealso	The patient resource containing this link is in use and valid, but points to another patient resource that is known to contain data about the same person. Data in this resource might overlap or contradict information found in the other patient resource. This link does not indicate any relative importance of the resources concerned, and both should be regarded as equally valid

Which of these link types is appropriate depends on how an institution manages patient records. This whole area of patient record overlaps and links, merges and unmerges are common sources of trouble for developers of healthcare applications. A full treatment of the subject of patient links and related identifier issues is outside the scope of this book (it is a whole book to itself!). The take home message at this point is: do not ignore the challenges patient registration problems create for downstream record keeping.

Bundle

The Bundle resource is a technical resource that is encountered in all environments where FHIR is used. Bundles are used to package a set of resources into a single package for transfer. Bundles are used in the following contexts:

- Returning a set of search results
- A server history
- Batches and Transactions
- Documents and Messages

Resources can be moved in and out of Bundles without altering them, and bundles never have any meaning in their own right. A bundle has the following features:

- An id, which is unique for this package and is never reused (UUIDs are strongly recommended)
- A type, which indicates the context in which the bundle is being used (document, message, transaction, transaction-response, batch, batch-response, history, searchset or collection)
- A set of links that provide context around the use of the bundle (e.g. search next and prev links)
- A set of entries, resources contained in the bundle. Note that bundles can be empty, and have no resources in them (e.g. an empty search result).

Each bundle entry has some combination of the following properties:

- Resource. Most entries have a resource, except in batch/transaction/history where some entries just have a request or a response
- fullUrl – an authoritative URL that identifies the resource
- Request or response – information about the request or response (batch/transaction/history only)
- Search Information – information about the resource's function in a search response.

The most difficult part of handling bundles is the relationship between the fullUrl, the entry resource id and the request URL when more than one of these is present. The specification makes a number of rules about how to interpret this and how to resolve references between the resources in a bundle.

List

The list resource links a set of resources together into a logical group. Here are some uses for List:

- Patient's current medication list
- List of Diagnostic Orders made as part of an admission
- List of patients currently admitted to a ward
- List of Care Plans for the patient

Superficially, the List and Bundle resources appear similar – both assemble resources in a sequence. However their function is very different. Bundles are a packaging mechanism that say nothing about meaning. List is a way of indicating meaning that has no packaging implications. A list has the same meaning whether the items it refers to are available on a RESTful interface, or packaged inside a Bundle with the list itself.

One important use of the List resource is for tracking 'current' lists. There are a number of common lists that track the current status of a patient:

- Current problem list
- Current medication list
- Current allergies

Note that for each of these lists, there are variations in scope, definition and workflow that make inter-operating between clinical systems challenging; at this time, clinical teams generally maintain their own lists, and use manual processes for migrating content.

Current lists present a challenge for the FHIR API: there is no inherent property of the list that marks it as 'current'. For example, there may be many 'medication lists' in an application, but only a few (or one) of them are the 'current' list. Further, lists may be superseded by another list without any change – or even any way to change – the list being superseded. This problem is even worse when applications are sharing and retaining digitally signed copies of past 'current' lists.

What makes a list current is its context, how it used, and not an inherent property of the list. So how does a client find a patient's current medication list? The FHIR API defines an operation that can be used to get a current list for a patient:

```
GET [base]/AllergyIntolerance?patient=42&_list=$current-allergies
```

The FHIR specification defines several list tokens for use with this operation[4] but there is a long way to go before these concepts are well understood and widely exchangeable.

Composition

The composition resource represents a set of information that was authored together as a single coherent statement of clinical meaning. There is no clinical information in a composition – just a whole lot of context information – who the patient was, the author, when it was made etc. Then the composition has a set of sections, each of which can contain links to other resources that contain the actual content of the composition or additional subsections.

This structure is the same logical content as the CDA content model (see Chap. 15), and serves a similar purpose. There is a key difference though: CDA, by definition, represents a frozen attested set of content; Composition, on the other hand, spans the authoring process. A composition resource is used to describe an ongoing editing session, which may be transiently in an incomplete state, and so it may be possible for the author to make further changes. The composition resource does not contain all the content in the way a CDA document does, it just references all the resources as they are found elsewhere

It is possible, though, to create a Bundle that contains the Composition resource, and all the other resources it links to. This is a frozen attested set of content that is functionally equivalent to a CDA document. A key advantage of this format over the current version of CDA is that the structured content in the section entries are the same common resource format used on the RESTful API, and can be extracted and archived in this way.

Some CDA implementers are examining what is involved in converting CDA to a FHIR document in order to make the content easier to handle. However, because of the way CDA templating and FHIR resources are defined, there is no general CDA $\rightarrow$ FHIR transform. The HL7 community is investigating whether it is possible to create common tools or mapping/transform artefacts for specific CDA implementation guides. In particular, CCDA is a focus for this work.

Provenance and AuditTrail

Provenance and AuditTrail are two important and related resources that play key roles in tracking data integrity. Provenance is a statement made by the initiator of an update to the data providing details about the data change action and AuditTrail is a statement made by the data custodian about a change made to the data. On a RESTful API, the client makes the provenance statement, and the Server creates the AuditTrail. In other contexts, the relationships may not be so simple.

OperationOutcome

OperationOutcome is one of two special resources – Parameters is the other – that are transient, and not meant to be stored. An OperationOutcome is the right thing to return whenever an operation fails or returns an error, and provides a human readable description of the problem, along with additional diagnostic information.

Sometimes there is some ambiguity whether a business error counts as a technical error or not. For example, when placing an order, and the order is rejected, is this an error? On a FHIR interface, this would be an error if there was no arrangement in the exchange details for the rejection to be represented explicitly [4].

References

1. http://www.w3.org/International/questions/qa-lang-why.en.php
2. http://www.hl7.org/fhir/terminologies-systems.html
3. http://unitsofmeasure.org
4. https://www.hl7.org/fhir/lifecycle.html#current

Chapter 21
Conformance and Terminology

Abstract Implementers have to decide how to use FHIR in their systems. The conformance layer helps with this by providing computable statements about how terminology is used, how the resource contents are used, and how the systems behave, which allows for code generation, and also for testing whether systems or resources conform to the claims made about them.

Keywords Conformance • Implementation guide • Terminology • Content • Profile • Behavior • Local adaptation • Terminology • Code system • Value set • Binding • Expansion • Element definition • Structure definition • Differential • Snapshot • Slice • Discriminator • TestScript

The conformance layer is an important part of the FHIR specification. The conformance resources are a set of resources that make statements about how a system works, or how it should work.

Functionally, these resources are used in several ways:

- To create implementation guides that document expected behavior
- To test that systems conform to a set of documented behavior
- To generate code for reading and writing resources, or search forms or input screens
- Comparing implementations to see if they are compatible

In order to create consistency of usage, the FHIR specification itself provides as much of its own content as possible using the conformance resources.

The conformance resources naturally fall into four categories:

- Terminology – these resources deal with describing how terminologies are used.
- Content – these resources describe the content of other resources. In other specifications, these are sometimes called 'profiles' or 'templates'.
- Behavior – these resources describe the system operations that can or will be used.
- Management – these resources deal with publishing an implementation guide (a set of conformance resources), and specifying testable behavior.

© Springer-Verlag London 2016   381
T. Benson, G. Grieve, *Principles of Health Interoperability*,
Health Information Technology Standards, DOI 10.1007/978-3-319-30370-3_21

Adapting the Platform Specification

The base FHIR specification describes a set of base resources, frameworks and APIs that are used in many different contexts in healthcare. However, there is wide variability between jurisdictions and across the healthcare ecosystem around practices, requirements, regulations, education and what actions are feasible and/or beneficial.

For this reason, the FHIR specification is a "platform specification" – it creates a common platform or foundation on which a variety of different solutions are implemented. This means that FHIR often requires further adaptation to particular contexts of use whenever it is used. Typically, these adaptations specify:

- Rules about which resource elements are or are not used, and what additional elements are added that are not part of the base specification
- Rules about which API features are used, and how
- Rules about which terminologies and used in particular elements
- Descriptions of how the Resource elements and API features map to local requirements and/or implementations

Note that because of the nature of the healthcare ecosystem, there may be multiple overlapping sets of adaptations – by healthcare domain, by country, by institution, and/or by vendor/implementation.

In the past, each country has taken HL7 specifications and further constrained them for their own use, describing how national agreed law, policy, and identifiers are represented in the HL7 messages or CDA documents. In addition, some business trading agreements are made at a national level. When vendors or institutions integrate applications, they start by combining national agreements, vendor capabilities, and institutional requirements – all these determine what content will be exchanged. This process is largely manual, and the outcome is different just about every time.

One goal of the FHIR conformance layer is to allow the results of the various inputs to be shared and compared in a computable fashion, to try and bring more order to the relative chaos that pervades this area now.

Common Conformance Metadata

All conformance resources have the same common set of metadata (See Table 21.1).

Conformance resources have two versions: the version above, and meta.versionId. The meta.versionId is maintained by the system, and is the record version. As a resource is copied from one server to another, the meta.versionId will change. The conformance resource version is under the control of the author – it is an explicit statement of how the resource life cycle should be understood.

Table 21.1 Common conformance resource metadata

url	The URL used to identify this resource, which is also the location where the master copy is to be found
identifier	Other identifiers by which this conformance resource, or the artefact it represents, are known by elsewhere (e.g. OID of a CDA template)
version	The current version. See below for discussion
name	A human readable name for the resource
status	Where the resource is in its status – draft, active retired. This is part of the decision about whether the conformance resource should be in use
experimental	True if the resource is for teaching, testing, or demonstration, rather than intended to ever be for real production use
publisher	The individual or organization that claims responsibility for publishing the resource
contact	Names, phone numbers, emails etc. by which authors or editors might be contacted
date	The date this version was published
useContext	A series of codes that indicate the scope in which this conformance resource is intended to be used. Two common uses of this are for country (e.g. this is for US usage), and for clinical specialty (this is for cardiology use)
requirements	An explanation of the rationale for creating this artefact
copyright	A statement of the copyright applicable to the content that this resource represents

The URL is the location at which the current version of the resource is to be found. When a new version is published, it replaces the existing version at that URL. Typically, then, breaking changes to a conformance resource mean a new URL, with a new name, while minor modifications that are consistent with existing usage mean a new version number.

Terminology Management

The terminology resources address the question of how terminologies are or should be used by a system. Figure 21.1 summarizes the important parts of the terminology system in FHIR.

Code Systems

A code system is any framework that publishes a list of codes, or a set of rules for composing codes, where the codes can be related to a set of definitions, so that they have a known meaning. This is a very loose definition – intentionally. Some code systems are very well known – SNOMED CT, LOINC, RxNorm, ICD-10, ICD-9,

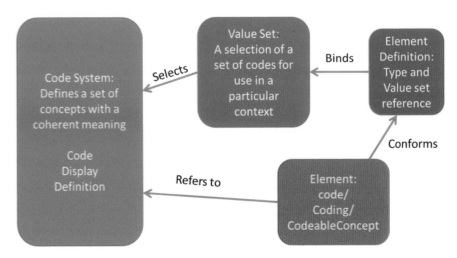

Fig. 21.1 Terminology concepts

but all sorts of other things are code systems too, such as configuration tables in applications, and enumeration values in programming code.

Each code system is identified by a 'system URI', which identifies the code system throughout the FHIR eco-system. In CDA, code systems are identified by OIDs (see Chap. 13/v3 data types). The same OIDs can be used as a URI by prepending "urn:oid:" to the front of it [1], but where possible, providing a real web URL that resolve to some content that describes the code system is much easier for an implementer to use.

Because code systems are always changing, they can be assigned a version as well. This is some string that the code system specifies to use for a particular version (e.g. for LOINC, 2.54). If nothing is specified, then the date of publication should be used.

Ideally, all code systems have the following properties:

- Uniqueness: each code has only on meaning (e.g. does not depend on context)
- Good definitions: each code has clear definition that establishes its meaning without references to inside knowledge, or other codes
- Concept Permanence – codes are not reused for other purposes, and the definitions are not changed significantly

Unfortunately, this is not always true of the more informal code systems, and this may make meaning degenerate to some degree.

From a FHIR perspective, a code system defines a list of possible codes, each with a display that is the preferred way to represent the code to a user.

Element: Coding/CodeableConcept

When a code is used in a resource, the element that contains the code makes a direct reference to the meaning as defined in the code system. This means that it has 4 properties: system, version, code, and display (as described in Chap. 20/Data Types). These refer uniquely to a defined meaning that an application processing the resource can use to display the meaning of the code, and make decisions based on the meaning. System and code are the minimally required elements for the element to carry a defined meaning. Version may be required for some code systems that do not have good concept permanence.

Element Definition: Binding

For elements that may have codes associated with them, the natural follow up question is, 'which codes are allowed in this element?'

In FHIR, every element has a definition that describes its meaning and allowable use (in fact, elements may have multiple definitions – this is discussed below). One of the features of the definition is to 'bind' the element to a value set. Binding an element to a value set means the code in the element must be one of the codes in the value set.

Unfortunately, due to the complexity of the terminology/content model interactions, bindings not only link an element to a value set, the binding also has a strength (See Table 21.2).

Table 21.2 Binding strength

Required	This is the most obvious binding strength – the code in the element has to come from the value set, and there are no exceptions
	This is mostly used for status codes, or workflow codes, where processes are dependent on the codes
Extensible	The code in the element has to come from the value set unless the value set does not contain the right code
	This is used for almost all clinical codes, since there are always new and exceptional cases
Preferred	The publisher believes that this is the best value set to use, but implementers can make their own choice
	HL7 uses this to recommend value sets, but where not all implementers agree to it. For instance, HL7 recommends that LOINC be used for all observables, but some HL7 stakeholders will only use SNOMED CT
Example	There is no recommended value set at all, but this value set serves as a good example of the kind of codes that can be used
	This is common in the specification; it is amazing how often everyone agrees that a coded element is essential for exchange, but there is no prospect of agreeing to a value set for said exchange

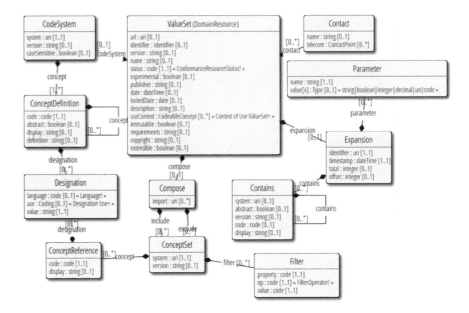

Fig. 21.2   UML class model for ValueSet

Value Sets

A Value Set is a selection of a set of codes defined by one or more code systems, for a particular use. Value sets have a number of uses:

- Defining what codes are valid for a particular element
- Presenting a user with a set of codes so they can pick the right one
- Describing a set of codes that trigger some workflow process, e.g. as part of a decision support process

Value Sets do not define their own codes – code systems define codes. Value Sets simply select codes from code systems.

ValueSet has three parts (See Fig. 21.2):

- A set of metadata (identity, who published it, why, etc.) as described above (*ValueSet* and *Contact*)
- A 'Content Logical Definition' that defines what codes are in the value set (*Compose* and *CodeSystem*)
- An 'Expansion' that explicitly represents the codes in the value set under a particular set of conditions (*Expansion*)

Note that ValueSets do not contain mappings between different code systems – there is a special resource called "ConceptMap" for this purpose.

Content Logical Definition The content logical definition is a series of statements to include or exclude content from a code system. Content from a code system can be selected for inclusion or exclusion by three different ways:

- Listing the codes that are being selected, or
- Selected a set of codes that meet the criteria specified in one or more filters, or
- If no codes are listed, and no filters are specified, then the entire code system is selected

A filter to select codes has three elements:

- Property Name – the name of the property on which the filter is based
- Operator – the kind of operation to make on the property for each code
- Value – the value for the operation

This is an abstract syntax that refers into the properties that are defined by the code system in question. Each code system defines a different set of properties. The FHIR specification defines a default set of properties for a code system, and defines specific properties for common important code systems. Other code systems may define their own properties.

In-line Code System The FHIR value set has an additional feature, a syntactical short cut for an extremely common pattern when working with ad-hoc code-systems such as application configuration tables:

- Publish a simple list of codes with displays and definitions
- Publish a value set that includes all the codes in the code system

The syntactical short cut is to define a value set that includes an in-line code system. When a value set includes an inline code system, it automatically includes all the code in the code system. Inlining a code system definition like this is only appropriate for simple ad-hoc code systems; it is not intended for use with formal code systems like LOINC, SNOMED CT, etc.

One important feature of the code system selection criteria is that they may specify a particular version of the code system – or they might not. In fact, it is more common for a value set not to specify the version of the code system it is based on, with the intention that it defaults to the current version of the code system in use.

Expansion If there are any code systems with unspecified versions, or if there is any filter based selection, then the content of the value set changes over time, and depends on context. Even if the context is fixed to specific versions, there is a lot of complexity here.

Because of this inherent complexity, the third part of value set is to represent the 'expansion' – that is the list of codes that are included in the value set under the current conditions. The expansion is different to the Content Logical Definition because:

- The expansion depends on a context – a set of code systems of particular versions
- A short definition (e.g. 'all SNOMED CT Clinical Findings') can result in a very large expansion

- The expansion changes as the context of use changes, and therefore as time passes
- It is possible to use the expansion without needing any additional knowledge about how the value set was constructed – it is just a list of system, version, code, display

One way to manage value sets is simply to store all the expansions, so they can be used when required. The problem with this is that storing expansions requires the storing system to be hooked into the underlying code system definitions so that it knows when a new version expansion is needed.

The $expand operation allows an application to request the current expansion in the local context. Following the RESTful API, reading a value set with id *123123* is

```
GET [base]/ValueSet/123123
```

An application that gets this value set may get a content logical definition, an expansion, or both. An application can ensure that the server returns the expansion using:

```
GET [base]/ValueSet/123123/$expand
```

In this case, the server figures out the expansion, and returns it. This moves all the logic around the content logical definition to the server. The client simply uses the list of returned codes. The expand operation is made especially useful because it takes a number of parameters:

- A text filter, to allow a user to enter some text, then search (e.g. "Ac Ap", to find Acute Appendicitis)
- Page parameters, to allow a client to page through an expansion in response to a user moving though it
- A profile reference to provide even more control over the expansion process

Terminology Service

The expand operation moves some of the challenge of dealing with code systems and value sets onto a dedicated server. This server is known as a terminology server. FHIR defines a terminology server, which is a server that provides the following functionality:

- Read/Search ValueSet and ConceptMap resources + potentially create/update for terminology maintenance
- Support the expansion operation, along with operations for validating codes and looking up additional information about them
- Support operations for translating between code systems, and for maintaining a closure table, to support search

Proper use of a terminology server allows an application to move all of the work (pain) of dealing with code system to a specialist service. A number of open source and commercial implementations exist [2].

Content Specification

Element Definition

Every element in a Resource or Data type has an 'Element Definition', which is the formal representation of the definition described in Chap. 20. An element definition has:

- The path, which is the identity of the element being defined (e.g. Patient. contact.name)
- Description, comments, requirements, usage notes, etc.
- Cardinality, Type, Usage flags
- Default/fixed/example values
- Invariants (rules that must be true about the element)
- Terminology Binding

These element definitions are packaged in a "StructureDefinition", which is a resource that contains all the element definitions for a particular resource or data type. The FHIR specification publishes a StructureDefinition for every Resource and Data Type.

Adapting the FHIR specification to local use means providing additional element definitions that make additional rules about the elements in a resource or data type. These additional definitions do not replace the base definition; instead, they add to the base definition. These additional definitions can not disagree with the base definitions. Adding these additional definitions, which are also packaged together using the StructureDefinition resource, is known as 'profiling' the FHIR specification.

Profiling FHIR: Structure Definition

Profiles constrain the use and meaning of the contents of resources or types. The FHIR profiling infrastructure has been built on all that has been learnt from constraining HL7 V3 and other specifications. Much of the complexity in healthcare application interoperability manifests in the FHIR eco-system in questions about how to define and use profiles.

Profiles are used for constraining existing resources or data types, or defining extensions (technically, these are profiles on the Extension data type), and are represented using the StructureDefinition resource.

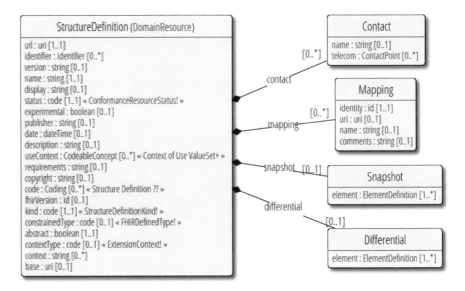

Fig. 21.3 UML class model for StructureDefinition

The StructureDefinition resource has the following parts (Fig. 21.3):

- Shared common metadata for all Conformance Resources (*StructureDefinition* and *Contact*)
- Codes to allow indexing/searching of profiles (*code*)
- Context – this establishes the intent of the profile (*kind, constrainedType, abstract, contextType, context,* and *base*) (Table 21.3)
- Mapping – defines mappings to other specifications/application constructs (*Mapping*)
- The revised definitions of the elements (*Differential* and *Snapshot*)

Each profile contains a set of rules on the element content in a resource. Profiles cannot define any new content the element is not allowed to have, or revoke any rules already established in the base resource. Note that profiles are allowed to add extensions because the base elements are allowed to have extensions.

Profiles can make the following kinds of statements:

- Restricting the cardinality of the element; e.g. the base might allow 0..*, and a particular application might support 1..2
- Ruling out use of an element by setting its maximum cardinality to 0
- Restricting the contents of an element to a single fixed value
- Making additional constraints on the content of nested elements within the resource (expressed as XPath statements)
- Restricting the types for an element that allows multiple types
- Requiring a typed element or the target of a resource reference to conform to another structure profile (declared in the same profile, or elsewhere)

Table 21.3 StructureDefinition context elements

Kind	Data Type, Resource, or Logical Model (not discussed in this book)
constrainedType	The underlying type being constrained e.g. Patient
Base	The structure definition this is a constraint on (this is different from constrainedType if the profile is derived from another profile)
Abstract	Whether this profile is defining a structure that can be used directly, or one that needs further definition in a derived structure
contextType	For extensions, where the extension can be used. Type can be:
Context	resource – the extension can be used in the resource paths listed in the context element (e.g. Patient.contact)
	datatype – the extension can be used in the nominated data type paths (e.g. ContactPoint.use)
	mapping – The extension can be used in any element mapped with this mapping (e.g. Any element mapped to a RIM Act)
	extension – The extension is defined to extend another definition

- Specifying a binding to a different terminology value set, as described above
- Providing alternative definitions, comments/usage notes and examples for the elements defined in a Resource to explain how they are used in the context of the Profile
- Providing more specific or additional mappings (e.g. to HL7 v2 or HL7 v3) for the resource when used in a particular context
- Declaring that one or more elements in the structure must be 'supported' (see below)

Differential and Snapshot

A profile is a series of statements about how a resource is used, made as element definitions. When implementers work with profiles, they generally want to know:

- What difference does this profile make compared to the base?
- What is the final set of cumulative rules this profile makes?

In general, a profile could simply list the differences that the profile makes, and leave it as an exercise for the readers – whether human or computer – to determine for themselves what the cumulative outcome of the rules is. But this is hard work for humans, and prone to creating error and confusion (as every CDA implementer knows all too well). For a computer, the chain of information to determine the cumulative outcome may not actually be available, and computing the correct outcome is not straightforward.

For this reason, FHIR profiles carry the information represented twice: once in the 'differential' which is just the details of what is being changed in this profile, and once in the 'snapshot', which is the computed outcome of applying the statements in the differential to the base.

Profiles can carry either a differential or a snapshot, or both. It is strongly recommended for a profile to represent both. The FHIR project provides tools for authoring profiles or generating snapshots – see Chap. 22/Tools for Implementers.

Slicing and Discrimination

One common pattern that arises when profiling resources comes where an element can repeat, and different repeats need to be profiled differently. For example, an Observation can have 0 or more component measures. The single most important usage of these components is a blood pressure measurement – this is a single observation with two components: Systolic and Diastolic pressure. In the profile, there are two different definitions, one for the Systolic component, and one for the Diastolic component, but in the base resource, there is only one component that repeats.

Applying multiple different definitions to a repeating element like this is called 'slicing' – the list has been sliced up into different sections, in order to give each section a different definition in the profile. In a profile, the first element describes how the slicing works, and then repeat elements with the same path describe the different slices. The slicing information on the first element defines how the slicing works (Table 21.4).

Must support

One flag that a profile can set for an element is whether it must be supported or not. This is used for optional elements, to specify that any application that conforms to the profile must 'support' the element. Typically, this means something like 'the

Table 21.4 Parameters that control how slicing works

ordered	Whether the slices must come in the order they are defined, or whether they can come in any order. (e.g. for Blood Pressure, it is universal practice that the Systolic is listed before the Diastolic, even though you can tell them apart by the component code)
Rules	Whether profiles that derive from this one are allowed to add additional slices or not – and if they are, and it is ordered, whether they can only come at the end
discriminator	An element (or list of elements) that can be used to tell which slice is which (e.g. for blood pressure, you can tell Systolic and Diastolic apart by the component code)
	The purpose of this element is to allow implementers to confidently write case/switch statements to handle the slices, rather than having to guess at how to quickly separate them. Note that some profiles do not have a way to easily tell the slices apart; these are hard for implementers to use

application must allow the user to see and/or choose a value for the element'. However what it exactly means to 'support' an element means is very dependent on the context and the task that the profile is describing. For this reason, the base FHIR specification never labels that an element must be supported, and any profile that labels an element 'must-support' must be explicit about what it means to support the element.

Using Profiles

A profile is a set of rules about the content of a resource. Profiles have several uses:

- Documenting how an application actually works
- Describing the set of acceptable content as part of an implementation guide
- Validating whether content conforms to the stated rules
- Testing whether content meets the pre-conditions for some business process

The FHIR project provides validation tooling – see Chap. 22/Tools for Implementers – which can be used to test whether a resource conforms to a nominated profile.

It is also possible to explicitly mark the resource with a claim that it conforms to the rules defined in a particular profile, by listing the canonical URL from the structure definition in the resource metadata:

```
<Patient>
  <meta>
  <profile        value="http://hl7.org/fhir/StructureDefinition/
  daf-patient"/>
  </meta>
</Patient>
```

This is a claim that this patient resource conforms to the DAF (Data Access Framework, a US national implementation guide) Patient profile. Note that it is just a claim; it is not necessarily true. Of course, you can test whether it is true using the validation tooling.

If you can find out whether a resource conforms to a set of expectations, and you can do that irrespective of its claims to conform to any resource, why explicitly mark a resource with a profile? There are 3 possible reasons for explicitly marking a resource with the profile it conforms to:

1. To make it easier to find resources that conform to a profile
2. To make it easier to process resources of a particular kind
3. Because knowing that a resource conforms to a profile conveys extra information not otherwise explicit in the resource

The first reason represents a performance denormalization – explicitly recording the conformance of a resource to a profile, in order to help support search and filtering, instead of having to test it every time. For this reason, any application is allowed to compare any resources it handles with profiles it knows, and update the list of conformance claims in the resource.

The second reason is another kind of denormalization – to use profiles as drivers of workflow. In essence, it means that instead of a system examining a resource to determine its relevant characteristics, and then deciding how to handle it, an application simply checks whether the resource claims to conform to some set of characteristics, and then treats it accordingly. Some application developers use this to externalize their costs, and by so doing create fragile applications that are easily broken by changing their eco-system. Other developers use this approach to develop robust high-performance applications. The difference between these two outcomes depends on where the profile claims come from; an application that applies the claims itself is robust compared to one that requires its trading partners to make the claim.

The third reason is a problem; if a profile is defined according to the rules described above, there is nothing in the profile that conveys secret knowledge that is not explicit in the resource, so it should never be the case that marking a resource with a profile conveys special knowledge.

As an example, consider an application that has a 'posture' element when recording a patient's height (since under some circumstances, posture affects the height measurement in clinically significant ways). The application could choose to add an extension for posture (or a component observation), and describe this in the profile. This is the recommended way to use profiles.

Alternatively, the application could try and define two different profiles, both of them identical other than that they have 'standing' or 'lying down' in the name of the profile. In this case, the only way to find out the patient's posture is to check the profile identifier the resource claims to conform to, and, in effect, the profile identifier conveys extra information. This information is secret, and potentially dangerous, since it is data that is only known to systems that explicitly know the profile. Profiles should not be used like that: removing the profile claim should not change the information that the resource contains.

Implicit Rules There are a few circumstances where, due to the existing business practices and exchange constraints, there is no choice but to have secret information like this, and this means that it is only safe for applications that are in the know to handle these resources. In these cases, the resource creators must mark the resource explicitly with an identifier for the secret knowledge, in the implicitRules element:

```
<Patient>
  <implicitRules value="http://acme.org/secret-agreement-url"/>
</Patient>
```

Applications that see a resource with an implicit rules identifier that they do not know are required to refuse to process the resource, on clinical safety grounds.

Obviously, the FHIR community highly recommends that this element not be used, as it represents a solid barrier to interoperability. Note that it does not create the barrier – it simply makes the existing barrier explicit.

Behavioral Specification

The behavior of a system is described by the 3 resources Conformance, OperationDefinition, and SearchParameter.

The center of the conformance sub-system is the Conformance resource. This is published by an application to describe what parts of the FHIR specification it supports, and how it uses the content of the resources and terminology. Conformance statements can be made for:

- Actual instances of a server system (e.g. Installed and configured, with known network address)
- Software (may be less precise, since not configured yet, and no known network address, but still has version, release date, etc.)
- Application – not necessarily existing, might be a statement of intent

Conformance statements can be published for both servers and clients. Servers are required to publish conformance statements, since the conformance resource can be fetched directly from the server itself (see Chap. 19/System Service). Clients are also encouraged to publish conformance statements – this allows tools to determine whether a client is compatible with a particular server. However, most clients do not have a conformance statement.

The response is a statement of how the system conforms to the FHIR specification by detailing all the functionality it provides, including:

- What resource types are supported
- What interactions and operations are supported for each resource type
- What search parameters are supported for each parameter

In addition, the conformance statement can specify the profiles that the system uses, and by so doing provide a really detailed statement of the functionality provided by the server.

The OperationDefinition and SearchParameter resources are adjuncts to the conformance statement resource; they define operations and search parameters that a server publishes, or that a client uses. The specification publishes all the operations and search parameters it defines as part of the specification.

Clients are able to use the behavioral resources to test server suitability and construct user interfaces. The FHIR community provides tools that allow for testing whether clients and servers can work together, and comparing profiles and behaviors to determine whether they are compatible, and the union or intersection of them.

Packaging and Testing

The last two resources in the Conformance sub-system are the ImplementationGuide and TestScript resources.

The ImplementationGuide resource serves to group together a set of conformance resources, examples, and other content to describe a particular scope of use for FHIR. The resource is used to gather all the parts of an implementation guide into a logical whole, and to publish a computable definition of all the parts.

The point of the logical whole is that applications can claim to conform to an implementation, or simple reference an implementation guide in the profile claim in a resource, or be assessed against an implementation guide. In addition to referencing all the profiles in the implementation guide, claiming conformance against an implementation guide can represent a closed implementation, in that applications may fail validation if they produce content not described by the implementation guide.

ImplementationGuide resources are also used by the FHIR tooling; the validation tooling uses it to locate the correct web address for the origin of a validation rule, and implementation guide publication tools also use the ImplementationGuide resource as the source for publishing implementation guides.

The TestScript resource contains a series of operations with rules about their expected outcomes that serve to test that a server conforms to an implementation guide. An application can execute the defined operations on a server in sequence, and test to see whether to server returns the correct results. Several servers exist to test whether test servers conform to test scripts [3].

References

1. RFC 3001, http://www.ietf.org/rfc/rfc3001.txt
2. http://wiki.hl7.org/index.php?title=FHIR_Tooling_Eco-system#Terminology_
   Tooling_Ecosystem
3. http://wiki.hl7.org/index.php?title=FHIR_Tooling_Eco-system#Conformance_.2F_Design_
   Tooling_Ecosystem

Chapter 22
Implementing FHIR

Abstract FHIR has strong focus on implementation. This is manifest by – amongst others – running regular connectathons, and publishing a safety checklist, providing open source tools and reference implementations. Implementers must make many decisions – e.g. XML or JSON – when implementing FHIR. Security is an important aspect of implementation. FHIR lays down a general architecture for security and defines some specific functionality for security labels. One really common and important method for handling security is to use SMART on FHIR.

Keywords Implementation • Connectathon • XML • JSON • Server • Safety • Tools • Reference implementation • Open source • Java • C# • Security • Smart on FHIR • Argonaut

From the start, FHIR has had a strong focus on implementation. The success of the specification is not judged by publication, but when – and how – it is implemented. This focus on outcome permeates the entire project and its priorities. Some examples of how implementation is a priority:

- The implementation friendly license on the core specification.
- The life cycle with its focus on adoption as a criterion for the maturity model.
- The core team's provision of libraries that run in many production systems.
- The presence of public testing servers.
- The connectathon process.

This chapter builds on the project's focus on implementation and discusses the implementation process from an implementer's perspective, considering:

- Why connectathons are so important, and how they can be used by implementers.
- General Implementation considerations, or how to think about getting started with FHIR, including how to organize information exchange, how to choose between XML and JSON, and some server considerations.
- How to use the Implementer's Safety Checklist to improve the implementation outcome.
- A review of useful tools for implementers, and the important reference implementations.
- How to implement secure systems, including Smart-On-FHIR.

© Springer-Verlag London 2016   397
T. Benson, G. Grieve, *Principles of Health Interoperability*,
Health Information Technology Standards, DOI 10.1007/978-3-319-30370-3_22

Connectathons

From early in the project, the FHIR team has held regular connectathons. The FHIR specification is driven by connectathon experiences, and the connectathon process is an explicit part of the maturity model.

A FHIR Connectathon is closely related to a 'Hackathon' event but the FHIR community chooses to use the word 'connectathon' because the focus has always been on connecting different systems together.

Typical FHIR connectathons work like this:

- A single room with a series of seats; one for each participant.
- Each participant gets a seat, table space, and power, along with wireless access.
- Each participant brings their own laptop with fully configured development environment and whatever other source code reflects their interest.
- The connectathon lasts one or two days (typical 2 day connectathons tend to go on extremely late on the first day).
- Flexible catering arrangements.
- An organizing team will specify one or more storyboards for systems to implement.
- A Google spreadsheet is used to record the outcomes. Typically, it has a tab for servers to register, a tab for client registration, a tab for client/server testing reports, and a tab for issues discovered.
- For the bulk of the connectathon, there is no formal program; the time is spent doing development, testing, debugging, and responding to issues discovered.
- Occasionally, issues that require group discussion arise; this may be handled as a small group discussion, or involve the whole group. Sometimes the discussion is amongst the connectathon organizers and the outcome is conveyed to entire group.
- At the end of the connectathon, there is a group presentation from the organizers that summarizes the outcomes, demonstrations, discussion of unresolved issues and next steps.

The organizational requirements for a connectathon are minimal. There often is not even a fee for participation, though charging is allowed. The important things to get right are the organizer roles and the storyboards.

Organizers are responsible for choosing the storyboards, interacting with the community to drive involvement, ensuring that there is the right mix of participants to allow meaningful participation in the event (no use having exclusively people with a client focus, or a server focus), resolving technical barriers to participation on the day, and conveying the outcomes of the event to the stakeholders and the wider FHIR community as appropriate.

Storyboards are key to the success of the connectathon. For some connectathons, the descriptions should be detailed and complete, and come with good testing data. For other connectathons, the description is best to be fairly loose, and focus on roles, capabilities, and the provision of testing data. There is no single formula for success.

The connectathons run by the FHIR team use a mix of these approaches. All HL7 connectathons feature multiple streams – different storyboards to involve different participants, though there is often overlap between them. Typically, the connectathons involve the following streams:

- Introductory Patient stream – search patients, add a patient, update a patient (using the public testing servers).
- Terminology Service – this has grown to cover fairly stringent testing of the terminology servers, and provision of services to support the other streams (again, using the public testing servers).
- Conformance Testing – an informal stream focused on providing testing services for other streams, and comparing the results of different testing software.
- Multiple other streams testing functionality under development by the various domain work groups at HL7.

Running multiple streams requires a depth of experience in the organizing team and the participants that only grows slowly.

The most important outcome from connectathons is the way that they foster the creation of a community that has an ongoing interest in connecting systems. Once a stream of connectathons is established, the community starts to organically migrate towards an ongoing virtual connectathon. Most of the servers and test tools live in the cloud anyway, so the community becomes the natural place to seek answers about how to interpret the FHIR specification or an implementation guide when these are not clear.

Anyone can run a FHIR connectathon using the description above as a template. No formal approval process is required, although there is a benefit in obtaining HL7 approval [1] for the connectathon, so that the outcomes of the connectathon can count towards the maturity process of the relevant HL7 artefacts. The HL7 approval process is primarily focused on ensuring that the organizers have enough technical depth and connection to the wider FHIR community to ensure continuity in the ecosystem. It also ensures that the maintainers of the public test servers do not take their servers down in the middle of a connectathon.

General Implementation Considerations

FHIR is a platform standard, intended to be implemented in multiple disparate architectures, for widely varying purposes. In addition, implementers have very different backgrounds, level of exposure to previous HL7 standards (or other interoperability standards), and lock-in to existing architectures. Because of this, there is no simple set of instructions for implementers. However, there are several overall principles to guide implementers:

- Use the public servers when testing basic functionality. If you are implementing a test server, compare your response to public test servers.
- Consider using one the reference implementations [2]. You do not have to, but they have a lot of useful stuff built into them.
- Avoid code generating from the schema (the schema has the least amount of useful information of all the expressions of the specification: just element name, cardinality, and type). If your architecture requires you to generate code from the schema, you can, but you will have to do extra work downstream.
- The FHIR specification makes writing a client easy. Servers are much harder work for many reasons (see Implementing a Server below).
- A number of clinical safety rules are written into the FHIR specification. Some of these represent hard won knowledge that is not obvious to inexperienced application developers. Check the implementers safety checklist (see Implementers Safety Checklist below).

The FHIR development team has striven to ensure that the FHIR specification meets the 5-5-5 rule:

- 5 s to find the specification on the web.
- 5 min to understand what it is.
- 5 h to have some working code.

Our experience is that this largely holds true. FHIR is widely regarded as allowing for anyone to develop healthcare applications without deep experience, but healthcare application developers should be aware of the final part of the -5 rule:

- 5 years to understand healthcare application development.

The FHIR team aims to make the specification itself as simple to use as possible, but FHIR cannot on its own simplify either healthcare applications or healthcare processes.

Push, Pull or Pub/Sub

One of the features of the RESTful interface is that there are multiple ways to share information on one system with information on another. Generally, there are three choices: push, pull, and pub/sub.

For this section, imagine a laboratory information system (LIS) is sharing generated data – DiagnosticReport and Observation resources – with a clinical data repository (EHR) (Table 22.1).

Each of these approaches achieves the same outcome, but with very different operational behavior, and differing consequences for custodianship and security. If a middleware server is introduced, these models can be combined, such as having the LIS post to a middleware server, and then the EHR queries the middleware server.

Table 22.1 Options for arranging data flow

Push	Whenever new or updated reports are available, the LIS as client uses the create and update operations to **push** the reports to the EHR as server
Pull	Whenever the EHR wants diagnostic reports, it uses the search and read operations as client on the LIS as server to **pull** them
Pub/Sub	The LIS makes the reports available as a server. The EHR as client uses the history operation to check for new reports every few minutes

Truly robust systems can operate in all three modes, but this has additional implementation costs. The base FHIR specification itself makes no requirement about which pattern should be used, but implementation guides can and do make these rules.

Choosing Between XML and JSON

The specifics of the XML and JSON formats are discussed in Chaps. 4 and 20. This section addresses making a choice between them.

For servers, the best strategy is to support both XML and JSON. This allows clients to choose between the two. Clients can also choose this approach too. In practice, this requires an object-based architecture, with a translation layer between the objects and the wire format (this is pretty much the default architecture for code generators). Standard code generators, such as XML schema based ones e.g. JAXB, can be used, but given the differences between the XML and JSON formats (some deliberate, and some inevitable due their fundamental difference), a standard code generator cannot produce a single object model that has both XML and JSON translators without a considerable degree of customization. As a consequence of this, most of the servers and clients that support both XML and JSON use one of the open source reference implementations.

If using a reference implementation in an application is not appropriate, implementations are often constrained already by their architecture to use either JSON or XML. This is the reason why both are supported by the specification. Client applications written to run in web browsers may use either, but JSON is generally a more natural fit internally.

In general, the difference between XML and JSON is that XML has a much deeper stack of specification and tooling. XML offers more for existing architectures – schema, validation libraries, translation tools, but also offers more ways to get tripped up by the format – namespaces, whitespace handling ambiguities, character set issues and mismatches between schema and document practices. JSON, as a simpler format, has neither of these advantages or disadvantages.

The specification itself has a noticeable preference for XML. The only reason for this preference is that XML includes an explicit comment syntax, which JSON does not. For the example resources in the specification, comments are a key part of their value. In production data exchange, however, there is no value to having comments in the resources, and since they have a potential for creating confusion they should be avoided in production data exchange, so this advantage of XML is only relevant for implementation during testing/learning/debugging.

Implementing a Server

The FHIR RESTful API naturally divides the world up into client and server, using the HTTP paradigm. Following the HTTP paradigm, clients are expected to be relatively light applications, while servers have to deal with security, threading, multiple representations, searching and indexing, and persistence. The server provides its services to make the clients life easier. In keeping with this philosophy, the FHIR specification moves work to the server whenever possible, and writing a server involves a lot more work than writing a client. This is mainly in the breadth of the services involved, though there are complex interrelationships between the services and security requirements.

For this reason, a key aspect of building a server is choosing what services to offer. Minimally, the only service a server has to offer is returning a conformance statement that says what services it provides. Beyond this, the services provided should be considered carefully and weighed against business benefits. Generally, for each resource, services should be judged against these considerations:

- Read: Get/Search should always be provided if a resource is supported:

  - Search parameters: reference parameters are the way joins are done, so supporting these is a priority.

- Write: create/update should generally be provided together if write access is needed:

  - Supporting 'upsert' – update to a new location – is best as a configuration choice.
  - Some resources are stored for audit only, and do not need to support update.

- Delete – support for delete is a policy question, but how to correct erroneous information should be clear.
- Versioning: recommended for legal reasons, but many systems do not:

  - Versioned updates (see Chap. 19/Version Tracking) and providing access to past versions are separate decisions.

The FHIR API is an external interface to a healthcare system. As long as the interface supports the functionality described in the base specification, and any applicable implementation guides, then it is a conformant application. How it is implemented – what happens behind the curtain, as it were – is irrelevant to the provision of the API.

Many different kinds of systems implement FHIR. The most obvious implementation is to develop a fresh server from ground up that is entirely based on FHIR, one that uses the resources as the 'lingua franca' of the system. The public test servers, the ones that implement 'all resources, all functions, all search parameters' – these are all built in order to serve FHIR resources.

However, such de-novo new implementations of FHIR servers are the exception. Thousands of pre-existing systems already provide the existing set of healthcare services, and no one is rushing out and re-writing them from scratch to be built around FHIR. Instead, FHIR is usually implemented as a bolt-on interface to existing legacy services and data.

Generally, the following strategies have been employed:

- Write an interface converter with some middleware that converts between FHIR operations and an internally supported protocol (v2, XDS for MHD – mobile access to health documents).
- Write a separate module that uses the application's internal data structures and programming libraries to make the data available directly using FHIR.
- Write a service that trickles data out from the system to a general purpose database (this has the advantage of putting the load on another server and separating public access from the production health system).
- Work a set of FHIR related functionality through the system, and gradually FHIR-ise the internal systems in whatever form is useful.

The last option is the most work, but any serious implementation of FHIR ends up heading down this path to some degree. One of the drivers for this is when application designers are starting to use FHIR as an internal modularization tool for their system design.

However the implementation is approached in these cases, the bottom line is that a system's ability to implement the features of the FHIR API, the contents of the resources, and the search parameters is constrained by its existing data model and storage implementation.

For this reason, the specification is clear that servers are not required to return a resource exactly as it was POSTED/PUT to it. Obviously, the more faithful a server is to the original content, the less impact this will have on the implementations it is supporting. But because systems are very often unable to do this, the specification does not mandate it. Some implementation guides may make such a requirement, however, either explicitly or implicitly.

One final note about storage: some implementations break resources and data types down to a SQL schema, and store the data atomically in relational data bases. Other implementations use a document database such as MongoDB [3]. Still others use a hybrid approach, storing resources as blobs in an SQL table, and extracting the content that is used for searching to a relational database structure that is normalized to some degree. There is no single answer to the question of which of these is superior – each has different sets of strengths and weaknesses.

Implementers Safety Checklist

The FHIR specification itself includes an implementer safety checklist [4]. All developers implementing FHIR should run through the content of this safety checklist before and after the implementation process.

Almost all interoperability developments occur in a limited context with one to a few trading partners, and relatively well controlled requirements. However, experience shows that over time, new participants and new requirements will creep into the eco-system and safety features that appeared unnecessary in a well-controlled system turn out to be necessary after all. *Ignore these checks at your peril.*

Production Exchange of Patient or Other Sensitive Data Will Always Use Some Form of Encryption on the Wire This is a fairly obvious thing to say in principle, but it is extremely common to find insecure exchange of healthcare data in practice. FHIR does not mandate that all exchange be encrypted, though many implementers have commented that it should. There are some valid use cases not to use encryption, such as terminology distribution etc. Implementers should check that their systems are secure. For further discussion, see the Security section below.

For Each Resource That My System Handles, I Have Reviewed the Modifier Elements In resource definitions, a number of elements are marked as modifying elements – see Chap. 20/Resource Definitions. Implementers are not required to support these elements in any meaningful fashion. Instead, implementers are required to ensure that their systems do not inappropriately ignore any of the possible values of the modifier elements. This may be achieved by:

- Ensuring that these values will never occur through proper use of the system (e.g. documenting that the system only handles human patients).
- Throwing an exception if an unsupported value is received.
- Ignoring the element that contains the modifier element (so that the value is irrelevant anyway).

Note that applications that store and echo or forward resources are not 'processing the resources'. Processing the resources means extracting data from them for display, conversion to some other format, or some form of automated processing.

My System Checks for ModifierExtension Elements Modifier Extensions (see Chap. 18/Extensions) are only seen rarely, but when they exist, they mean that an implementer has extended an element with something that changes the meaning of the element, and it is not safe to ignore the extension. For safety purposes, implementers should routinely add some kind of code instruction like this:

```
Assert(object.hasNoModifiers, "Object at path %p has Unknown modi-
fier extensions")
```

This should be done for each object processed. Of course, the exact manifestation of this instruction will vary depending on the language. Performing these checks is a chore, so it is frequently not done, but it should be done for safety purposes. Note that one cheap way to achieve this is to write a statement in the documentation of the application: "Do not send this application any modifier extensions". Like all cheap ways, this is likely to not be as effective as actually checking for extensions throughout an application.

My System Supports Elements Labeled as "Must-Support" in the Profiles That Apply to My System Implementation Guides are able to mark particular elements as 'must-support' (see Chap. 21/Profiling Resources). This means that although the element is optional, an application must be able to populate or read the element correctly. What precisely it means to do this correctly varies widely, so Implementation Guides must indicate exactly what they mean when marking an element as 'must-support', and applications that claim to conform need to do whatever is prescribed.

For Each Resource That My System Handles, My System Handles the Full Life Cycle (Status Codes, Record Currency Issues, and Erroneous Entry Status) Many resources have a life cycle tied to some business process. Applications are not required to implement the full business life cycle – they should implement what is needed. But systems need to fail explicitly if the life cycle they expect does not match the content of the resources they are receiving

A common and important area where applications fail to interoperate correctly is when records are created in error, or linked to the wrong context, and then must be retracted. For instance, when a diagnostic report is sent to an EHR linked to the wrong patient. There are a variety of ways to handle this, with different implications for the record keeping outcomes. Failure to get this right is a well-known area of clinical safety failure.

The FHIR specification makes some rules around how erroneous entry of resources is indicated. Applications should ensure that they handle these correctly.

My System Can Render Narratives Properly (Where They Are Used) The general theory of text vs data is discussed in Chap. 18 /The importance of Human Display. Resources can contain text, data or both. Systems are not obliged to be able to display the narrative; they can always choose to process the data. But in many cases, it is a good idea to offer the user a choice to see the original narrative of the resource (or resources, in many cases), particularly for clinical resources. This might be described as 'see original document' in a user relevant language.

The FHIR specification makes no explicit requirements in this regard, since the correct behavior is so variable. Implementers should judge for themselves what is appropriate in this regard.

My System Has Documented How Distributed Resource Identification Works in Its Relevant Contexts of Use, and Where (and Why) Contained Resources Are Used Many of the clinical safety issues that arise in practice arise from misalignment between systems around how identification and identifiers work. In the FHIR context, this risk is particularly acute given how easy it is to develop interfaces and connect systems together. Any applications that assign identifiers or create resources with an explicit identity should document their assumptions and processes around this. This is particularly important where there is the prospect of more than two trading partners.

The same applies to contained resources (see Chap. 18/References between Resources): a system should refrain from using contained resources as much as possible, and where it necessary, document the usage.

My System Manages Lists of Current Resources Correctly See Chap. 20/ Important Resources for a discussion of current lists. If the system has current lists, it must be clear how to get the correct current list.

My System Makes the Right Provenance Statements and AuditEvent Logs, and Uses the Right Security Labels Where Appropriate See Chap. 20 for a brief discussion of Provenance and AuditEvent, and see below for a discussion of Security Labels. The record keeping requirements have been reviewed, and Provenance and AuditTrail are used properly to meet those requirements.

My System Checks That the Right Patient Consent Has Been Granted (Where Applicable) Patient consent requirements vary around the world. FHIR includes the ability to track and exchange patient consent explicitly, which is a relatively new integration capability. Various jurisdictions are still feeling out how to exchange consent to meet legislative and cultural requirements.

When Other Systems Return Http Errors from the RESTful API and Operations (Perhaps Using Operation Outcome), My System Checks for Them and Handles Them Appropriately Ignoring errors, or not handling them properly, is a common operational problem when integrating systems. FHIR implementers should audit their system explicitly to be sure that the http status code is always checked, and errors in OperationOutcomes are handled correctly

My System Publishes a Conformance Statement with StructureDefinitions, ValueSets, and OperationDefinitions, etc., So Other Implementers Know How the System Functions This is discussed in depth in Chap. 21. While servers have no choice but to publish a conformance statement, the degree of detail is up to the implementer. The more detail published, the easier it will be for systems to integrate.

My System Produces Valid Resources It is common to encounter production systems that generate invalid v2 messages or CDA documents. All sorts of invalid content can be encountered, including invalid syntax due to not escaping properly, wrong codes (see Chap. 11), and disagreement between narrative and data.

In the FHIR ecosystem, some public servers scrupulously validate all resources, while others do not. It is common to hear implementers announce at connectathon that their implementation is complete, because it works against a non-validating server, and not worry about the fact it does not work against the validating servers.

Use the validation services (see below) to check that your resources really are valid, and make sure that you use a DOM (document object model) or are very careful to escape all your strings.

Check for implicitRules All resources can carry an implicitRules pointer – see Chap. 21. While this is discouraged, there are cases where it is needed. If a resource has an implicitRules reference, you must refuse to process it unless you know the reference. Remember to check for this.

Tools for Implementers

As part of the implementation focus, the FHIR team provides a number of useful tools to implementers. The most important of these are the public test servers:

- http://fhir2.healthintersections.com.au/ (all resource types, all API features, XML + JSON)
- http://spark.furore.com (all resource types, all API features, XML + JSON)
- http://fhirtest.uhn.ca/ (all resource types, all API features, XML + JSON)
- http://sqlonfhir-dstu2.azurewebsites.net/fhir(all resource types, but not all API features, XML + JSON)

There are many other test servers, including ones from Oridashi, Epic, Cerner, Health Samurai and others, and the current list of test servers is maintained at the hl7 Wiki [5].

Any implementer can connect to these servers and use the API. All of them support non-authenticated access, and many also support Smart-on-FHIR (see below). Since any implementer can use the API, it is possible for implementers to run into each other – e.g. both perform updates on the same resource. The maintainers of the public servers occasionally wipe the database and start afresh (usually by preloading all the example resources out of the specification).

Another important set of tools for implementers is validation tooling. As discussed above, encountering invalid content is a common experience in v2 and CDA. One way that the FHIR project has addressed this is to provide a set of services built on the conformance framework described in the previous chapter that are easily consumed. Implementers can validate resources using any of the following methods:

- Use the Java Validation Jar provided as part of the specification [6].
- Ask one of the public test servers to validate resources [7], using the $validate operation described at http://hl7.org/fhir/resource-operations.html#validate, or using a web interface if they provide one.
- Open the resource in Notepad++ and ask the FHIR Notepad++ plug-in to validate it.
- Validate it against the XML Schema + Schematron.

The schema/schematron method is the least capable of these methods – the others provide a much deeper validation, particularly with regard to coded content.

Beyond this, there are a wide variety of other useful tools for implementers to use:

- Open source reference implementations (see below).
- Validation tools [8] as discussed above [8].
- Terminology Services.
- A Notepad++ Plug-in that has a set of useful features for developers.
- A Profile editor [9].
- A ValueSet Editor [10].
- A public registry for conformance resources.

A curated list of tools for implementers can be found at http://wiki.hl7.org/index.php?title=FHIR_Tooling_Eco-system.

Open Source Reference Implementations

The FHIR project provides a number of open source reference implementations. These are provided free for implementers to take advantage of. There are several reference implementations targeting different platforms:

JAVA/HAPI For Java, there are two reference implementations, the *Java Reference Implementation* [11] and HAPI-FHIR [12]. These two APIs exist for historical reasons. The Java reference implementation includes an object model, parsers, and all the validation, generation and terminology tooling used during the build. The Java Reference implementation is the basis of the tooling that produces the specification, so it is completely conformant to the spec; or perhaps since the Java Reference Implementation is used to build the specification, bugs in the Java reference implementation become bugs in the specification, and vice versa. The public server http://ontoserver.csiro.au/fhir is implemented using the Java Reference Implementation.

The Java reference implementation was always maintained with the needs of the build process as a priority, but it is suitable to be used in production systems, and has been used by many systems, including systems in production. The Java Reference Implementation is not integrated with any client and server frameworks, in order to keep the build tool dependencies down. For this reason, and others, the HAPI team developed their own implementation (Fig. 22.1).

The HAPI (HL7 Application Programming Interface, pronounced "happy") project is a team of programmers with a long and well-recognized track record of producing open source software for HL7 related integration projects. The HAPI framework does not include all the various support utilities that the Java Reference Implementation does, but it includes many framework and design features that are important for either client or server development. Use of HAPI is generally the preferred approach for implementers, and has been used in a wide range of implementations, from mobile phone solutions all the up to implementations by the

Fig. 22.1 James Agnew
(University Health
Network, Toronto),
program lead for the
HAPI-FHIR project.

largest cloud service providers. The public server http://fhirtest.uhn.ca/ is implemented using FHR-HAPI

The maintainers of the Java reference implementation and the FHIR specification tooling are working with the HAPI team to bring the two implementations together; as of going to press, this process is a nearing completion.

DotNet/C# (Fig. 22.2)

The DotNet reference implementation is written in C#, and supports DotNet 4+. It provides an object model, parsers, and various utilities. It has been used in production on a wide range of systems from mobile phones through to cloud servers, including http://spark.furore.com and http://sqlonfhir-dstu2.azurewebsites.net/fhir.

Pascal (Delphi) The Pascal implementation is the basis for the http://fhir2.health-intersections.com.au server (and others at healthintersections.com.au). It has an object model, parsers, validation and generation utilities, and a Smart-on-FHIR implementation. The server built on the reference implementation [13] includes a full blown terminology service. The notepad++ plug-in is also built using the Pascal reference implementation.

Other Open-Source Reference Implementations In addition to these 3 primary reference implementations, there is ongoing work to maintain a Swift implementation [14] and various JavaScript and XML Utilities [11]. There are other open source implementations available around the Internet, but these have not been widely used, or are not maintained on a formal basis.

Fig. 22.2 Maintainers of the DotNet reference implementation, Ewout Kramer (Furore) and Brian Postlethwaite (Telstra) (credit: Ken Rubin http://kenrubinphotography.com/)

Security

FHIR is not a security specification, nor does it require any particular security implementation. However, most implementations require some kind of security, such as:

- Communications Security: unauthorized systems should not be able to eavesdrop on the exchanges, nor tamper with them.
- Authentication / Authorization / Access Control: Users and/or clients usually need to be authenticated, and then any actions need to be subject to thorough access control. Security Labels may be used to assist the access control system.
- Audit/Provenance / Digital Signatures – applications will require some degree of this.
- Narrative and Attachments - FHIR includes XHTML narrative, and allows for binary resources and attachments. These have their own security concerns.
- Consent and Data Management Policies - FHIR defines a set of capabilities to support data exchange. Not all the capabilities that FHIR enables may be appropriate or legal for use in some combinations of context and jurisdiction (e.g. HIPAA includes many rules for exchange between institutions). It is the responsibility of implementers to ensure that relevant regulations and other requirements are met.

FHIR provides some building blocks for implementing a proper security system, and the web security protocols provide others. Particular implementation guides may require specific security approaches for particular solutions.

Access Control and Security Labels Most FHIR servers will need to implement access control on incoming requests using some kind of access control engine. Generally, there are three main architectures to consider (Fig. 22.3):

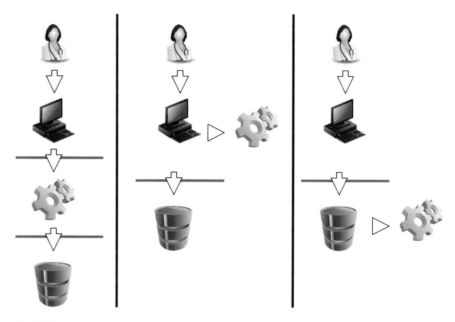

Fig. 22.3 Access control architectures
Legend:

 The consumer that is using a healthcare related system.

 The client application the user is using (application, mobile app, website, etc.)

 The security system (authentication and access control).

 The server – a clinical/healthcare repository.
——— RESTful API.

In most contexts, there has to be some layer of security between the end-user and the server repository. The security engine – some kind of role-based or access-based decision engine – can either included in three different places (Table 22.2).

Note that it is possible – and even common practice – to include security engines in all 3 places, though the complex interplay between the security engines is a primary source of security flaws.

One factor to keep in mind is that when security is needed, the processing engine that responds to RESTful requests needs to use the access control engine deep inside its process. Here are some reasons why:

- Clients may have access to some resources and not others, based on their properties, in ways that are not obvious in the request or the response (e.g. a client that has access to a patient's records and any medications linked from the patient's MedicationStatements).

Table 22.2 Security architectures

Integrated into Client	The client makes security decisions, and the server does whatever the client says. This is very convenient for the server, but prone to failure the client's protections are circumvented, which is often easy in a web framework
Facade on Server	The security engine inspects requests and responses between the client and the server to apply security. This allows implementers to modularize applications, such that the security portion is re-usable across multiple servers (a Cloud API provider advantage), and also allows for easier testing of the security functionality
Integrated into Server	The security engine is integrated inside the server, and the server uses the security sub-system as part of the process of responding to requests

- A request to read a resource type may lead to returning a contained resource type that the client is not authorized to access.
- A response might contain content that client has no access to, and so must be selectively removed.
- Requests to create and update content may propose links to content that the client has no access to.
- Search chain parameters and _includes can reach deeply into the underlying content.

For these reasons, all servers must have some aspect of option #3 in their design, to implement the kind of decisions above. A common cloud solution when using a primary architecture based round option #2 is for a generic cloud provider to provide surface security, and to implement the simpler front-line rules on the cloud façade, and for the cloud provider to pass an OAuth/OpenID token (e.g. the Smart on FHIR token – see below) through to the application server for it to implement the more specific content specific security rules, such as those just listed.

Security Labels

All resources can carry one or more security labels. These are tags that convey additional information about the resource for use by the access control system, so it can make the right decisions when granting or denying access to a resource. Typical security labels indicate:

- A general confidentiality level (Low → Very restricted).
- That a patient is a celebrity or a staff member.
- Information is not to be shared with the patient, or with their family.
- Information relates to sensitive medical diagnoses (typically, STDs, and mental health issues).
- Special agreements pertain to disclosure.
- That information is not to be stored by the receiving application.

This list is far from complete; the complete HL7 catalogue of security labels has more than 500 labels, and institutions can define their own.

Security labels are used to make explicit information that is often implicit, and so represent a form of denormalization for convenience and consistency. This is especially useful because what makes a particular resource sensitive may be information in its relationships (e.g. this admission was to a psych treatment ward), or the information may be in narrative, and only a human would know to mark it as sensitive (though there are prototype FHIR client processes that search through resources applying security labels based on natural language and terminological processing of the resource contents). Making the security labels explicit removes any ambiguity and uncertainty.

The FHIR specification does not require any particular consequences in the API for any security labels; they are always regarded as standard inputs to the access control engine, which is expected to enforce the agreed organizational policy. HL7 is currently working on what forms of standardizing access control that can be achieved.

Smart on FHIR

Smart on FHIR is a set of open specifications to integrate apps with healthcare data provider systems (e.g. Electronic Health Records, HIEs, etc.) It builds on top of the FHIR interface to add:

- A set of scopes and permissions agreed between client and server that are specific to the app context.
- A simple reliable and secure login process using OAuth2 and OpenID Connect.
- Consistent UI integration between client and server.
- Interactive decision support between user, server, and decision support systems.

The Smart on FHIR login process can be used for all sorts of purposes, just as apps can be used for all sorts of purposes [15]. The combination of the FHIR API, and the Smart on FHIR integration layer creates the capability for all sorts of powerful integrations in a way that has not been possible before.

The foundation of Smart on FHIR is a pair of launch sequences, EHR-Launch sequence (Fig. 22.4), and the standalone launch sequence (Fig. 22.5). They differ because of which system initiates the exchange. Note: the Smart on FHIR launch sequences are described in terms of "EHR" systems, but there is nothing specific about EHR applications; any data provider can play this role.

In the EHR launch system, the EHR decides, based on application logic or configuration, that an application is to launched.

In the standalone launch sequence, the initiation details differ a little, but the overall pattern is the same. The main difference is in the control – generally, the EHR initiated launch sequence starts with a context integrated with the EHR workflow, while the standalone launch sequence allows for integrating the EHR services into some other workflow.

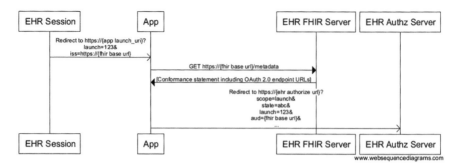

Fig. 22.4 EHR Launch sequence

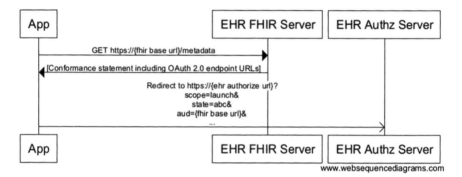

Fig. 22.5 Standalone launch sequence

This section does not describe the exact details of the Smart on FHIR protocol; they are out of scope and changing in response to implementation experience and security review. For further details, see http://docs.smarthealthit.org/.

The Smart on FHIR protocol uses web methods and the authorization process uses a web interface – e.g. a browser. It is not necessary for the app to be a web app – it can be a native OS program, or a mobile app – but it must switch to a browser or bring up a browser control to carry out the authorization process. It is important to understand the difference between authorization and authentication here.

The launch process is focused on authorization – that is, getting a user to authorize an application to perform a set of actions that the user has the right to approve. The user cannot approve things they do not have the right to do in the first place, and they do not need to authorize the application to do everything it wants to do, although some parts of the expected workflow might not work if they do not. Note that it is up to the server how much it asks the user whether they trust the application; they may not get asked anything, and authorization is granted based on institutional policy only.

Although the focus of this process is *authorization*, in order to let the user authorize an application, the server must first identify the user by authenticating them.

One of the by-products of the launch process may be an OpenID Connect token that carries signed statements about the user identity. This allows the client to delegate authentication of the user to the server, as well as authorization. However, this is optional, in that the server does not have to provide the user identity to the client; it might decline to do so, or the user may not authorize the application to have their identity (although most healthcare clients would probably decline to serve the user in that case). The authorization server can delegate authentication of the user onwards to yet another server by acting as a client and repeating the OAuth process [16].

At the conclusion of the launch process – if successful – the outcome is an application that:

- Has a bearer token that identifies its session to the server.
- Has a set of scope tokens that describe what operations it is allowed to perform on the server.
- May have information about the user identity.
- Has a defined scope of operations for this session – which may be limited to a particular subset of information on the server, such as all records for a particular patient, or it may have the entire record set in scope.

The application can now perform any authorized action.

This is a powerful tool to add on top of the **FHIR RESTful API**, which has broad support for the healthcare process. The combination of authorized applications, operating against open servers that implement the FHIR API has the potential to transform the healthcare eco-system, and it is exactly this kind of the transformation that is the focus of the Argonaut project.

The Argonaut project is a joint collaboration by the major US based EHR system vendors to provide a patient's summary health information to either patients or other healthcare providers. This is to be achieved by implementing Smart on FHIR on top a set of FHIR services that cover the meaningful use minimum data set (for readers outside USA, think of this as basic clinical summary information).

The goal of the Argonaut program is to create genuine interoperability that just works. This has been the goal of the FHIR project from the start. We need to move healthcare IT systems to the point where interoperability just works, and users – whether they are patients or providers – can easily get the information they need, when they need. This is needed to provide better clinical systems, and better healthcare for all of us.

References

1. Contact your local HL7 affiliate, or email fmgcontact@hl7.org
2. https://www.hl7.org/fhir/downloads.html
3. https://www.mongodb.org/
4. https://www.hl7.org/fhir/implementation.html#2.0.0.1
5. http://wiki.hl7.org/index.php?title=Publicly_Available_FHIR_Servers_for_testing

6. http://hl7.org/fhir/validation.html#jar
7. http://fhir2.healthintersections.com.au
8. http://hl7.org/fhir/validation.html
9. http://fhir.furore.com/
10. http://www.healthintersections.com.au/FhirServer/
11. http://hl7.org/fhir/downloads.html
12. http://jamesagnew.github.io/hapi-fhir/
13. https://github.com/grahamegrieve/fhirserver
14. https://github.com/smart-on-fhir/Swift-FHIR
15. Mandl KD, Mandel JC, Kohane IS. Driving innovation in health systems through an apps-based information economy. Cell Syst. 2015;1(1):8–13.
16. http://www.healthintersections.com.au/?p=2108

Glossary[1]

Abstract message The basic level definition of an HL7 V2 message associated with a particular trigger event. It includes the data fields that will be sent within a message, the valid response messages and the treatment of application level errors.

Access control Means of ensuring that the resources of a data processing system can be accessed only by authorized entities in authorized ways

Accountability Property that ensures that the actions of an entity may be traced uniquely to that entity

ACK Acknowledgement message

ACR American College of Radiology

Acronym An abbreviation formed by using the initial components in a phrase or name.

Act Any action of interest. Something that has happened or may happen.

Actor An abstraction for entities outside a system that interact directly with the system. An actor participates in a use case or a coherent set of use cases to accomplish an overall purpose.

ActRelationship A relationship between two Acts

ADT Admission Discharge and Transfer

AMIA American Medical Informatics Association

ANSI American National Standards Institute. ANSI represents US interests on International standards organizations such as ISO.

API Application Program Interface. A set of rules and specifications that enable communication between software programs in much the same way that a user interface facilitates interaction between humans and computers.

Application A software program or set of related programs that provide some useful healthcare capability or functionality.

---

[1] The definitions in this Glossary are derived from the HL7 Glossary, SNOMED CT User Guide, 2008, CEN EN 13606, HIMSS Dictionary of Healthcare Information Technology Terms, Acronyms and Organizations and other sources.

© Springer-Verlag London 2016

T. Benson, G. Grieve, *Principles of Health Interoperability*,
Health Information Technology Standards, DOI 10.1007/978-3-319-30370-3

Application layer The seventh and highest layer of the OSI model. Provides resources for the interaction that takes place between a user and an application.

Application role An abstraction that expresses a portion of the messaging behaviour of an information system.

ARRA American Recovery and Reinvestment Act of 2009. It includes the HITECH bill, which provides incentives to providers and hospitals to adopt Health Information Technology.

Archetype Reusable, structured models of clinical information concepts that appear in EHRs such as test result, physical examination and medication order, expressed in terms of constraints on a reference model.

Artefact Any deliverable resulting from the discovery, analysis and design activities leading to the creation of HL7 message specifications.

Architecture A framework from which computer system components can be developed in a coherent manner and in which every part fits together without containing a mass of design detail

ASCII American Standard Code for Information Interchange

Association A reference from one class to another class or to itself, or a connection between two objects (instances of classes).

ASTM American Society for the Testing of Materials

Attestation Process of certifying and recording legal responsibility for a particular unit of information.

Attribute-value pair The combination of an attribute with a value that is appropriate for that attribute. The attribute name identifies the type of information and the attribute value provides a value. Example: FINDING SITE=Lung structure

Audit trail Chronological record of activities of information system users, which enables prior states of the information to be faithfully reconstructed

Authentication Process of reliably identifying security subjects by securely associating an identifier and its authenticator.

Authorization Authorization is the process of giving someone permission to do or have something. Authorization is sometimes seen as both the preliminary setting up of permissions by a system administrator and the actual checking of the permission values that have been set up when a user is getting access.

BCS British Computer Society

Binding Indicates how an element content is taken from a value set.

Browser A tool for exploring and searching the terminology content. A browser can display hierarchy sections and concept details (relationships between concepts, descriptions and Ids, etc.).

BSI British Standards Institute. BSI represents British interests on International standards organizations such as CEN and ISO.

caBIG Cancer Biomedical Informatics Grid

CAP College of American Pathologists

Cardinality A measure of the number of elements in a set.

Care Plan A care plan is an ordered assembly of expected or planned activities, including observations, goals, services, appointments and procedures, usually organized in phases or sessions, which have the objective of organizing and managing healthcare activity for the patient, often focused upon one or more of the patient's healthcare problems. Care plans may include order sets as actionable elements, usually supporting a single session or phase. Also known as Treatment Plan.

CCDA Consolidated CDA, an XML-based implementation guide that specifies the encoding, structure, and semantics for a document that summarizes a single patient's clinical information

CCHIT Certification Commission for Health Information Technology

CCITT Comité Consultatif International Télégraphique et Téléphonique

CCOW Clinical Context Object Workgroup; HL7 standard for single sign on.

CCR ASTM E2369 – 05 Standard Specification for Continuity of Care Record. An XML-based standard that specifies a way to create a clinical summary of a patient's information

CD Concept descriptor data type

CDA Clinical Document Architecture

CDC Centers for Disease Control

CDISC Clinical Data Interchange Standards Consortium. CDISC mission is to develop and support global, platform-independent data standards that enable information system interoperability to improve medical research and related areas of healthcare.

CEN Comité Européen de Normalisation (European Committee for Standardization)

CENELEC Comité Européen de Normalisation Electrotechnique

CEN/TC 251 CEN Technical Committee 251 responsible for standards within health informatics in Europe

Certificate Authority (CA) Issues digital certificates in a public key infrastructure environment

Choice A message construct that includes alternative portions of the message. For a choice due to specialization, the sender picks one of the alternatives and sends it along with a flag.

CIM Constrained Information Model

Class An abstraction of a thing or concept in a particular application domain.

Class A class represents a concept within the system being modeled.

Classification The systematic placement of things or concepts into categories, which share some common attribute, quality or property.

Clinical Decision Support (CDS) Clinical Decision Support (CDS) refers broadly to providing clinicians or patients with clinical knowledge and patient-related information, intelligently filtered or presented at appropriate times, to enhance patient care. Clinical knowledge of interest could range from simple facts and relationships to best practices for managing patients with specific disease states, new medical knowledge from clinical research and other types of information.

Clone A class from the Reference Information Model (RIM) that has been used in a specialized context and whose name differs from the RIM class from which it was replicated. This makes it possible to represent specialized uses of more general classes to support the needs of messaging.

CMS Centers for Medicare and Medicaid Services

CNE Coded No Exceptions

CMET Common message element type (CMET) is a specialised message type in a Hierarchical Message Description (HMD) that MAY be included by reference in other HMD's.

Code A fixed sequence of signs or symbols, alphabetic or numeric characters, serving to designate an object or concept.

CodeSystem A FHIR resource that presents information about a coding system.

Coding Scheme A system of classifying objects or entities such as diseases, procedures or symptoms, using a finite set of numeric or alphanumeric identifiers.

Component An identifiable item in the main body of SNOMED CT, or in an authorized Extension. Each component is a uniquely identifiable instance of one of the following: Concept, Description, Relationship, Subset, Subset Member, Cross Map Set, Cross Map Target, History Component.

ComponentID A general term used to refer to the primary identifier of any SNOMED CT Component. All ComponentIDs follow the form of the SCTID specification.

Composite data type A data type assigned to a message element type that contains one or more components, each of which is represented by an assigned data type.

Composition The set of information committed to one EHR by one agent, as a result of a single clinical encounter or record documentation session.

Concept A clinical idea to which a unique ConceptID has been assigned in SNOMED CT.

Concept equivalence When two SNOMED CT concepts or post-coordinated expressions have the same meaning. Concept equivalence can occur when a post-coordinated expression has the same meaning as a pre-coordinated Concept; or when two different post-coordinated expressions have the same meaning.

Concept Model The SNOMED CT Concept Model is the complete set of rules that govern the ways in which concepts are permitted to be modeled using relationships to other concepts.

ConceptID The unique identifier (code) for each SNOMED CT concept. Refer to the SNOMED Technical Reference Guide for a full explanation of how this identifier is structured.

Example: For the concept Pneumonia (disorder), the ConceptID is 233604007

Concepts Table A table that includes all SNOMED CT concepts. Each concept is represented by a row.

Confidentiality Property that information is not made available or disclosed to unauthorized individuals, entities, or processes.

Conformance (FHIR) A set of statements about how the FHIR API and resources are used by a actual or possible system.

Conformance Profile A conformance profile is a constraint to either an underlying standard or another conformance profile. Normally, it specifies a single message or document.

Connectathon (FHIR) A gathering of implementers to test out how well systems can exchange data based on the FHIR specification.

Constraint Narrowing down of the possible values for an attribute; a suggestion of legal values for an attribute (by indicating the data type that applies, by restriction of the data type, or by definition of the domain of an attribute as a subset of the domain of its data type). MAY also include providing restrictions on data types. A constraint imposed on an association MAY limit the cardinality of the association or alter the navigability of the association (direction in which the association can be navigated). A Refined Message Information Model (R-MIM) class MAY be constrained by choosing a subset of its Reference Information Model (RIM) properties (ie, classes and attributes) or by cloning, in which the class' name is changed.

Context Model A model that specifies relationships relating to semantic context that has been defined outside of the SNOMED CT Concept Model.

Continua Alliance Continua Health Alliance is a non-profit, open industry coalition of healthcare and technology companies joining together in collaboration to improve the quality of personal healthcare, such as those used in the home.

Control event wrapper A wrapper that contains domain specific administrative information related to the "controlled event" which is being communicated as a messaging interaction. The control event wrapper is used only in messages that convey status, or in commands for logical operations being coordinated between applications (eg, the coordination of query specification/query response interactions).

CPOE Computerized practitioner order entry

CPT-4 Current Procedural Terminology. Coding system used in the US as a guide to services for which patients may be billed.

CRE Care record element

CRS Care Record Service (NHS)

CTS Common Terminology Services. The CTS defines the minimum set of functions required for terminology interoperability within the scope of HL7's messaging and vocabulary browsing requirements.

CTV3 Clinical Terms Version 3 (Read Codes Version 3)

CTV3ID A five-character code allocated to a concept or term in CTV3. For data compatibility and mapping purposes, SNOMED CT concepts include a record of the corresponding concept codes from the Clinical Terms Version 3 (CTV3, previously known as Read Codes) and SNOMED RT.

CUI Microsoft Health/NHS CFH Common User Interface (CUI) provides user interface design guidance and toolkit controls that address a wide range of patient safety concerns for healthcare organizations worldwide, enabling a new generation of safer, more usable and compelling health applications to be quickly and easily created [http://www.mscui.net]

CWE Coded With Exceptions

DAM Domain Analysis Model

Database A collection of stored data typically organized into fields records and files and an associated description (schema)

Data type The structural format of the data carried in an attribute. It MAY constrain the set of values an attribute may assume.

Defining relationship A relationship used to define the meaning of a concept.

Delimit To mark or set off. For example the day, month and year in a string such as 2/5/2009 are delimited by the "/" symbol.

Description Each Description is assigned a unique DescriptionID and connects a Term and a Concept.

DescriptionID An SCTID that uniquely identifies a Description.

Dialect A language modified by the vocabulary and grammatical conventions applied in a particular geographical or cultural environment.

DICOM Digital Imaging and Communications in Medicine

Digital Representation of an entity based on binary (on/off) signals.

DIN Deutsches Institut fur Normung – the German national standards organization.

dm+d Dictionary of medicines and devices, containing unique identifiers and associated textual descriptions for medicines and medical devices used in the NHS.

DMIM Domain Message Information Model

DNS Domain Name System, an Internet system to translate human-readable names into Internet addresses

Domain expert Individual who is knowledgeable about the concepts in a particular problem area within the healthcare arena and/or is experienced with using or providing the functionality of that area.

Domain Message Information Model (D-MIM) A form of Refined Message Information Model (R-MIM) constructed to represent the totality of concepts embodied in the individual R-MIMs needed to support the communication requirements of a particular HL7 domain.

DRG Diagnosis Related Group

DSTU Draft Standard for Trial Use

DTD Document Type Definition (XML)

EAI Enterprise Application Integration

ED Encapsulated Data Type

EDI Electronic Data Interchange – based on electronic sending and receiving of messages

EDIFACT Electronic Data Interchange For Administration, Commerce, and Transport – a set of rules and syntax for EDI maintained by the UN.

EHR Electronic Health Record. A comprehensive, structured set of clinical, demographic, environmental, social, and financial data and information in electronic form, documenting the healthcare given to a single individual.

EHR-S FM EHR System Functional Model–provides a reference list of over 160 functions that may be present in an Electronic Health Record System (EHR-S)

EHR System The set of components that form the mechanism by which patient records are created, used, stored, and retrieved.

Element A FHIR resource is composed of a tree of elements, which are the root type for all other elements.

ElementDefinition The description of an element and it's possible content.

EMR Electronic medical record

EN Norme Europeene (European Standard) approved by CEN and which normally takes precedence over local or national standards.

Encounter Encounter serves as a focal point linking clinical, administrative, and financial information. Encounters occur in many different settings – ambulatory care, inpatient care, emergency care, home healthcare, field and virtual (telemedicine).

Entity A person, animal, organization or thing. Something that has separate and distinct existence and objective or conceptual reality. Something that exists as a particular and discrete unit. An organization (as a business or governmental unit) that has an identity separate from those of its members.

ENV Europaische Vornorm (European Pre-standard) – a standard that has yet to be put into a final and definitive form for approval as an EN.

EOM End of Message

Eponym The use of a person's name to describe an entity.

EP Eligible Provider

EPHI Electronic protected health information

EPR Electronic Patient Record (owned by the patient)

ESC Escape

ETP Electronic Transfer of Prescriptions

Expansion A list of the codes in a value set under current conditions.

Expression A collection of references to one or more concepts used to express an instance of a clinical idea. An expression containing a single concept identifier is referred to as a pre-coordinated expression. An expression that contains two or more concept identifiers is a post-coordinated expression. The concept identifiers within a post-coordinated expression are related to one another in accordance with rules expressed in the SNOMED CT Concept Model.

Extension (SNOMED CT) Extensions are complements to a released version of SNOMED CT. Extensions are components that are created in accordance with the data structures and authoring guidelines applicable to SNOMED CT.

Extension (FHIR) Additional data added to an element as a pair of (url, value) as allowed by the FHIR specification.

FDA Food and Drug Administration

FHIR Fast Healthcare Interoperability Resources

Field The smallest named unit of data in a database. Fields are grouped together to form records.

File A collection of electronic data. A file has a name by which it is known to the computer and may contain, for example, data, records, text, image etc.

Folder The high level organisation within an EHR, dividing it into compartments relating to care provided for a single condition, by a clinical team or institution, or over a fixed time period such as an episode of care.

FMM FHIR Maturity Model - an estimate of how far through the development process a resource has progressed.

FTP File Transfer Protocol

Fully defined concept Fully defined concepts can be differentiated from their parent and sibling concepts by virtue of their relationships.

Fully Specified Name (FSN) A phrase that describes a concept uniquely and in a manner that is intended to be unambiguous.

Generalization An association between two classes, referred to as superclass and subclass, in which the subclass is derived from the superclass. The subclass inherits all properties from the superclass, including attributes, relationships, and states, but also adds new ones to extend the capabilities of the parent class. Essentially, a specialization from the point-of-view of the subclass.

GP General Medical Practitioner

GP2GP GP to GP record transfer service (NHS)

Graphical expression A visual representation of a model that uses graphic symbols to represent the components of the model and the relationships that exist between those components.

GUI Graphical user interface

HAI Hospital acquired infection

HCO Healthcare organization

HDF HL7 Development Framework

HES Hospital Episode Statistics (NHS)

Healthcare agent Person, device, or software that performs a role in a healthcare activity

Healthcare organization Organisation involved in the direct or indirect provision of healthcare services to an individual or to a population. NOTE Groupings or subdivisions of an organisation, such as departments, may also be considered as organisations where there is a need to identify them.

HealthCare Party Person involved in the direct or indirect provision of healthcare services to an individual or to a population.

HealthCare Professional. A person who is authorized by a nationally recognized body to be qualified to perform certain health duties.

HealthCare Provider A HealthCare Provider is a person licensed, certified or otherwise authorized or permitted to administer healthcare in the ordinary course of business or practice of a profession, including a healthcare facility.

HealthCare Service Service provided with the intention of directly or indirectly improving the health of the person or populations to whom it is provided.

HIE Health Information Exchange

Hierarchical Message Description A specification of the exact fields of a message and their grouping, sequence, optionality, and cardinality. This specification contains message types for one or more interactions, or that represent one or more common message element types. This is the primary normative structure for HL7 messages.

Hierarchy An ordered organization of concepts. General concepts are at the top of the hierarchy; at each level down the hierarchy, concepts become increasingly specialized.

HIMSS Healthcare Information and Management Systems Society

HIPAA Health Insurance Portability and Accountability Act, 1996

HIO Health Information Organization, an organization that holds patient information and/or provides services to allow members of the organization to exchange health information

HIS Health (or Hospital) Information System

HISP Health Information Service Provider, the entity that is responsible for delivering health information as messages between senders and receivers over the Internet

HITECH Health Information Technology for Economic and Clinical Health Act, a bill that, as a part of the American Recovery and Reinvestment Act of 2009, aims to advance the use of health information technology such as electronic health records

HITPC Healthcare IT Policy Committee, a federal advisory committee charged with making recommendations to the National Coordinator for Health IT surrounding standards implementation specifications, and certifications criteria in order to shape a nationwide infrastructure for the adoption of healthcare information technology and the exchange of meaningful patient medical information

HITSC Healthcare IT Standards Committee, a federal advisory committee charged with providing standards guidance and testing infrastructure to support the recommendations of the HIT Policy Committee

HITSP Health Information Technology Standards Panel

HL7 Health Level Seven is an international standards-development organization, whose mission is to provide standards for the exchange, integration, sharing, and retrieval of electronic health information; support clinical practice; and support the management, delivery and evaluation of health services

HMD Hierarchical Message Description

Homonym One term having two or more independent meanings

HTML Hypertext Markup Language

HTTP Hypertext Transfer Protocol

ICD International Classification of Diseases

ICP Integrated Care Pathway

ICPC International Classification of Primary Care

ICPM International Classification of Procedures in Medicine

ICT Information and Communication Technology

Identifier A piece of data that uniquely identifies an item, information, or a person as the subject of this identity within a given context

IDN Integrated Delivery Network, a network of healthcare organizations organized under a parent holding company that provides a continuum of healthcare services

IEC International Electrotechnical Commission

IEEE Institute of Electrical and Electronics Engineers

IHE Integrating the Health Environment. IHE (Integrating the Healthcare Enterprise) is an industry-led initiative to improve the way computer systems in healthcare share information. IHE promotes the coordinates use of established standards such as HL7 and DICOM to address specific clinical needs. http://www.ihe.net/

IHTSDO International Health Terminology Standards Development Organization

IM&T Information Management and Technology

IMIA International Medical Informatics Association

Implementation Guide A specification that describes how to use a general platform specification (v2, CDA, FHIR) in a particular context to solve a specific problem.

Implementation Technology A technology selected for use in encoding and sending HL7 messages. For example, XML is being used as an implementation technology for Version 3.

Implementation Technology Specification (ITS) A specification that describes how HL7 messages are sent using a specific implementation technology. It includes, but is not limited to, specifications of the method of encoding the messages, rules for the establishment of connections and transmission timing and procedures for dealing with errors.

Information Model A structured specification, expressed graphically and/or in narrative, of the information requirements of a domain. An information model describes the classes of information required and the properties of those classes, including attributes, relationships, and states. Examples in HL7 are the Domain Reference Information Model, Reference Information Model, and Refined Message Information Model.

Integration Profile An integration profile describes the workflow for a specific use case. It combines actors and interactions.

Interaction A single, one-way information flow that supports a communication requirement expressed in a scenario.

Interface A common boundary between two associated systems across which information may flow. The interface may filter or modify data as it passes across the boundary.

Interface Terminology Systematic collections of clinically oriented phrases or terms aggregated to support clinicians' entry of patient information directly into computer programs, such as clinical documentation systems or decision support tools. They may mediate between a user's colloquial conceptualizations of concept descriptions and an underlying reference terminology.

International Release The required international components of the SNOMED CT terminology, along with related works and resources, maintained and distributed by the IHTSDO.

Internet The International network of computers providing support for data exchange, Email and the World-wide Web.

IOM Institute of Medicine

ISB Information Standards Board (NHS)

ISO International Organization for Standardization – the body overseeing endorsement and publication of international standards.

ISO/TC 215 International Standards Organization/Technical Committee 215 (Health Informatics)

ISP International Standardized Profile

ITS Implementation Technology Specification

ITU International Telecommunications Union

IVR Interactive Voice Response

JSON Java Script Object Notation

LAN Local Area Network

Language Subset SNOMED CT can be translated into any language or dialect. These translations use existing SNOMED CT concepts, along with new language-specific descriptions. A language subset is a set of references to the descriptions that are members of a language edition of SNOMED CT. Additionally, this subset specifies the type of description (FSN, Preferred Term or synonym).

LOINC Logical Observation Identifiers Names and Codes

LR Legitimate Relationship

LSP Local Service Provider (NHS)

Mandatory If an attribute is designated as mandatory, all message elements, which make use of this attribute, SHALL contain a non-null value or they SHALL have a default that is not null.

Markup Computer-processable annotations within a document. Markup encodes a description of a document's storage layout and logical structure. In the context of HL7 Version 3, markup syntax is according to the XML Recommendation.

Master file Common lookup table used by one or more application systems.

May The conformance verb MAY is used to indicate a possibility.

MBDS Minimum Basic Data Set

Meaningful Use Often abbreviated as MU, defined in the Final Rule from CMS published in July, 2010 under the ARRA HITECH provisions

MeSH Medical Subject Headings

Message A package of information communicated from one application to another. See also message type and message instance.

Message element A unit of structure within a message type.

Message element type A portion of a message type that describes one of the elements of the message.

Message instance A message, populated with data values, and formatted for a specific transmission based on a particular message type.

Message payload Data carried in a message.

Message type A set of rules for constructing a message given a specific set of instance data. As such, it also serves as a guide for parsing a message to recover the instance data.

Meta-model A model used to specify other models. For example, the meta-model for a relational database system might specify elements of type 'Table', 'Record', and 'Field.'

MIB Medical Information Bus

MIM Message Implementation Manual published by NHS Connecting for Health.

MIME Multipurpose Internet Mail Extensions, an Internet standard that extends e-mail to support content beyond simple ASCII plaintext data.

MPI Master Patient Index

MT Message Type

Model A semantically complete abstraction of a system

Multiplicity In the information model, multiplicity is a specification of the minimum and maximum number of objects from each class that can participate in an association. Multiplicity is specified for each end of the association.

Narrative XHTML in a FHIR resource added to provide human readability.

Nationwide Health Information Network A set of standards, services and policies that enable secure health information exchange over the Internet

Navigability Direction in which an association can be navigated (either one way or both ways).

NCI National Cancer Institute

NCPDC National Council for Prescription Drug Program

NDC National Drug Code

NHS National Health Service

NHSCR NHS Central Register

NIST National Institute for Science and Technology

NLM National Library of Medicine

Nomenclature A set or system of names or terms, as those used in a particular science or art.

NPfIT National Programme for Information Technology (NHS)

Null A value for a data element that indicates the absence of data. A number of "flavors" of null are possible.

Object An instance of a class. A part of an information system containing a collection of related data (in the form of attributes) and procedures (methods) for operating on that data

Object identifier A scheme to provide globally unique identifiers. This object identifier (OID) scheme is an ISO standard (ISO 8824:1990).

ODA Open Document Architecture

ODP Open Distributed Processing (ISO/IEC 10746, used for describing distributed systems)

OHT Open Health Tools is a community of open source developers, health professionals, and an ecosystem that brings together members from the health and IT professions to create a common health interoperability framework, exemplary tools and reference applications to support health information interoperability. The fact that this software framework is available under a commercially friendly open source license means that anyone, any company, and any hospital, whether or not they are a member, can build applications using this framework – without any payment required for the software.

OID Object Identifier

OMG Object Management Group

ONC Office of the National Coordinator for Health Information Technology in the Department of Health and Human Services, the principal Federal entity charged with coordinating nationwide efforts to promote the use of health information technology.

Ontology The hierarchical structuring of knowledge about things by subcategorizing them according to their essential (or at least relevant and/or cognitive) qualities.

OpenEHR OpenEHR is a not-for-profit foundation to make EHRs "adaptable and future-proof" through the use of a technology independent architecture.

OSI Open Systems Interconnection

OWL Web Ontology Language

P2P Peer-to-peer

P4P Pay for performance

PACS Picture Archiving and Communication System

PAP Policy Administration Point creates and manages policies and consent directives.

Participation The involvement of a Role in an Act

PAS Patient Administration System

Patient One who is suffering from any disease or behavioral disorder and is under treatment for it.

PCAST President's Advisory Council on Science and Technology

PCP Primary Care Provider

PDP Policy Decision Point evaluates and issues authorization decision.

PDS Personal Demographics Service (NHS)

PEP Policy Enforcement Point intercepts user's access request to a resource and enforces PDP's decision

PHIN Public health information network

PHR Personal Health Record, an electronic health record managed by a patient. A PHR may be "connected", opening a patient-friendly portal to information ultimately owned by a healthcare organization, care provider, or insurance company; or it may be "unconnected," providing a patient-owned space for storing and editing personal medical information.

PICS Protocol Implementation Conformance Statement

PIM Platform Independent Model

PIN Personal Identification Number

PKI Public Key Infrastructure

PN Person Name data type

POC Point of Care

POMR Problem oriented medical record, originally developed by Dr Larry Weed.

Post-coordination Representation of a clinical idea using a combination of two or more concept identifiers. A combination of concept identifiers used to represent a single clinical idea is referred to as a post-coordinated expression (see expression). Many clinical ideas can also be represented using a single SNOMED CT concept identifier (see pre-coordination). Some clinical ideas may be represented in several different ways. SNOMED CT technical specifications include guidance of logical transformations that reduce equivalent expressions to a common canonical form.

PQRI Physician Qualify Reporting Initiative

Pre-coordination Representation of a clinical idea using a single concept identifier. A single concept identifier used to represent a specific meaning is referred to as a pre-coordinated expression (see expression). SNOMED CT also allows the use of post-coordinated expressions (see post-coordination) to represent a meaning using a combination of two or more concept identifiers. However, including commonly used concepts in a pre-coordinated form makes the terminology easier to use.

Preferred Term The Term that is deemed to be the most clinically appropriate way of expressing a SNOMED CT Concept in a clinical record.

Primitive Concept A concept is primitive when its modeling (attributes and parents) does not fully express its meaning. A concept definition is the list of its relationships to other concepts. Primitive concepts do not have the unique relationships needed to distinguish them from their parent or sibling concepts.

Privacy Freedom from intrusion into the private life or affairs of an individual when that intrusion results from undue or illegal gathering and use of data about that individual.

PRP Policy Retrieval Point where consent policy is stored.

Problem List The problem list of a given individual can be described by formal diagnosis coding systems (such as ICD-10) or by other professional descriptions of healthcare issues affecting an individual. Problems can be short or long term in nature, chronic or acute, and have a status. In a longitudinal record, all problems may be of importance in the overall long term care of an individual, and may undergo changes in status repeatedly. Problems are identified during patient visits, and may span multiple visits, encounters, or episodes of care.

Profile A set of functions required in a particular setting or available as part of a particular system or component

Profile (FHIR) A set of rules about how a resource is used in a particular context (e.g. part of an Implementation Guide).

PSIS Personal Spine Information Service (NHS)

PSM Platform Specific Model

QMAS Quality Management and Analysis System (NHS)

QMR Quick Medical Reference

QOF Quality and Outcomes Framework (NHS)

QoS Quality of Service

Realization The relationship between a specification and its implementation.

Realm A sphere of authority, expertise, or preference that influences the range of Components required, or the frequency with which they are used. A Realm may be a nation, an organization, a professional discipline, a specialty, or an individual user.

Receiver The application fulfilling the Receiving Application role in an interaction

Receiver responsibility An obligation on an application role that receives an interaction as defined in the interaction model.

Record A writing by which some act or event, or a number of acts or events are recorded.

Recursion An association that leads from a class directly or indirectly back to that class.

Reference A url that links from one resource to another.

Reference Implementation A software library or application that implements FHIR functionality, suitable for re-use in production systems.

Reference Information Model (RIM) The HL7 information model from which all other V3 information models (eg, R-MIMs) and messages are derived.

Reference Terminology A reference terminology is a terminology in which every concept designation has a formal, machine-usable definition supporting data aggregation and retrieval. Reference terminologies are designed to provide exact and complete representations of a given domain's knowledge, including its entities and ideas, and their interrelationships, and are typically optimized to support the storage, retrieval, and classification of clinical data.

Refined Message Information Model (R-MIM) An information structure that represents the requirements for a set of messages. A constrained subset of the Reference Information Model (RIM), which MAY contain additional classes that are cloned from RIM classes. Contains those classes, attributes, associations, and data types that are needed to support one or more Hierarchical Message Descriptions (HMD). A single message can be shown as a particular pathway through the classes within an R-MIM.

Relationship An association between two Concepts. A Relationship Type indicates the nature of the association.

Relationship Type The nature of a Relationship between two Concepts. The RelationshipType field indicates the ConceptID for the concept in SNOMED that forms the relationship between two other concepts (ConceptID1 and ConceptID2)

RelationshipID A SCTID that uniquely identifies a Relationship between three concepts: a source concept (ConceptID1), a target concept (ConceptID2), and a relationship type.
Each row in the Relationships Table represents a relationship "triplet" (ConceptID1 – RelationshipType – ConceptID2) identified by a RelationshipID.

Relationships Table A table consisting of rows, each of which represents a Relationship.

Release Version A version of SNOMED CT released on a particular date. Except for the initial release of SNOMED CT that was called "SNOMED CT First Release," subsequent releases use the release data. Example: "SNOMED CT July 2008 Release"

Required One of the allowed values in conformance requirements, it means that the message elements SHALL appear every time that particular message type is used for an interaction. If the data is available, the element SHALL carry the data, otherwise a null value MAY be sent.

Requirement A desired feature, property or behaviour of a system.

Resource An package of data with a known location (URL), of one of the types defined in the FHIR specification.

REST REpresentational State Transfer - see http://www.ics.uci.edu/~fielding/pubs/dissertation/rest_arch_style.htm.

RESTful A style of interaction that follows the approach defined as "REST".

RFID Radio frequency identification (RFID) is a generic term that is used to describe a system that transmits the identity (in the form of a unique serial number) of an object or person wirelessly, using radio waves.

RIM HL7 Reference Information Model

RHIO Regional Health Information Organization

RMIM HL7 Refined Message Information Model

Role A part played by or the responsibility of an Entity

RoleLink A relationship between two Roles.

Root Concept The single Concept "SNOMED CT Concept" that is at the top of the entire SNOMED CT hierarchy of concepts.

Rubric The title or name of a class or category.

SaaS Software as a service

SAEAF Services Aware Enterprise Architecture Framework. HL7's SAEAF defines the artefacts and specification semantics needed to support interoperability in healthcare, life sciences, and clinical research.

Sanctioned relationships Relationships between SNOMED CT concepts that are sanctioned by the SNOMED CT Concept Model. Sanctioned relationships are specified in a row in the SNOMED CT Relationships table, as opposed to 'Allowable' relationships, which are a pattern in the Concept Model.

Scenario A sequence of actions that illustrates behaviour. A scenario may be used to illustrate an interaction or the execution of a use case instance.

Schematron Schematron is an XML structure validation language for making assertions about the presence or absence of patterns in trees. It is a simple and powerful structural schema language.

SCR Summary Care Record

SCT SNOMED Clinical Terms

SCT Enabled Application A software application designed to support the use of SNOMED CT.

SCTID SNOMED Clinical Terms Identifier

SDO Standards Development Organization

SDS Spine Directory Service (NHS)

Section EHR data within a composition that belongs under one clinical heading, usually reflecting the flow of information gathering during a clinical encounter, or structured for the benefit of future human readership.

Semantics Meaning of symbols and codes

Semantic interoperability Ability for data shared by systems to be understood at the level of fully defined domain concepts.

Sender The application fulfilling the Sending Application role in an interaction.

Service A consultation, diagnosis, treatment or intervention performed for a person and/or other activity performed for a person. Includes health, goods and support services.

Set A form of collection, which contains an unordered list of unique elements of a single type.

SGML Standardized General Markup Language

Shall The conformance verb SHALL is used to indicate a requirement.

Should The conformance verb SHOULD is used to indicate a recommendation.

SIG Special Interest Group

S/MIME Secure/Multipurpose Internet Mail Extensions, an Internet standard for securing MIME data. S/MIME provides privacy and data security through encryption; and authentication, integrity assurance, and non-repudiation of origin through signing.

SMTP Simple Mail Transport Protocol, an industry standard for transporting e-mail.

SNOMED An acronym for the Systematized Nomenclature of Human and Veterinary Medicine originally developed by the College of American Pathologists.

SNOMED Clinical Terms (SNOMED CT) The clinical terminology maintained and distributed by the IHTSDO. The First Release of SNOMED Clinical Terms was the result of the merger of the CTV3 and SNOMED RT.

SNOMED Clinical Terms Identifier (SCTID) A unique identifier applied to each SNOMED CT component (Concept, Description, Relationship, Subset, etc.).

SOA Service Oriented Architecture provides methods for systems development and integration where systems package functionality as interoperable services. A SOA infrastructure allows different applications to exchange data with one another. Service-orientation aims at a loose coupling of services with operating systems, programming languages and other technologies that underlie applications. SOA separates functions into distinct units, or services, which developers make accessible over a network in order that users can combine and reuse them in the production of applications. These services communicate with each other by passing data from one service to another, or by coordinating an activity between two or more services.

SOAP Simple object access protocol

Specialization An association between two classes (designated superclass and subclass), in which the subclass is derived from the superclass. The subclass inherits all properties from the superclass, including attributes, relationships, and states, but also adds new ones to extend the capabilities of the superclass.

Specification A detailed description of the required characteristics of a product.

SQL Structured query language

Standard A document, established by consensus and approved by a recognized body, that provides, for common and repeated use, rules, guidelines or characteristics for activities or their results, aimed at the achievement of the optimum degree of order in a given context.

Storyboard Defines what happens from the users point of view. A narrative of relevant events defined using interaction or activity diagrams or use cases. The storyboard provides one set of interactions that will typically occur in the domain.

String A sequence of text characters.

StructureDefinition A definition of a structure in FHIR - either an underlying resource or data type, or a Profile.

Stylesheet A file that describes how to display an XML document of a given type.

Subclass A class that is the specialization of another class (superclass).

Subset A group of Components (eg Concepts, Descriptions or Relationships) that share a specified common characteristic or common type of characteristic. Example: UK English Subset

Superclass A class that is the generalization of one or more other classes (subclasses).

Swimlane A partition on activity graphs for organizing responsibilities for activities, often corresponding to the organizational units in a business model.

Synonym A term that is an acceptable alternative to the preferred term as a way of expressing a concept. Synonyms allow representations of the various ways a concept may be described.

Syntax Rules for structuring words into sentences or computer commands or electronic messages.

System A collection of connected units organized to accomplish a purpose.

Table view An expression of the Hierarchical Message Description (HMD) common and message type definition condensed in size to fit on a printed page.

TAG Technical advisory group

Taxonomy The science or technique of classification

TC Technical Committee

TCP/IP Transmission Control; Protocol / Internet Protocol. A protocol for communication between computers, used as a standard for transmitting data over networks and as the basis for standard Internet protocols.

Template A template is an RMIM, which is used to constrain another model

Term A text string that represents a concept. The Term is part of the Description. There are multiple descriptions per Concept.

Terminology A set of concepts designated by terms belonging to a special domain of knowledge, or subject field.

Terminology Binding An instance of a link between a terminology component and an information model artefact.

Terminology server Software that provides access to SNOMED CT (and/or to other terminologies). A Terminology server typically supports searches and Navigation through Concepts. A server may provide a user interface (eg a browser or set of screen controls) or may provide low-level software services to support access to the terminology by other applications.

Transaction A complete set of messages for a particular trigger event, eg a message and a response.

Transitive closure table A table that lists all of the ancestor codes of each concept, used in fast subsumption testing

Transport wrapper A wrapper that contains information needed by a sending application or message handling service to route the message payload to the designated receiver. All HL7 Version 3 messages require an appropriately configured transport wrapper.

Trigger Event Defines what causes a message to be sent. An event which, when recorded or recognized by an application, indicates the need for an information flow to one or more other applications, resulting in one or more interactions.

TRUD Terminology Reference Data Update Distribution Service (NHS)

TSC Technical Steering Committee (HL7)

TTP Trusted third party

UMDNS Universal medical device nomenclature system

UML Unified Modeling Language

UMLS Unified Medical Language System

UN/CEFACT United Nations Centre for Trade Facilitation and Electronic Business

UKTC UK Terminology Centre (NHS)

UPI Unique Patient Identifier

URL Uniform resource locator

Use case The specification of sequences of actions, including variant sequences and error sequences, which a system can perform by interacting with outside actors.

VA Veterans Administration

Valid document A document that meets all of the validity constraints in the XML specification.

ValueSet A FHIR Resource the defines a set of codes selected from one of more Coding Systems.

Value set A vocabulary domain that has been constrained to a particular realm and coding system.

View Specific information displayed on a computer monitor after it has been filtered for a different user or purpose.

Vocabulary The set of all concepts that can be taken as valid values in an instance of a coded attribute or field.

W3C World Wide Web Consortium

WAN Wide Area Network

WEDI Workgroup on Electronic data Interchange

WHO World Health Organization

Wrapper The control or envelope information in which the message payload resides.

WWW World Wide Web

X.509 Digital Certificate A standard for asserting that an entity is who it purports to be.

XDM The IHE Cross-Enterprise Document Media Interchange integration profile, a specification for the exchange of electronic health record documents on portable media. XDM provides an option for zipped file transfer over e-mail.

XDR The IHE Cross-Enterprise Document Reliable Interchange integration profile, a specification for the interchange of electronic health record documents through reliable point-to-point network communication, based on a push of information.

XDS The IHE Cross-Enterprise Documenting Sharing integration profile, a specification for managing the sharing, finding, and retrieval of electronic health
record documents among a defined group of healthcare enterprises.

XML Extensible Mark-up Language

XSL Extensible Style sheet Language. The XSL family comprises three languages:

- XSL Transformations (XSLT): an XML language for transforming XML
  documents
- XSL Formatting Objects (XSL-FO): an XML language for specifying the
  visual formatting of an XML document
- XML Path Language (XPath): used to address the parts of an XML
  document.

XSLT Extensible Stylesheet Language Transformations (XSLT) is an XML-based
language used for the transformation of XML documents into other XML or
"human-readable" documents. The original document is not changed; rather, a
new document is created based on the content of an existing one. The new document may be serialized (output) by the processor in standard XML syntax or in
another format, such as HTML or plain text. XSLT is most often used to convert
data between different XML schemas or to convert XML data into HTML or
XHTML documents for web pages, creating a dynamic web page, or into an
intermediate XML format that can be converted to PDF documents.

References and Further Reading

1. Ainsworth J, Buchan I. Combining health data uses to ignite health system learning. Methods Inf Med. 2015;54:479–87.
2. Alderwick H, Roberton R, Appleby J, Dunn P, Maguire D. Better value in the NHS: the role of changes in clinical practice. London: The Kings Fund; 2015.
3. Ambler S. The elements of UML 2.0 style. Cambridge: Cambridge University Press; 2005.
4. Armour F, Miller G. Advanced use case modelling. Boston: Addison Wesley; 2000.
5. ASTM E1238.88. 1988.
6. ASTM. Specification for continuity of care record, E2369-05. 2006.
7. Beale T, Heard S, editors. OpenEHR architecture overview. Release 1.0.1. London: The openEHR Foundation; 2007.
8. Beck K. Extreme programming explained: embrace change, vol. 2. Reading: Addison-Wesley; 2005.
9. Beeler W. HL7 version 3 – an object-oriented methodology for collaborative standards development. Int J Med Inform. 1998;48:151–61.
10. Benson T, Neame R. Healthcare computing: a guide to health information management and systems. Harlow: Longman; 1994.
11. Benson T. Medical informatics: a report for managers and clinicians. Harlow: Longman Health Services Management; 1991.
12. Benson T. Prevention of errors and user alienation in healthcare IT integration programmes. Inform Prim Care. 2007;15:1–7.
13. Benson T. The history of the Read Codes: the inaugural James Read Memorial Lecture 2011. Inform Prim Care. 2011;19(3):173–82.
14. Benson T. Why general practitioners use computers and hospital doctors do not – part 1: incentives. BMJ. 2002;325:1086–9.
15. Benson T. Why general practitioners use computers and hospital doctors do not – part 2: scalability. BMJ. 2002;325:1090–3.
16. Benson T. Why industry is not embracing standards. Int J Med Inform. 1998;48:133–6.
17. Berwick DM. Medical associations: guilds or leaders. BMJ. 1997;314:1564. (http://www.bmj.com/cgi/content/full/314/7094/1564).
18. Bleich HL, Lawrence L. Weed and the problem-oriented medical record. MD Comput. 1993;10(2):70.
19. Blois MS. Information and medicine: the nature of medical descriptions. Berkeley: University of California Press; 1984.
20. Blum BI, Duncan K, editors. A history of medical informatics. Reading: Addison Wesley; 1990.

© Springer-Verlag London 2016

437

T. Benson, G. Grieve, *Principles of Health Interoperability*,
Health Information Technology Standards, DOI 10.1007/978-3-319-30370-3

21. Blumenthal D. Stimulating the adoption of health information technology. N Engl J Med. 2009;360:15.
22. Booch G, Rumbaugh J, Jacobson I. The unified modeling language user guide. Reading: Addison-Wesley; 1999.
23. Boone KW. The CDA™ book. London: Springer; 2011.
24. Bos B. XML in 10 Points. W3C. 1999. www.w3c.org/XML/1999/XML-in-10-points
25. Braunstein ML. Practitioner's guide to health informatics. Cham: Springer; 2015.
26. Bray T, Paoli J, Sperberg-McQueen CM, Maler E, Yergeau F. Extensible markup language (XML). World Wide Web Consortium. 1998; Recommendation REC-xml-19980210.
27. Bray T. The JavaScript Object Notation (JSON) Data Interchange Format. Internet Engineering Task Force (IETF). 2014; RFC 7157.
28. Brennan S. The NHS IT project: the biggest computer programme in the world – ever! Oxford: Radcliffe; 2005.
29. Business Process Modeling Notation. www.bpmn.org/
30. Caldicott F. Information: to share or not to share: the information governance review. London: Department of Health; 2013.
31. Canada Standards Collaborative. www.infoway-inforoute.ca/lang-en/standards-collaborative
32. Carrol L. Through the looking glass and what alice found there. London: Macmillan; 1871.
33. CEN CR 12587:1996. Medical informatics – methodology for the development of healthcare messages. CEN Report. 1996.
34. CEN CR 1350:1993. Investigation of syntaxes for existing interchange formats to be used in health care. CEN Report. 1993.
35. CEN EN 13606-1. Health informatics—Electronic health record communication—Part 1: Reference Model.
36. CEN EN 13606-2. Health informatics—Electronic health record communication—Part 2: Archetype Interchange Specification.
37. CEN EN 13606-3. Health informatics—Electronic health record communication—Part 3: Reference Archetypes and Term Lists.
38. CEN EN 13606-4. Health informatics—Electronic health record communication—Part 4: Security Features.
39. CEN EN 13606-5. Health informatics—Electronic health record communication—Part 5: Exchange Models.
40. Chisholm J. The read clinical classification. BMJ. 1990;300:1092.
41. Christensen CM, Grossman JH, Hwang J. The innovator's prescription: a disruptive solution for health care. New York: McGraw Hill; 2009.
42. Cimino JJ. Desiderata for controlled medical vocabularies in the twenty-first century. Methods Inform Med. 1998;37:394–403.
43. Cockburn A. Agile software development. Boston: Addison-Wesley; 2002.
44. Cockburn A. Writing effective use cases. Boston: Addison-Wesley; 2001.
45. Coiera E. Four rules for the reinvention of health care. BMJ. 2004;328:1197–9.
46. Coiera E. Guide to health informatics, 3rd ed. Boca Raton: CRC Press; 2015.
47. Collins T. Crash: ten easy ways to avoid a computer disaster. London: Simon and Schuster; 1997.
48. Common User Interface Project. www.mscui.net
49. Continua Alliance. www.continuaalliance.org
50. Cooper D et al. Internet X.509 public key infrastructure certificate and Certificate Revocation List (CRL) profile. IETF Network Working Group RFC 5280. 2008. http://www.ietf.org/rfc/rfc5280.txt
51. Coplien JO, Harrison NB. Organizational patterns of agile software development. Upper Saddle River: Pearson Prentice Hall; 2005.
52. Coulter A. Engaging patients in healthcare. Maidenhead: McGraw HIll; 2011.

53. Daschle T. Critical: what we can do about the health-care crisis. New York: Thomas Dunne Books; 2008.
54. De Dombal FT. Surgical decision making. Oxford/Boston: Butterworth Heinemann; 1993.
55. Department of Health. The good practice guidelines for GP electronic patient records v4. 2011.
56. DesRoches CM, Campbell EG, Rao SR, et al. Electronic health records in ambulatory care – a national survey of physicians. N Engl J Med. 2008;359:50–60.
57. DH. Delivering 21st century IT support for the NHS: National Strategic Programme. Leeds: Department of Health; 2002.
58. Dolin RH, Alschuler L, Beebe C, Biron PV, Boyer SL, Essin D, Kimber E, et al. The HL7 clinical document architecture. J Am Med Inform Assoc. 2001;8:552–69.
59. Dolin RH, Alschuler L, Boyer S, Beebe C, Behlen FM, Biron PV, Shabo A. HL7 clinical document architecture release 2. J Am Med Inform Assoc. 2006;13:30–9.
60. Dolin RH, Alschuler L. Approaching semantic interoperability in Health Level Seven. J Am Med Inform Assoc. 2011;18:99–103.
61. Ellis D. Medical computing and applications. Chichester: Ellis-Horwood; 1987.
62. Eriksson H-E, Penker M. Business modeling with UML: business patterns at work. New York: Wiley; 2000.
63. EU. ICT standards in the health sector: current situation and prospects. A Sectoral e-Business Watch study by Empirica. Special Study No. 1. 2008.
64. Evans E. Domain-driven design: tackling complexity in the heart of software. Boston: Addison-Wesley; 2004.
65. eXtensible Access Control Markup Language (XACML) Version 2.0. OASIS Standard, 1 Feb 2005. oasis-access_control-xacml-2.0-core-spec-os.
66. Forrey AF, McDonald CJ, DeMoor G, Huff SM, Leavelle D, Leleand D, et al. Logical Observation Identifier Names and Codes (LOINC) database, A public use set of codes and names for electronic reporting of clinical laboratory test results. Clin Chem. 1996;42:81–90.
67. Fowler M. UML distilled: a brief guide to the standard object modeling language. 3rd ed. Boston: Addison-Wesley; 2003.
68. Fox J, Das S. Safe and sound. Artificial intelligence in hazardous applications. Menlo Park: MIT Press; 2000.
69. Gibbons P et al. Coming to terms: scoping interoperability in health care. Final. Health level Seven EHR Interoperability Work Group. 2007.
70. Glushko RJ, McGrath T. Document engineering: analyszing and designing documents for business informatics and web services. Cambridge: MIT Press; 2005.
71. Google Health. http://code.google.com/apis/health/
72. Greenhalgh T. How to read a paper: the basics of evidence based medicine. 2nd ed. London: BMJ Books; 2001.
73. Guyatt G, Cook D, Haynes B. Evidence based medicine has come a long way. BMJ. 2004;329:990–1.
74. Hammond WE, Cimino JJ. Standards in biomedical informatics. In: Shortliffe EH, Cimino JJ, editors. Biomedical informatics: computer applications in health care and biomedicine. 4th ed. London: Springer; 2014. p. 211–54.
75. Hayes G, Barnett D, editors. UK health computing: recollections and reflections. Swindon: BCS; 2008.
76. Health Insurance Reform: Security Standards; Final Rule. Department of Health and Human Services. Federal Register Vol. 68, No. 34 February 20 2003.
77. Heitmann K, Bloebel B, Dudeck J. HL7 Communication standard in medicine, short introduction and information. Cologne: Verlag Alexander Mönch; 1999.
78. Henderson M. HL7 messaging, version 2. 2nd ed. Aubrey: O'Tech Inc; 2007.
79. HIMSS. Dictionary of healthcare information technology terms, acronyms and organizations. Chicago: HiMSS; 2006.
80. Hinchley A. Understanding version 3: a primer on th eHL7 version 3 healthcare interoperability standard – Normative edition. 4th ed. Cologne: Verlag Alexander Mönch; 2007.

81. HL7 Implementation Guide for Clinical Document Architecture, Release 2: Consent Directives, Release 1. HL7 Draft Standard for Trial Use, CDAR2_IG_CONSENTDIR_R1_DSTU_2011JAN, January 2011.
82. HL7 V3 TERMINFO. HL7 Version 3 Implementation Guide: Using SNOMED CT, Release 1.5. http://www.hl7.org/v3ballot/html/infrastructure/terminfo/terminfo.html
83. HL7. EHR system functional model and standard draft Standard for trial use (DSTU). 2004.
84. HL7. Implementation Guide: CDA Release 2 – Continuity of Care Document (CCD). 2007.
85. HL7. Using SNOMED CT in HL7 Version 3; Implementation Guide. 2008.
86. Hodge MH. History of the TDS medical information system. In: Blum BI, Duncan K, editors. A history of medical informatics. New York: ACM Press; 1990. p. 328–44.
87. Hohpe G, Woolf B. Enterprise integration patterns: designing, building and deploying messaging solutions. Boston: Addison-Wesley; 2004.
88. Høy A. Guidelines for translation of SNOMED CT. Version 1. Copenhagen: IHTSDO; 2009. www.ihtsdo.org/fileadmin/user_upload/Docs_01/SNOMED_CT/SNOMED_CT_Publications/IHTSDO_Translation_Guidelines_20090309_v1-00.pdf.
89. Hoyt RE, Yoshihashi AK, editors. Health informatics: practical guide for healthcare and information technology professionals. 6th ed. Raleigh: Informatics Education; 2014.
90. IHE (Integrating the Healthcare Enterprise). www.ihe.net
91. IHE. IT infrastructure technical framework volume 1 (ITI TF-1). www.ihe.net/Technical_Framework/upload/ihe_iti_tf_2.0_vol1_FT_2005-08-15.pdf
92. IHTSDO. Compositional grammar for SNOMED CT expressions in HL7 version 3. External Draft for Trial Use. Version 0.06. 2008.
93. Institute of Electrical and Electronics Engineers. IEEE Standard computer dictionary: a compilation of IEEE standard computer glossaries. New York: Institute of Electrical and Electronics Engineers; 1990.
94. Institute of Medicine. Crossing the quality chasm: a new health system for the 21st century. Washington, DC: National Academy Press; 2001.
95. Institute of Medicine. Patient safety: achieving a new standard for care. Washington, DC: National Academy Press; 2004.
96. Institute of Medicine. The computer-based patient record: an essential technology for health care. Revised Edition. Washington, DC: National Academy Press; 1997.
97. International Health Terminology Standards Development Organisation (IHTSDO). www.ihtsdo.org
98. ISO 13606-1:2008. Health informatics – Electronic health record communication – Part 1: Reference Model.
99. ISO. ISO Strategic plan 2005–2010: standards for a sustainable world. Geneva: ISO; 2004.
100. ISO/IEC 27001. Information technology: security techniques: information security management systems: requirements. Geneva: International Organization for Standardization; 2013.
101. ISO/IEC. Guide 2:2004, definition 3.2.
102. ISO/IEC. Open Distributed Processing – Reference Model: Overview. ISO/IEC 10746-1. 1998.
103. ISO/TS 13606-4. Health informatics – Electronic health record communication – Part 4: Security. 2009.
104. Jha AK, DesRoches CM, Campbell EG, et al. Use of electronic health records in US hospitals. N Engl J Med. 2009;360:1628–38.
105. Johansen I, Henriksen G, Demkjær K, Bjerregaard Jensen H, Jørgensen L. Quality assurance and certification of health IT-systems communicating data in primary and secondary health sector. Presentation at MIE 2003, 7 May 2003, St Malo.
106. Kleppe A, Warmer J, Bast W. MDA explained: the model driven architecture: practice and promise. Boston: Addison-Wesley; 2003.
107. Kruchten P. The rational unified process: an introduction. 3rd ed. Reading: Addison Wesley; 2003.
108. Lamberts H, Woods M, editors. ICPC: International Classification of Primary Care. Oxford/New York: Oxford University Press; 1987.

109. Larman C. Applying UML, and patterns: an introduction to object-oriented analysis and design and iterative development. 3rd ed. Upper Saddle River: Prentice Hall; 2005.

110. Lindberg DA, Humphreys BL, McCray AT. The unified medical language system. Methods Inf Med. 1993;32(4):281–91.

111. Mandl KD, Szolovits P, Kohane IS. Public standards and patients' control: how to keep electronic medical records accessible but private. BMJ. 2001;322:283–7.

112. Marshall C. Enterprise modeling with UML: designing successful software through business analysis. Reading: Addison Wesley; 2000.

113. McLuhan M. The gutenberg galaxy: the making of typographic man. Toronto: University of Toronto Press; 1962.

114. MedCom – IT brings the Danish health sector together. November 2008.

115. MedCom – the Danish Health Care Data Network. A Danish health care network in two years. Odense: Danish Centre for Health Telematics; 1996.

116. Mellor SJ, Scott K, Uhl A, Weise D. MDA distilled: principles of model-driven architecture. Boston: Addison Wesley; 2004.

117. Microsoft Health Vault. http://msdn.microsoft.com/en-us/healthvault/default.aspx

118. Microsoft. Connected health framework architecture and design blueprint: a stable foundation for Agile health and social care, 2nd ed. March 2009.

119. Negroponte N. Being digital. London: Coronet Books; 1995.

120. NHS CUI Design Guide Workstream – Design Guide Entry – Terminology – Display Standards for Coded Information. 2007.

121. NHS CUI Design Guide Workstream – Design Guide Entry – Terminology – Elaboration. 2007.

122. NHS CUI Design Guide Workstream – Design Guide Entry – Terminology – Matching. 2007.

123. NHS CUI Design Guide Workstream – Design Guide Entry – Terminology – Postcoordination. 2007.

124. NICE. Colorectal cancer: the diagnosis and management of colorectal cancer, Clinical guideline. London: National Institute for Clinical Excellence; 2011.

125. NICE. Improving outcomes in colorectal cancers: manual update. London: National Institute for Clinical Excellence; 2004.

126. O'Neil MJ, Payne C, Read JD. Read codes version 3: a user led terminology. Methods Inform Med. 1995;34:187–92.

127. OECD. OECD guidelines on the protection of privacy and transborder flows of personal data. Paris: OECD; 1980.

128. Open Health Tools. www.openhealthtools.org

129. OpenEHR Foundation. www.openehr.org

130. Palfrey J, Gasser U. Interop: the promise and perils of highly interconnected systems. New York: Basic Books; 2012.

131. Perry J. OXMIS problem codes for primary medical care. Oxford: Oxford Community Health Project; 1978.

132. Pianykh OS. Digital imaging and communications in medicine (DICOM): a practical introduction and survival guide. 2nd ed. Berlin: Springer; 2012.

133. Pilone D. UML pocket reference. Farnham: O'Reilly; 2003.

134. Preece J. The use of computers in general practice. 4th ed. Edinburgh: Churchill Livingstone; 2000.

135. Protti D, Johansen I. Further lessons from Denmark about computer systems in physician offices. Electron Healthcare. 2003;2(2):36–43. http://www.itacontario.com/health care/2003/03ElectronicHC.pdf.

136. Ramsdell B, editor. Secure/Multipurpose Internet Mail Extensions (S/MIME) Version 3.1 Message specification. IETF Network Working Group RFC 3851. July 2004. http://www.ietf. org/rfc/rfc3851.

137. Read J, Benson T. Comprehensive coding. Br J Healthcare Comput. 1986;3(2):22–5.

138. Rector A, Nowlan W, Kay S. Foundations for an electronic medical record. Methods Inf Med. 1991;30:179–86.

139. Rector A, Solomon W, Nowlan A, Rush T, Claassen A, Zanstra P. A terminology server for medical language and medical information systems. Methods Inf Med. 1994;34:147–57.

140. Rector AL, Qamar R, Marley T. Binding ontologies and coding systems to electronic health records and messages. KR-MED 2006 biomedical technology in action, Baltimore; November 2006.

141. Rosenberg D, Scott K. Applying use case driven object modelling with UML: an annotated e-Commerce example. Boston: Addison-Wesley; 2001.

142. Rothstein MA. HIPAA privacy rule 2.0. J Law Med Ethics. 2013;41(2):525–8.

143. Royal Society. Digital healthcare: the impact of information and communication technologies on health and healthcare. London: The Royal Society; 2006.

144. Rumbaugh J, Jacobson I, Booch G. The unified modeling language reference manual. Reading: Addison-Wesley; 1999.

145. Sager N, Friedman C, Lyman MS. Medical language processing: computer management of narrative data. Reading: Addison-Wesley; 1987.

146. Schadow G, Russler D, Mead C, Case J, McDonald C. The unified service action model: documentation for the clinical area of the HL7 reference information model. Revision 2.6. Regenstrief Institute. May 2000. http://aurora.rg.iupui.edu/RIM/USAM-2.6.PDF

147. Schultz JR. A history of the PROMIS technology: an effective human interface. In: Goldberg A, editor. A history of personal workstations. Reading: Addison Wesley; 1988.

148. Scott JT, Rundall TG, Vogt TM, Hsu J. Kaiser Permanente's experience of implementing an electronic medical record: a qualitative study. BMJ. 2005;331:1313–6.

149. Scott T, Rundall TG, Vogt TM, Hsu J. Implementing an electronic medical record system: successes, failures, lessons. Oxford: Radcliffe Publishing; 2006.

150. Brown SJ, Duguid P. The social life of information. Boston: Harvard Business School Press; 2000.

151. Selvachandran SN, Hodder RJ, Ballal MS, Jones P, Cade D. Prediction of colorectal cancer by a patient consultation questionnaire and scoring system: a prospective study. Lancet. 2002;360:278–83.

152. Shannon CE. A mathematical theory of communication. Bell Syst Tech J. 1948:27: 379–423, 623–56.

153. Shapiro C, Varian HR. Information rules: a strategic guide to the network economy. Boston: Harvard Business School Press; 1999.

154. Shortliffe EH, Cimino JJ, editors. Biomedical informatics: computer applications in health care and biomedicine. 4th ed. New York: Springer; 2014.

155. Slack WV. Cybermedicine: how computing empowers doctors and patients for better health care. San Francisco: Jossey-Bass; 1997.

156. Slee V, Slee D, Schmidt HJ. The endangered medical record: ensuring its integrity in the age of informatics. St. Paul: Tringa; 2000.

157. SNOMED Clinical Terms Technical Implementation Guide. IHTSDO. January 2015.

158. SNOMED CT Compositional Grammar Specification and Guide. The International Health Terminology Standards Development Organisation, Copenhagen. Version 2.03. July 2015.

159. SNOMED CT Editorial Guide. IHTSDO. January 2016. https://confluence.ihtsdotools.org/pages/releaseview.action?pageId=5505533

160. SNOMED CT IHTSDO Glossary. IHTSDO. January 2015.

161. SNOMED CT Starter Guide. IHTSDO. December 2014.

162. Spackman K, Campbell K, Cote R. SNOMED RT: a reference terminology for health care. Proc AMIA Symp. 1997: 640–4.

163. Stokes AV. OSI standards and acronyms. 3rd ed. Manchester: NCC Blackwell; 1991.

164. Straus SE, Richardson WS, Glasziou P, Haynes RB. Evidence-based medicine: how to practice and teach EBM. 3rd ed. Edinburgh: Churchill Livingstone; 2005.

165. Sullivan F, Wyatt JC. ABC of health informatics. Malden: Blackwell Publishing; 2006.

166. Tanenbaum A, Wetherall D. Computer networks. 5th ed. Boston: Pearson; 2010.

167. Taylor P. From patient data to medical knowledge: the principles and practice of health informatics. Malden/Oxford: Blackwell Publishing; 2006.
168. Topol E. The creative destruction of medicine: how the digital revolution will create better health care. New York: Basic Books; 2012.
169. Topol E. The patient will see you now: the future of medicine is in your hands. New York: Basic Books; 2015.
170. UML Resource Page. http://www.uml.org/
171. Van de Velde R, Degoulet P. Clinical information systems: a component-based approach. New York: Springer; 2003.
172. van der Vlist E. Using W3C XML Schema. October 21 2001. http://www.xml.com/pub/a/2000/11/29/schemas/part1.html
173. W3C World Wide Web Consortium. www.w3.org/
174. Wachter R. The digital doctor: hope, hype, and harm at the dawn of medicine's computer age. New York: McGraw-Hill; 2015.
175. Wager KA, Lee FW, Glaser JP. Health care information systems: a practical approach for health care management. 3rd ed. San Francisco: Jossey-Bass; 2013.
176. Walker J, Pan E, Johnston J, Adler-Milstein J, Bates DW, Middleton B. The value of health information exchange and interoperability. Health Aff (Millwood). 2005 Jan-Jun; Suppl Web Exclusives: W5-10-W5-18.
177. Wanless D. Securing our future health: taking a long-term view. Final report. London: HM Treasury; 2002.
178. Warmer J, Kleppe A. The object constraint language: getting your models ready for MDA. 2nd ed. Boston: Addison-Wesley; 2003.
179. Weed LL, Weed L. Medicine in Denial. Charleston: Createspace; 2011.
180. Weed LL. Medical records that guide and teach. NEJM. 1968; 278: 593–9 and 652–7.
181. Weed LL. Knowledge coupling: new premises and new tools for medical care and education. New York: Springer; 1991.
182. Wenger E, McDermot R, Snyder WM. Cultivating communities of practice. Boston: Harvard Business School Press; 2002.
183. Westbrook J, Ampt A, Kearney L, Rob MI. All in a day's work: an observational study to quantify how and with whom doctors on hospital wards spend their time. MJA. 2008;188(9):506–9.
184. White SA, Miers D. BPMN modeling and reference guide: understanding and using BPMN. Lighthouse Point: Future Strategies; 2008.
185. www.ebusiness-watch.org/studies/special_topics/2007/documents/Special-study_01-2008_ICT_health_standards.pdf
186. Wyatt JC. Clinical knowledge and practice in the information age: a handbook for health professionals. London/Lake Forest: RSM Press; 2001.
187. Zielinski K, Duplaga M, Ingram D, editors. Information technology solutions for healthcare. London: Springer; 2006.

.

Index

© Springer-Verlag London 2016
T. Benson, G. Grieve, *Principles of Health Interoperability*,
Health Information Technology Standards, DOI 10.1007/978-3-319-30370-3